Texts in Computer Science

Editors
David Gries
Fred B. Schneider

For further volumes:
http://www.springer.com/series/3191

Michael R. Berthold · Christian Borgelt ·
Frank Höppner · Frank Klawonn

Guide to Intelligent
Data Analysis

How to Intelligently Make
Sense of Real Data

 Springer

Prof. Dr. Michael R. Berthold
FB Informatik und
Informationswissenschaft
Universität Konstanz
78457 Konstanz
Germany
Michael.Berthold@uni-konstanz.de

Dr. Christian Borgelt
Intelligent Data Analysis &
Graphical Models Research Unit
European Centre for Soft Computing
C/ Gonzalo Gutiérrez Quirós s/n
Edificio Científico-Technológico
Campus Mieres, 3ª Planta
33600 Mieres, Asturias
Spain
christian.borgelt@softcomputing.es

Prof. Dr. Frank Höppner
FB Wirtschaft
Ostfalia University of Applied Sciences
Robert-Koch-Platz 10-14
38440 Wolfsburg
Germany
f.hoeppner@ostfalia.de

Prof. Dr. Frank Klawonn
FB Informatik
Ostfalia University of Applied Sciences
Salzdahlumer Str. 46/48
38302 Wolfenbüttel
Germany
f.klawonn@ostfalia.de

Series Editors
David Gries
Department of Computer Science
Upson Hall
Cornell University
Ithaca, NY 14853-7501, USA

Fred B. Schneider
Department of Computer Science
Upson Hall
Cornell University
Ithaca, NY 14853-7501, USA

ISBN 978-1-4471-2572-3   ISBN 978-1-84882-260-3 (eBook)
DOI 10.1007/978-1-84882-260-3
Springer London Dordrecht Heidelberg New York

British Library Cataloguing in Publication Data
A catalogue record for this book is available from the British Library

Cover design: VTeX, Vilnius

Printed on acid-free paper

Springer is part of Springer Science+Business Media (www.springer.com)

Preface

The main motivation to write this book came from all our problems to find suitable material for a textbook that would really help us to teach the practical aspects of data analysis together with the needed theoretical underpinnings. Many books out there tackle either one or the other of these aspects (and, especially for the latter, there are some fantastic text books out there), but a book providing a good combination was nowhere to be found.

The idea to write our own book to address this shortcoming arose in two different places at the same time—when one of the authors was asked to review the book proposal of the others, we quickly realized that it would be much better to join forces instead of independently pursuing our individual projects.

We hope that this book helps others to learn what kind of challenges data analysts face in the real world and at the same time provides them with solid knowledge about the processes, algorithms, and theories to successfully tackle these problems. We have put a lot of effort into balancing the practical aspects of applying and using data analysis techniques while making sure at the same time that we did not forget to also explain the statistical and mathematical underpinnings behind the algorithms beneath all of this.

There are many people to be thanked, and we will not attempt to list them all. However, we do want to single out Iris Adä who has been a tremendous help with the generation of the data sets used in this book. She and Martin Horn also deserve our thanks for an intense last minute round of proof reading.

Konstanz, Germany   Michael R. Berthold
Oviedo, Spain   Christian Borgelt
Braunschweig, Germany   Frank Höppner and Frank Klawonn

Contents

Symbols

A, A_i	attribute, variable [e.g., $A_1 = color$, $A_2 = price$, $A_3 = category$]
ω	a possible value of an attribute [e.g., $\omega = red$]
$\Omega, \mathrm{dom}(\cdot)$	set of possible values of an attribute [e.g., $\Omega_1 = \Omega_{\mathrm{color}} = \mathrm{dom}(A_i) = \{red, blue, green\}$]
$\mathcal{A}$	set of all attributes [e.g., $\mathcal{A} = \{color, price, category\}$]
m	number of considered attributes [e.g., 3]
x	a specific value of an attribute [e.g., $x_2 = x_{\mathrm{price}} = 4000$]
$\mathcal{X}$	space of possible data records [e.g., $\mathcal{X} = \Omega_{A_1} \times \cdots \times \Omega_{A_m}$]
$\mathcal{D}$	set of all records, data set, $\mathcal{D} \subseteq \mathcal{X}$ [e.g., $\mathcal{D} = \{\mathbf{x}_1, \mathbf{x}_2, \ldots, \mathbf{x}_n\}$]
n	number of records in data set
$\mathbf{x}$	record in database [e.g., $\mathbf{x} = (x_1, x_2, x_3) = (red, 4000, luxury)$]
$\mathbf{x}_A$	attribute A of record $\mathbf{x}$ [e.g., $\mathbf{x}_{\mathrm{price}} = 4000$]
$\mathbf{x}_{2,A}$	attribute A of record $\mathbf{x}_2$
$\mathcal{D}_{A=v}$	set of all records $\mathbf{x} \in \mathcal{D}$ with $\mathbf{x}_A = v$
C	a selected categorical target attribute [e.g., $C = A_3 = category$]
Ω_C	set of all possible classes [e.g., $\Omega_C = \{quits, stays, unknown\}$]
Y	a selected continuous target attribute [e.g., $Y = A_2 = price$]
$\mathcal{C}$	cluster (set of associated data objects) [e.g., $\mathcal{C} \subseteq \mathcal{D}$]
c	number of clusters
$\mathcal{P}$	partition, set of clusters $\{\mathcal{C}_1, \ldots, \mathcal{C}_c\}$
$p_{i|j}$	membership degree of data #j to cluster #i
$[p_{i|j}]$	membership matrix
$d.$	distance function, metric (d_E: Euclidean)
$[d_{i,j}]$	distance matrix

Chapter 1
Introduction

In this introductory chapter we provide a brief overview over some core ideas of intelligent data analysis and their motivation. In a first step we carefully distinguish between "data" and "knowledge" in order to obtain clear notions that help us to work out why it is usually not enough to simply collect data and why we have to strive to turn them into knowledge. As an illustration, we consider a well-known example from the history of science. In a second step we characterize the data analysis process, also often referred to as the knowledge discovery process, in which so-called "data mining" is one important step. We characterize standard data analysis tasks and provide a brief catalog of methods and tools to tackle them.

1.1 Motivation

Every year that passes brings us more powerful computers, faster and cheaper storage media, and higher bandwidth data connections. Due to these groundbreaking technological advancements, it is possible nowadays to collect and store enormous amounts of data with amazingly little effort and at impressively low costs. As a consequence, more and more companies, research centers, and governmental institutions create huge archives of tables, documents, images, and sounds in electronic form. Since for centuries lack of data has been a core hindrance to scientific and economic progress, we feel compelled to think that we can solve—at least in principle—basically any problem we are faced with if only we have enough data.

However, a closer examination of the matter reveals that this is an illusion. Data alone, regardless of how voluminous they are, are not enough. Even though large databases allow us to retrieve many different single pieces of information and to compute (simple) aggregations (like average monthly sales in Berlin), general patterns, structures, and regularities often go undetected. We may say that in the vast amount of data stored in some databases we cannot see the wood (the patterns) for the trees (the individual data records). However, it is most often exactly these patterns, regularities, and trends that are particularly valuable if one desires, for example, to increase the turnover of a supermarket. Suppose, for instance, that a

M.R. Berthold et al., *Guide to Intelligent Data Analysis,*
Texts in Computer Science 42,
DOI 10.1007/978-1-84882-260-3_1, © Springer-Verlag London Limited 2010

supermarket manager discovers, by analyzing the sales and customer records, that certain products are frequently bought together. In such a case sales can sometimes be stimulated by cleverly arranging these products on the shelves of the market (they may, for example, be placed close to each other, or may be offered as a bundle, in order to invite even more customers to buy them together).

Unfortunately, it turns out to be harder than may be expected at first sight to actually discover such patterns and regularities and thus to exploit a larger part of the information that is contained in the available data. In contrast to the overwhelming flood of data there was, at least at the beginning, a lack of tools by which raw data could be transformed into useful information. Almost fifteen years ago John Naisbett aptly characterized the situation by saying [3]: "We are drowning in information, but starving for knowledge." As a consequence, a new research area has been developed, which has become known under the name of *data mining*. The goal of this area was to meet the challenge to develop tools that can help humans to find potentially useful patterns in their data and to solve the problems they are facing by making better use of the data they have. Today, about fifteen years later, a lot of progress has been made, and a considerable number of methods and implementations of these techniques in software tools have been developed. Still it is not the tools alone, but the *intelligent composition* of human intuition with the computational power, of sound background knowledge with computer-aided modeling, of critical reflection with convenient automatic model construction, that leads *intelligent data analysis* projects to success [1]. In this book we try to provide a hands-on approach to many basic data analysis techniques and how they are used to solve data analysis problems if relevant data is available.

1.1.1 Data and Knowledge

In this book we distinguish carefully between *data* and *knowledge*. Statements like "Columbus discovered America in 1492" or "Mister Smith owns a VW Beetle" are **data**. Note that we ignore whether we already know these statements or whether we have any concrete use for them at the moment. The essential property of these statements we focus on here is that they refer to single events, objects, people, points in time, etc. That is, they generally refer to single instances or individual cases. As a consequence, their domain of application and thus their utility is necessarily limited.

In contrast to this, **knowledge** consists of statements like "All masses attract each other" or "Every day at 7:30 AM a flight with destination New York departs from Frankfurt Airport." Again, we neglect the relevance of these statements for our current situation and whether we already know them. Rather, we focus on the essential property that they do *not* refer to single instances or individual cases but are general rules or (physical) laws. Hence, if they are true, they have a large domain of application. Even more importantly, though, they allow us to make predictions and are thus highly useful (at least if they are relevant to us).

We have to admit, though, that in daily life we also call statements like "Columbus discovered America in 1492" knowledge (actually, this particular statement is used as a kind of prototypical example of knowledge). However, we neglect here this vernacular and rather fuzzy use of the notion "knowledge" and express our regrets that it is not possible to find a terminology that is completely consistent with everyday speech. Neither single statements about individual cases nor collections of such statements qualify, in our use of the term, as knowledge.

Summarizing, we can characterize data and knowledge as follows:

data
- refer to single instances
  (single objects, people, events, points in time, etc.)
- describe individual properties
- are often available in large amounts
  (databases, archives)
- are often easy to collect or to obtain
  (e.g., scanner cashiers in supermarkets, Internet)
- do not allow us to make predictions or forecasts

knowledge
- refers to *classes* of instances
  (*sets* of objects, people, events, points in time, etc.)
- describes general patterns, structures, laws, principles, etc.
- consists of as few statements as possible
  (this is actually an explicit goal, see below)
- is often difficult and time-consuming to find or to obtain
  (e.g., natural laws, education)
- allows us to make predictions and forecasts

These characterizations make it very clear that generally knowledge is much more valuable than (raw) data. Its generality and the possibility to make predictions about the properties of new cases are the main reasons for this superiority.

It is obvious, though, that not all kinds of knowledge are equally valuable as any other. Not all general statements are equally important, equally substantial, equally significant, or equally useful. Therefore knowledge has to be assessed, so that we do not drown in a sea of irrelevant knowledge. The following list (which we do not claim to be complete) lists some of the most important criteria:

criteria to assess knowledge
- correctness (probability, success in tests)
- generality (domain and conditions of validity)
- usefulness (relevance, predictive power)
- comprehensibility (simplicity, clarity, parsimony)
- novelty (previously unknown, unexpected)

In the domain of science, the focus is on correctness, generality, and simplicity (parsimony) are in the focus: one way of characterizing science is to say that it is the search for a minimal correct description of the world. In economy and industry, however, the emphasis is placed on usefulness, comprehensibility, and novelty: the main goal is to gain a competitive edge and thus to increase revenues. Nevertheless, neither of the two areas can afford to neglect the other criteria.

1.1.2 Tycho Brahe and Johannes Kepler

We illustrate the considerations of the previous section with an (at least partially) well-known example from the history of science. In the sixteenth century studying the stars and the planetary motions was one of the core areas of research. Among its proponents was Tycho Brahe (1546–1601), a Danish nobleman and astronomer, who in 1576 and 1584, with the financial help of King Frederic II, built two observatories on the island of Ven, about 32 km north-east of Copenhagen. He had access to the best astronomical instruments of his time (but no telescopes, which were used only later by Galileo Galilei (1564–1642) and Johannes Kepler (see below) to observe celestial bodies), which he used to determine the positions of the sun, the moon, and the planets with a precision of less than one angle minute. With this precision he managed to surpass all measurements that had been carried out before and to actually reach the theoretical limit for observations with the unaided eye (that is, without the help of telescopes). Working carefully and persistently, he recorded the motions of the celestial bodies over several years.

Stated plainly, Tycho Brahe collected data about our planetary system, fairly large amounts of data, at least from the point of view of the sixteenth century. However, he failed to find a consistent scheme to combine them, could not discern a clear underlying pattern—partially because he stuck too closely to the geocentric system (the earth is in the center, and all planets, the sun, and the moon revolve around the earth). He could tell the precise location of Mars on any given day of the year 1582, but he could not connect its locations on different days by a clear and consistent theory. All hypotheses he tried did not fit his highly precise data. For example, he developed the so-called Tychonic planetary system (the earth is in the center, the sun and the moon revolve around the earth, and the other planets revolve around the sun on circular orbits). Although temporarily popular in the seventeenth century, this system did not stand the test of time. From a modern point of view we may say that Tycho Brahe had a "data analysis problem" (or "knowledge discovery problem"). He had obtained the necessary data but could not extract the hidden knowledge.

This problem was solved later by Johannes Kepler (1571–1630), a German astronomer and mathematician, who worked as an assistant of Tycho Brahe. Contrary to Brahe, he advocated the Copernican planetary system (the sun is in the center, the earth and all other planets revolve around the sun in circular orbits) and tried all his life to reveal the laws that govern the motions of the celestial bodies. His approach was almost radical for his time, because he strove to find a mathematical description. He started his investigations with the data Tycho Brahe had collected and which he extended in later years. After several fruitless trials and searches and long and cumbersome calculations (imagine: no pocket calculators), Kepler finally succeeded. He managed to combine Tycho Brahe's data into three simple laws, which nowadays bear his name: **Kepler's laws**. After having realized in 1604 already that the course of Mars is an ellipse, he published the first two of these laws in his work

"Astronomia Nova" in 1609 [6] and the third law ten years later in his magnum opus "Harmonices Mundi" [4, 7]:

1. The orbit of every planet (including the earth) is an ellipse,
   with the sun at a focal point.
2. A line from the sun to the planet sweeps out equal areas
   during equal intervals of time.
3. The squares of the orbital periods of any two planets relate to each other
   like the cubes of the semimajor axes of their respective orbits:
   $T_1^2/T_2^2 = a_1^3/a_2^3$, and therefore generally $T \sim a^{\frac{3}{2}}$.

Tycho Brahe had collected a large amount of astronomical data, and Johannes Kepler found the underlying laws that can explain them. He discovered the hidden knowledge and thus became one of the most famous "data miners" in history.

Today the works of Tycho Brahe are almost forgotten—few have even heard his name. His catalogs of celestial data are merely of historical interest. No textbook on astronomy contains excerpts from his measurements—and this is only partially due to the better measurement technology we have available today. His observations and precise measurements are raw data and thus suffer from a decisive drawback: they do not provide any insight into the underlying mechanisms and thus do not allow us to make predictions. Kepler's laws, on the other hand, are treated in basically all astronomy and physics textbooks, because they state the principles according to which planets and comets move. They combine all of Brahe's observations and measurements in three simple statements. In addition, they permit us to make predictions: if we know the location and the speed of a planet relative to the sun at any given moment, we can compute its future course by drawing on Kepler's laws.

How did Johannes Kepler find the simple astronomical laws that bear his name? How did he discover them in Tycho Brahe's long tables and voluminous catalogs, thus revolutionizing astronomy? We know fairly little about his searches and efforts. He must have tried a large number of hypotheses, most of them failing. He must have carried out long and cumbersome computations, repeating some of them several times to eliminate errors. It is likely that exceptional mathematical talent, hard and tenacious work, and a significant amount of good luck finally led him to success. What we can be sure of is that he did not possess a universally applicable procedure or method to discover physical or astronomical laws.

Even today we are not much further: there is still no silver bullet to hit on the right solution. It is still much easier to collect data, with which we are virtually swamped in today's "information society" (whatever this popular term actually means) than to discover knowledge. Automatic measurement instruments and scanners, digital cameras and computers, and an abundance of other automatic and semiautomatic devices have even relieved us of the burden of manual data collection. In addition, database and data warehouse technology allows us to store ever increasing amounts of data and to retrieve and to sample them easily. John Naisbett was perfectly right: "We are drowning in information, but starving for knowledge."

It took a distinguished researcher like Johannes Kepler several years (actually half a lifetime) to evaluate the data that Tycho Brahe had collected—data that from a modern point of view are negligibly few and of which Kepler actually analyzed closely only those about the orbit of Mars. Given this, how can we hope today to cope with the enormous amounts of data we are faced with every day? "Manual" analyses (like Kepler's) have long ceased to be feasible. Simple aids, like the visualization of data in charts and diagrams, even though highly useful and certainly a first and important step, quickly reach their limits. Thus, if we refuse to surrender to the flood of data, we are forced to develop and employ computer-aided techniques, with which data analysis can be simplified or even automated to some degree. These are the methods that have been and still are developed in the research areas of intelligent data analysis, knowledge discovery in databases and data mining. Even though these methods are far from replacing human beings like Johannes Kepler, especially since a mindless application can produce artifacts and misleading results, it is not entirely implausible to assume that Kepler, if he had been supported by these methods and tools, could have reached his goal a little earlier.

1.1.3 Intelligent Data Analysis

Many people associate any kind of data analysis with **statistics** (see also Appendix A, which provides a brief review). Statistics has a long history and originated from collecting and analyzing data about the population and the state in general.

Statistics can be divided into *descriptive* and *inferential statistics*. **Descriptive statistics** summarizes data without making specific assumptions about the data, often by characteristic values like the (empirical) mean or by diagrams like histograms. **Inferential statistics** provides more rigorous methods than descriptive statistics that are based on certain assumptions about the data generating random process. The conclusions drawn in inferential statistics are only valid if these assumptions are satisfied.

Typically, in statistics the first step of the data analysis process is to *design the experiment* that defines how data should be collected in order to be able to carry out a reliable analysis based on the obtained data. To capture this important issue, we distinguish between *experimental* and *observational studies*. In an **experimental study** one can control and manipulate the data generating process. For instance, if we are interested in the effects of certain diets on the health status of a person, we might ask different groups of people to stick to different diets. Thus we have a certain control over the data generating process. In this experimental study, we can decide which and how many people should be assigned to a certain diet.

In an **observational study** one cannot control the data generating process. For the same dietary study as above, we might simply ask people on the street what they normally eat. Then we have no control about which kinds of diets we get data and how many people we will have for each diet in our data.

No matter whether the study is experimental or observational, there are usually independence assumptions involved, and the data we collect should be representative. The main reason is that inferential statistics is often applied to *hypothesis testing* where, based on the collected data, we desire to either confirm or reject some hypothesis about the considered domain. In this case representative data and certain independencies are required in order to ensure that the test decisions are valid.

In contrast to hypothesis testing, **exploratory data analysis** is concerned with *generating hypotheses* from the collected data. In exploratory data analysis there are no or at least considerably weaker model assumptions about the data generating process. Most of the methods presented in this book fall into this category, since they are mostly universal methods designed to achieve a certain goal but are not based on a rigorous model as in inferential statistics.

The typical situation we assume in this book is that we already have the data. They might not have been collected in the best way, or in the way we would have collected them had we been able to design the experiment in advance. Therefore, it is often difficult to make specific assumptions about the data generating process. We are also mostly goal-oriented—that is, we ask questions like "Which customers will yield the highest profit"?—and search for methods that can help us to answer such questions or to solve our problems.

The opportunity of analyzing large business databases that were initially collected for completely different purposes came with the availability of powerful tools and technologies that can process and analyze massive amounts of data, so-called **data mining** techniques. A few years ago some people seemed to believe that with just the right data mining tool at hand any kind of desired knowledge could be squeezed out of a given database *automatically* with no or only little human interference. However, practical experience demonstrates that every problem is different and a full automatization of the data analysis process is simply impossible. Today we understand by **knowledge discovery in databases** (KDD) an interactive "process of identifying valid, novel, potentially useful, and ultimately understandable patterns in data" [3]. This process consists of multiple phases, and the data mining or modeling step became just a single step in it. That is, after a period of time where powerful tools were (sometimes) naively applied to the data, the "intelligent analyst" is brought back into the loop. As a consequence, the KDD process differs not so much anymore from classical statistical data analysis (except where the lacking principled data acquisition takes its toll). To emphasize that every project is different and therefore intelligence is required to make the most out of the already gathered data, we use the term **intelligent data analysis**, which was coined by David Hand [1, 5] (and is used today almost synonymously with the KDD process).

In this book we strove to provide a comprehensive guide to intelligent data analysis, outlining the process and its phases, presenting methods and algorithms for various tasks and purposes, and illustrating them with two freely available software tools. In this way we hope to offer a good starting point for anyone who wishes to become more familiar with the area of intelligent data analysis.

1.2 The Data Analysis Process

There are at least two typical situations in which intelligent data analysis may help us to find solutions to certain problems or provide answers to questions that arise.

In the first case, the problem at hand is by no means new, but it is already solved as a matter of routine (e.g., approval of credit card applications, technical inspection during quality assurance, machine control by a plant operator, etc.). If data has been collected for the past cases together with the result that was finally achieved (such as poor customer performance, malfunction of parts, etc.), such historical data may be used to revise and optimize the presently used strategy to reach a decision. In the second case, a certain question arises for the first time, and only little experience is available, or the experience is not directly applicable to this new question (e.g., starting with a new product, preventing abuse of servers, evaluating a large experiment or survey). In such cases, it is supposed that data from related situations may be helpful to generalize the new problem or that unknown relationships can be discovered from the data to gain insights into this unfamiliar area.

What if we have no data at all? This situation does not occur literally in practice, since in most cases there is always *some* data. Especially in businesses huge amounts of data have been collected and stored for operational reasons in the past (e.g., billing, logistics, warranty claims) that may now be used to optimize various decisions or offer new options (e.g., predicting customer performance, reducing stock on hand, tracking causes of defects). So the right question should be: How do we know if we have enough *relevant* data? This question is not answered easily. If it actually turns out that the data is not sufficient, one option is to acquire new data to solve the problem. However, as already pointed out in the preceding section, the experimental design of data acquisition is beyond the scope of this book.

There are several proposals about what the intelligent data analysis process should look like, such as SEMMA (an acronym for *sample, explore, modify, model, assess* used by SAS Institute Inc.), CRISP-DM (an acronym for *CRoss Industry Standard Process for Data Mining* as defined by the CRISP-DM consortium) [2], or the KDD-process [3] (see [8] for a detailed comparison). In this book, we are going to follow the CRISP-DM process, which has been developed by a consortium of large companies, such as NCR, Daimler, and SPSS, and appears to be the most widely used process model for intelligent data analysis today.

CRISP-DM consists of six phases as shown in Fig. 1.1. Most of these phases are usually executed more than once, and the most frequent phase transitions are shown by arrows. The main objective of the first **project understanding** step (see Chap. 3) is to identify the potential benefit as well as the risks and efforts of a successful project, such that a deliberate decision on conducting the full project can be made. The envisaged solution is also transferred from the project domain to a more technical, data-centered notion. This first phase is usually called business understanding, but we stick to the more general term *project understanding* to emphasize that our problem at hand may as well be purely technical in nature or a research project rather than economically motivated.

Next we need to make sure that we will have sufficient data at hand to tackle the problem. While we cannot know this for sure until the end of the project, we at least have to convince ourselves that there is enough relevant data. To achieve this, we proceed in the **data understanding** phase (see Chap. 4) with a review of the available databases and the information contained in the database fields, a visual

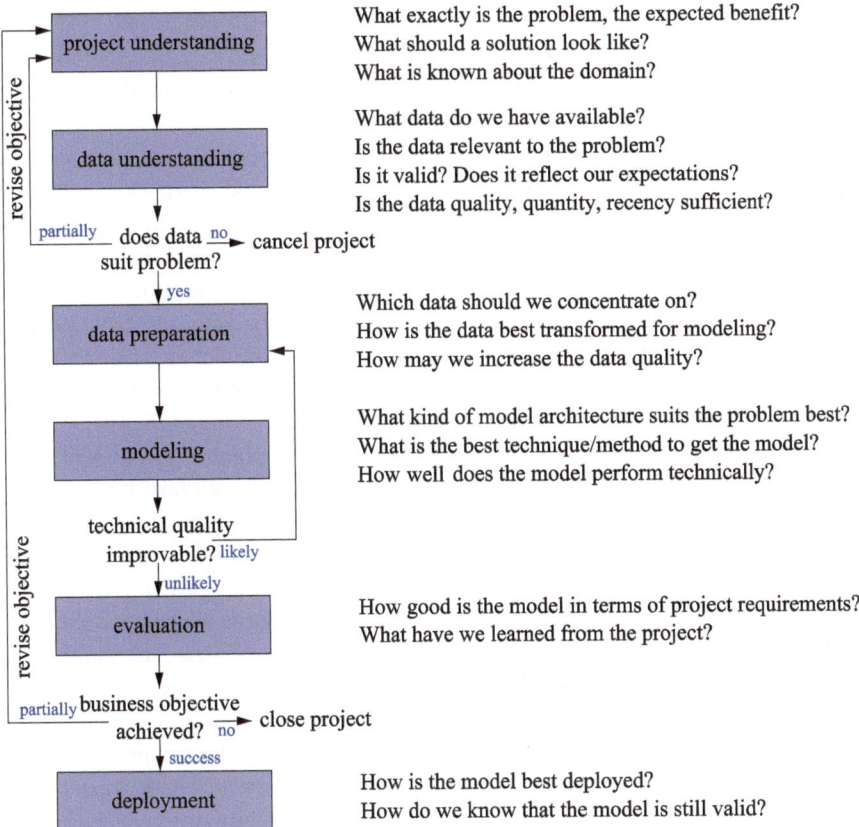

Fig. 1.1 Overview of the CRISP-DM process together with typical questions to be asked in the respective phases

assessment of the basic relationships between attributes, a data quality audit, an inspection of abnormal cases (outliers), etc. For instance, outliers appear to be abnormal in some sense and are often caused by faulty insertion, but sometimes they give surprising insights on closer inspection. Some techniques respond very sensitively to outliers, which is why they should be treated with special care. Another aspect is empty fields which may occur in the database for various reasons—ignoring them may introduce a systematic error in the results. By getting familiar with the data, typically first insights and hypotheses are gained. If we do not believe that the data suffices to solve the problem, it may be necessary to revise the project's objective.

So far, we have not changed any field of our database. However, this will be required to get the data into a shape that enables us to apply modeling tools. In the **data preparation** phase (Chap. 6) the data is selected, corrected, modified, even new attributes are generated, such that the prepared data set best suits the problem and the envisaged modeling technique. Basically all deficiencies that have been identified in the data understanding phase require special actions. Often the outliers and missing values are replaced by estimated values or true values obtained from other sources. We may restrict the further analysis to certain variables and to a selection of the records from the full data set. Redundant and irrelevant data can give many techniques an unnecessarily hard time.

Once the data is prepared, we select and apply **modeling** tools to extract *knowledge* out of the data in the form of a *model* (Chaps. 5 and 7–9). Depending on what we want to do with the model, we may choose techniques that are easily interpretable (to gain insights) or less demonstrative black-box models, which may perform better. If we are not pleased with the results but are confident that the model can be improved, we step back to the data preparation phase and, say, generate new attributes from the existing ones, to support the modeling technique or to apply different techniques. Background knowledge may provide hints on useful *transformations* that simplify the representation of the solution.

Compared to the modeling itself, which is typically supported by efficient tools and algorithms, the data understanding and preparation phases take considerable part of the overall project time as they require a close manual inspection of the data, investigations into the relationships between different data sources, often even the analysis of the process that generated the data. New insights promote new ideas for feature generation or alter the subset of selected data, in which case the data preparation and modeling phases are carried out multiple times. The number of steps is not predetermined but influenced by the process and findings itself.

When the technical benchmarks cannot be improved anymore, the obtained results are analyzed in the **evaluation** phase (Chap. 10) from the perspective of the problem owner. At this point, the project may stop due to unsatisfactory results, the objectives may be revised in order to succeed under a slightly different setting, or the found and optimized model may be deployed.

After **deployment**, which ranges from writing a report to the creation of a software system that applies the model automatically to aid or make decisions, the project is not necessarily finished. If the project results are used continuously over time, an additional monitoring phase is necessary: during the analysis, a number of assumptions will be made, and the correctness of the derived model (and the decisions that rely on the model) depends on them. So we better verify from time to time that these assumption still hold to prevent decision-making on outdated information.

In the literature one can find attempts to create cost models that estimate the costs associated with a data analysis project. Without going into the details, the major key factors that remained in a reduced cost model derived from 40 projects were [9]:

- the number of tables and attributes,
- the dispersion of the attributes (only a few vs. many values),
- the number of external data sources,
- the type of the model (prediction being the most expensive),
- the attribute type mixture (mixture of numeric and nonnumeric), and
- the familiarity of the staff with data analysis projects in general,
  the project domain in particular, and the software suites.

While there is not much we can do about the problem size, the goal of this book is to increase the familiarity with data analysis projects by going through each of the phases and providing first instructions to get along with the software suites.

1.3 Methods, Tasks, and Tools

Problem Categories   Every data analysis problem is different. To avoid the effort of inventing a completely new solution for each problem, it is helpful to think of different problem categories and consider them as building blocks from which a solution may be composed. These categories also help to categorize the large number of different tools and algorithms that solve specific tasks. Over the years, the following set of method categories has been established [3]:

- **classification**
  Predict the outcome of an experiment with a finite number of possible results (like *yes/no* or *unacceptable/acceptable/good/very good*). We may be interested in a prediction because the true result will emerge in the future or because it is expensive, difficult, or cumbersome to determine it.
  Typical questions: *Is this customer credit-worthy? Will this customer respond to our mailing? Will the technical quality be acceptable?*
- **regression**
  Regression is, just like classification, also a prediction task, but this time the value of interest is numerical in nature.
  Typical questions: *How will the EUR/USD exchange rate develop? How much money will the customer spend for vacation next year? How much will the machine's temperature change within the next cycle?*
- **clustering, segmentation**
  Summarize the data to get a better overview by forming groups of similar cases (called clusters or segments). Instead of examining a large number of similar records, we need to inspect the group summary only. We may also obtain some insight into the structure of the whole data set. Cases that do not belong to any group may be considered as abnormal or outliers.
  Typical questions: *Do my customers divide into different groups? How many operating points does the machine have, and what do they look like?*
- **association analysis**
  Find any correlations or associations to better understand or describe the interdependencies of all the attributes. The focus is on *relationships* between all attributes rather than focusing on a single target variable or the cases (full record).
  Typical questions: *Which optional equipment of a car often goes together? How do the various qualities influence each other?*
- **deviation analysis**
  Knowing already the major trends or structures, find any exceptional subgroup that behaves differently with respect to some target attribute.
  Typical questions: *Under which circumstances does the system behave differently? Which properties do those customers share who do not follow the crowd?*

The most frequent categories are *classification* and *regression*, because decision making always becomes much easier if reliable predictions of the near future are available. When a completely new area or domain is explored, cluster analysis and association analysis may help to identify relationships among attributes or records. Once the major relationships are understood (e.g., by a domain expert), a deviation analysis can help to focus on *exceptional situations* that deviate from regularity.

Catalog of Methods   There are various methods in each of these categories to find reliable answers to the questions raised above. However, there is no such thing as a *single gold method* that works perfectly for all problems. To convey some idea which method may be best suited for a given problem, we will discuss various methods in Chaps. 7–9. However, in order to organize these chapters, we did not rely on the problem categories collected above, as some methods can be used likewise for more than one problem type. We rather used the intended task of the data analysis as a grouping criterion:

- **finding patterns**   (Chap. 7)
  If the domain (and therefore the data) is new to us or if we expect to find interesting relationships, we explore the data for new, previously unknown patterns. We want to get a full picture and do not concentrate on a single target attribute yet. We may apply methods from, for instance, segmentation, clustering, association analysis, or deviation analysis.
- **finding explanations**   (Chap. 8)
  We have a special interest in some target variable and wonder why and how it varies from case to case. The primary goal is to gain new insights (knowledge) that may influence our decision making, but we do not necessarily intend automation. We may apply methods from, for instance, classification, regression, association analysis, or deviation analysis.
- **finding predictors**   (Chap. 9)
  We have a special interest in the prediction of some target variable, but it (possibly) represents only one building block of our full problem, so we do not really care about the *how* and *why* but are just interested in the best-possible prediction. We may apply methods from, for instance, classification or regression.

Available Tools   As already mentioned, the key to success is often the proper combination of data preparation and modeling techniques. Data analysis software suites are of great help as they reduce data formatting efforts and ease method linking. There is a long list of commercial and free software suites and tools, including the following *classical* products:

- IBM SPSS PASW Modeler (formerly Clementine)
  Clementine was the first commercial data mining workbench in 1994 and is a commercial product from SPSS, now IBM.
  http://www.spss.com/
- SAS Enterprise Miner
  A commercial data mining solution from SAS.
  http://www.sas.com/

- The R-project
  R is a free software environment for statistical computing and graphics.
  http://www.r-project.org/
- Weka
  Weka is a popular open-source collection of machine learning algorithms, initially developed by the University of Waikato, New Zealand.
  http://www.cs.waikato.ac.nz/ml/weka/

For an up-to-date list of software suites see, for instance,

http://www.kdnuggets.com/software/suites.html

Although the choice of the software suite has considerable impact on the project time (usability) and can help to avoid errors (because some of them are easily spotted using powerful visualization capabilities), the suites cannot take over the full analysis process. They provide at best an initial starting point (by means of analysis templates or project wizards), but in most cases the key factor is the *intelligent combination* of tools and background knowledge (regarding the project domain and the utilized tools). The suites exhibit different strengths, some focus on supporting the human data analyst by sophisticated graphical user interfaces, graphical configuration and reporting, while others are better suited for batch processing and automatization.

In this book, we will use R, which is particularly powerful in statistical techniques, and KNIME (the *Konstanz Information Miner*[1]), which is an open-source data analysis tool that is growing in popularity due to its graphical workflow editor and its ability to integrate other well-known toolkits.

1.4 How to Read This Book

In the next chapter we will take a glimpse at the intelligent data analysis process by looking over the shoulder of Stan and Laura as they analyze their data (while only one of them actually follows CRISP-DM). The chapter is intended to give an impression of what will be discussed in much greater detail throughout the book. The subsequent chapters roughly follow the CRISP-DM stages: we analyze the problem first in Chap. 3 (project understanding) and then investigate whether the available data suits our purposes in terms of size, representativeness, and quality in Chap. 4 (data understanding). If we are confident that the data is worth carrying out the analysis, we discuss the data preparation (Chap. 6) as the last step before we enter the modeling phase (Chaps. 7–9). As already mentioned, data preparation is already tailored to the methods we are going to use for modeling; therefore, we have to introduce the principles of modeling already in Chap. 5. Deployment and monitoring is briefly addressed in Chap. 10. Readers who, over the years, have lost some of their statistical knowledge can (partially) recover it in Appendix A. The statistics

---

[1] Available for download at http://www.knime.org/.

appendix is not just a glossary of terms to quickly look up details but also serves as a book within the book for a few preparative lessons on statistics before delving into the chapters about intelligent data analysis.

Most chapters contain a section that equips the reader with the necessary information for some first hands-on experience using either R or KNIME. We have settled on R and KNIME because they can be seen as extremes on the range of possible software suites: R is a statistical tool, which is (mostly) command-line oriented and is particularly useful for scripting and automatization. KNIME, on the other hand, supports the composition of complex workflows in a graphical user interface.[2] Appendices B and C provide a brief introduction into both systems.

References

1. Berthold, M., Hand, D.: Intelligent Data Analysis. Springer, Berlin (2009)
2. Chapman, P., Clinton, J., Kerber, R., Khabaza, T., Reinartz, T., Shearer, C., Wirth, R.: Cross Industry Standard Process for Data Mining 1.0, Step-by-step Data Mining Guide. CRISP-DM consortium (2000)
3. Fayyad, U.M., Piatetsky-Shapiro, G., Smyth, P., Uthurusamy, R. (eds.): Advances in Knowledge Discovery and Data Mining. AAAI Press/MIT Press, Menlo Park/Cambridge (1996)
4. Feynman, R.P., Leighton, R.B., Sands, M.: The Feynman Lectures on Physics. Mechanics, Radiation, and Heat, vol. 1. Addison-Wesley, Reading (1963)
5. Hand, D.: Intelligent data analysis: issues and opportunities. In: Proc. 2nd Int. Symp. on Advances in Intelligent Data Analysis, pp. 1–14. Springer, Berlin (1997)
6. Kepler, J.: Astronomia Nova, aitiologetos seu physica coelestis, tradita commentariis de motibus stellae martis, ex observationibus Tychonis Brahe. (New Astronomy, Based upon Causes, or Celestial Physics, Treated by Means of Commentaries on the Motions of the Star Mars, from the Observations of Tycho Brahe) (1609); English edition: New Astronomy. Cambridge University Press, Cambridge (1992)
7. Kepler, J.: Harmonices Mundi (1619); English edition: The Harmony of the World. American Philosophical Society, Philadelphia (1997)
8. Kurgan, L.A., Musilek, P.: A survey of knowledge discovery and data mining process models. Knowl. Eng. Rev. **21**(1), 1–24 (2006)
9. Marban, O., Menasalvas, E., Fernandez-Baizan, C.: A cost model to estimate the effort of data mining process (DMCoMo). Inf. Syst. **33**, 133–150 (2008)

---

[2]The workflows discussed in this book are available for download at the book's website.

Chapter 2
Practical Data Analysis: An Example

Before talking about the full-fledged data analysis process and diving into the details of individual methods, this chapter demonstrates some typical pitfalls one encounters when analyzing real-world data. We start our journey through the data analysis process by looking over the shoulders of two (pseudo) data analysts, Stan and Laura, working on some hypothetical data analysis problems in a sales environment. Being differently skilled, they show how things should and should not be done. Throughout the chapter, a number of typical problems that data analysts meet in real work situations are demonstrated as well. We will skip algorithmic and other details here and only briefly mention the intention behind applying some of the processes and methods. They will be discussed in depth in subsequent chapters.

2.1 The Setup

Disclaimer   The data and the application scenario used in this chapter are fictional. However, the underlying problems are motivated by actual problems which are encountered in real-world data analysis scenarios. Explaining particular applicational setups would have been entirely out of the scope of this book, since in order to understand the actual issue, a bit of domain knowledge is often helpful if not required. Please keep this in mind when reading the following. The goal of this chapter is to show (and sometimes slightly exaggerate) pitfalls encountered in real-world data analysis setups and not the reality in a supermarket chain. We are painfully aware that people familiar with this domain will find some of the encountered problems strange, to say the least. Have fun.

The Data   For the following examples, we will use an artificial set of data sources from a hypothetical supermarket chain. The data set consists of a few tables, which have already been extracted from an in-house database:[1]

---

[1]Often just getting the data is a problem of its own. Data analysis assumes that you have access to the data you need—an assumption which is, unfortunately, frequently not true.

M.R. Berthold et al., *Guide to Intelligent Data Analysis,*
Texts in Computer Science 42,
DOI 10.1007/978-1-84882-260-3_2, © Springer-Verlag London Limited 2010

- **Customers**: data about customers, stemming mostly from information collected when these customers signed up for frequent shopper cards.
- **Products**: A list of products with their categories and prices.
- **Purchases**: A list of products together with the date they were purchased and the customer card ID used during checkout.

The Analysts   Stan and Laura are responsible for the analytics of the southern and northern parts, respectively, of a large supermarket chain. They were recently hired to help better understand customer groups and behavior and try to increase revenue in the local stores. As is unfortunately all too common, over the years the stores have already begun all sorts of data acquisition operations, but in recent years quite a lot of this data has been merged—however, still without a clear picture in mind. Many other stores had started to issue frequent shopping cards, so the directors of marketing of the southern and northern markets decided to launch a similar program. Lots of data have been recorded, and Stan and Laura now face the challenge to fit existing data to the questions posed. Together with their managers, they have sat down and defined three data analysis questions to be addressed in the following year:

- differentiate the different customer groups and their behavior to better understand their impact on the overall revenue,
- identify connections between products to allow for cross selling campaigns, and
- help design a marketing campaign to attract core customers to increase their purchases.

Stan is a representative of the typical self-taught data analysis newbie with little experience on the job and some more applied knowledge about the different techniques, whereas Laura has some training in statistics, data processing, and data analysis process planning.

2.2 Data Understanding and Pattern Finding

The first analysis task is a standard data analysis setup: customer segmentation—find out which types of customers exist in your database and try to link them to the revenue they create. This can be used later to care for clientele that are responsible for the largest revenue source or foster groups of customers who are under-represented. Grouping (or *clustering*) records in a database is the predominant method to find such customer segments: the data is partitioned into smaller subsets, each forming a more coherent group than the overall database contains. We will go into much more detail on this type of data analysis methods in Chap. 7. For now it suffices to know that some of the most prominent clustering methods return one typical example for each cluster. This essentially allows us to reduce a large data set to a small number of representative examples for the subgroups contained in the database.

Table 2.1 Stan's clustering result

Cluster-id	Age	Customer revenue
1	46.5	€ 1,922.07
2	39.4	€ 11,162.20
3	39.1	€ 7,279.59
4	46.3	€ 419.23
5	39.0	€ 4,459.30

The Naive Approach   Stan quickly jumps onto the challenge, creates a dump of the database containing customer purchases and their birth date, and computes the age of the customers based on their birth date and the current day. He realizes that he is interested in customer clusters and therefore needs to somehow aggregate the individual purchases to their respective "owner." He uses an aggregating operator in his database to compute the total price of the shopping baskets for each customer. Stan then applies a well-known clustering algorithm which results in five prototypical examples, as shown in Table 2.1.

Stan is puzzled—he was expecting the clustering algorithm to return reasonably meaningful groups, but this result looks as if all shoppers are around 40–50 years old but spend vastly different amount of money on products. He looks into some of the customers' data in some of these clusters but cannot seem to find any interesting relations or any reason why some seem to buy substantially more than others. He changes some of the algorithm's settings, such as the number of clusters created, but the results are similarly uninteresting.

The Sound Approach   Laura takes a different approach. Routinely she first tries to understand the available data and validates that some basic assumptions are in fact true. She uses a basis data summarization tool to report the different values for the string attributes. The distribution of first names seems to match the frequencies she would expect. Names such as "Michael" and "Maria" are most frequent, and "Rosemarie" and "Anneliese" appear a lot less often. The frequencies of the occupations also roughly match her expectations: the majority of the customers are employees, while the second and third groups are students and freelancers, respectively. She proceeds to checking the attributes holding numbers. In order to check the age of the customers, she also computes the customers' ages from their birth date and checks minimum and maximum. She spots a number of customers who obviously reported a wrong birthday, because they are unbelievably young. As a consequence, she decides to filter the data to only include people between the ages of 18 and 100. In order to explore the data more quickly, she reduces the overall customer data set to 5,000 records by random sampling and then plots a so-called histogram, which shows different ranges of the attribute *age* and how many customers fall into that range. Figure 2.1 shows the result of this analysis.

This view confirms Laura's assumptions—the majority of shoppers is middle aged, and the number of shoppers continuously declines toward higher age groups.

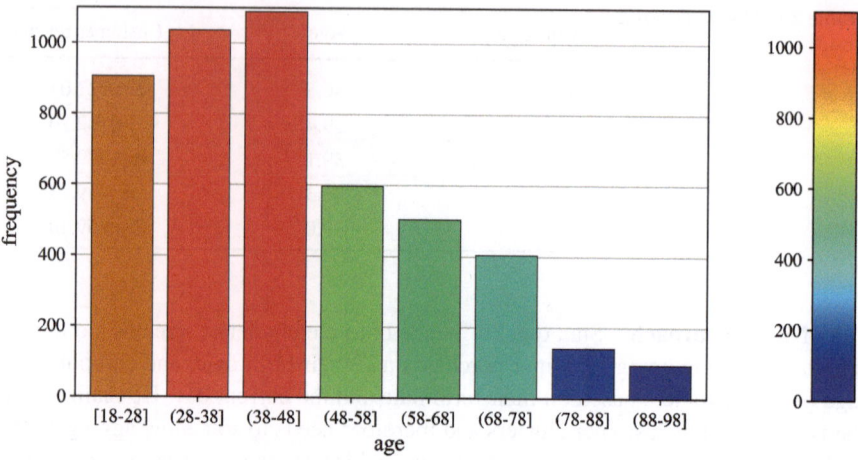

Fig. 2.1 A histogram for the distribution of the value of attribute *age* using 8 bins

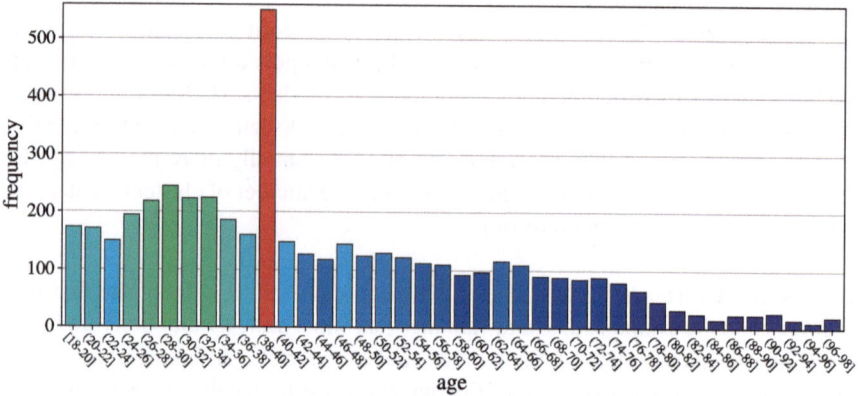

Fig. 2.2 A histogram for the distribution of the value of attribute *age* using 40 bins

She creates a second histogram to better inspect the subtle but strange cliff at around age 48 using finer setting for the bins. Figure 2.2 shows the result of this analysis.

Surprised, she notices the huge peak in the bin of ages 38–40. She discusses this observation with colleagues and the administrator of the shopping card database. They have no explanation for this odd concentration of 40-year-old people either. After a few other investigations, a colleague of the person who—before his retirement—designed the data entry forms suspects that this may have to do with the coding of missing birth dates. And, as it turns out, this is in fact the case: forms where people entered no or obviously nonsensical birth dates were entered into the form as zero values. For technical reasons, these zeros were then converted into the Java 0-date which turns out to be January 1, 1970. So these people all turn up with the same birth date in the customer database and in turn have the same age after the

Table 2.2 Laura's clustering result

Cluster	Age	Avg. cart price	Avg. purchases/month
1	75.3	€ 19.-	5.6
2	42.1	€ 78.-	7.8
3	38.1	€ 112.-	9.3
4	30.6	€ 16.-	4.8
5	44.7	€ 45.-	3.7

conversion Laura performed initially. Laura marks those entries in her database as "missing" in order to be able to distinguish them in future analyses.

Similarly, she inspects the shopping basket and product database and cleans up a number of other outliers and oddities. She then proceeds with the customer segmentation task. As in her previous data analysis projects, Laura first writes down her domain knowledge in form of a cognitive map, indicating relationships and dependencies between the attributes of her database. Having thus recalled the interactions between the variables of interest, she is well aware that the length of customer's history and the number of overall shopping trips affect the overall basket price, and so she settles on the average basket price as a better estimator for the value of a particular customer. She considers also distinguishing the different product categories, realizing that those, of course, also potentially affect the average price. For the first step, she adds the average number of purchases per month, another indicator for the revenue a customer brings in. Data aggregation is now a bit more complex, but the modern data analysis tool she is using allows her to do the required joining and pivoting operations effortlessly. Laura knows that clustering algorithms are very sensitive to attributes with very different magnitudes, so she normalizes the three attributes to make sure they all three contribute equally to the clustering result. Running the same clustering algorithm that Stan was using, with the same setting for the number of clusters to be found, she gets the result shown in Table 2.2.

Obviously, there is a cluster (#1) of older customers who have a relatively small average basket price. There is also another group of customers (#4) which seems to correlate to younger shoppers, also purchasing smaller baskets. The middle-aged group varies wildly in price, however. Laura realizes that this matches her assumption about family status—people with families will likely buy more products and hence combine more products into more expensive baskets, which seems to explain the difference between clusters #2/#3 and cluster #5. The latter also seem to shop significantly less often. She goes back and validates some of these assumptions by looking at shopping frequency and average basket size as well and also determines the overall impact on store revenues for these different groups. She finally discusses these results with her marketing and campaign specialists to develop strategies to foster the customer groups which bring in the largest chunk of revenue and develop the ones which seem to be under-represented.

2.3 Explanation Finding

The second analysis goal is another standard shopping basket analysis problem: find product dependencies in order to better plan campaigns.

The Naive Approach   Stan recently read in a book on practical data analysis how association rules can find arbitrary such connections in market basket data. He runs the association rule mining algorithm in his favorite data analysis tool with the default settings and inspects the results. Among the top-ranked generated rules, sorted by their confidence, Stan finds the following output:

```
'foie gras' (p1231) <- 'champagne Don Huberto' (p2149),
      'truffle oil de Rossini' (p578) [s=1E-5, c=75%]
'Tortellini De Cecco 500g' (p3456)'
   <- 'De Cecco Sugo Siciliana' (p8764) [s=1E-5, c=60%]
```

He quickly infers that this representation must mean that foie gras is bought whenever champagne and truffle oil are bought together and similarly for the other rule. Stan knows that the confidence measure c is important, as it indicates the strength of the dependency (the first rule holds in 3 out of 4 cases). He considers the second measure of frequency s to be less important and deliberately ignores its fairly small value. The two rules shown above are followed by a set of other, similarly luxury/culinary product-oriented rules. Stan concludes that luxury products are clearly the most important products on the shelf and recommends to his marketing manager to launch a campaign to advertise some of the products on the right side of these rules (champagne, truffle oil) to increase the sales of the left side (foie gras). In parallel, he increases orders for these products, expecting a recognizable increase in sales. He proudly sends the results of his analysis to Laura.

The Sound Approach   Laura is puzzled by those nonintuitive results. She reruns the analysis and notices the support values of the rules extracted by Stan—some of the rules Stan extracted have indeed a remarkably high confidence, and some do almost forecast shopping behavior. However, they have very low support values, meaning that only a small number of shopping baskets containing the products were ever observed. The rules that Stan found are not representative at all for his customer base. To confirm this, she runs a quick query on her database and sees that, indeed, there is essentially no influence on the overall revenue.

She notices that the problem of low support is caused by the fact that Stan ran the analysis on product IDs, so in effect he was forcing the rules to differentiate between brands of champagne and truffle oil. She reruns the analysis based on the product categories instead, ranks them by a mix of support and confidence, and finds a number of association rules with substantially higher support:

```
tomatoes <- capers, pasta  [s=0.007, c=32%]
tomatoes <- apples  [s=0.013, c=22%]
```

Laura focuses on rules with a much higher support measure s than before and also realizes that the confidence measure c is significantly higher than one would expect by chance. The first rule seems to be triggered by a recent fashion of Italian cooking, whereas the apple/tomato-rule is a known aspect.

However, she is still irritated by one of the rules discovered by Stan, which has a higher than suspected confidence despite a relatively low support. Are there some gourmets among the customers who prefer a very specific set of products? Rerunning this analysis on the shopping card owners yields almost the same results, so the (potential) gourmets appear among their regular customers. Just to be sure, she inspects how many different customers (resp. shopping cards) occur for baskets that support this rule. As she had conjectured, there is a very limited number of customers that seem to have a strong affection for these products. Those few customers have bought this combination frequently, thus inflating the overall support measure (which refers to shopping baskets). This means that the support in terms of the *number of customers* is even smaller than the support in terms of *number of shopping baskets*. The response to any kind of special promotion would fall even shorter than expected from Stan's rule.

Apparently the time period in which the analyzed data has been collected influences the results. Thinking about it, she develops an idea how to learn about changes in the customers shopping behavior: She identifies a few rules, some rather promising other well-known facts, and decides to monitor those combinations on a regular basis (say quarterly). She got to know that a chain of liquor stores will soon open a number of shops close to the own markets, so she picks some rules with beverages in their conclusion part to see if the opening has any impact on the established shopping patterns of the own customers. As she fears a loss of potential sales, she plans a comparison of rules obtained not only over time but also among markets in the vicinity of such stores versus the other markets. She wonders whether promoting the products in the rule's antecedent may help to bring back the customer and decides to discuss this with the marketing&sales team to determine if and where appropriate campaigns should be launched, once she has the results of her analysis.

2.4 Predicting the Future

The third and final analysis goal we consider in this brief overview is a forecasting or prediction problem. The idea is to find some relationship in our existing data that can help us to predict if and how customers will react to coupon mailings and how this will affect our future revenue.

The Naive Approach Stan believes that no detailed analysis is required for this problem and notices that it is fairly straightforward to monitor success. He has seen at a competitor how discount coupons attract customers to purchase additional products. So he suggests launching a coupon campaign that gives customers a discount of 10% if they purchase products for more than €50. This coupon is mailed to all customers on record. Throughout the course of the next month, he carefully monitors his database and is positively surprised when he sees that his campaign is obviously working: the average price of shopping baskets is going up in comparison with previous months. However, at the end of the quarter he is shocked to see that overall revenues for the past quarter actually fell. His management is finally fed up with the lack of performance and fires Stan.

The Sound Approach   Laura, who is promoted to head of analytics for the northern and southern super market chain first cancels Stan's campaign and looks into the underlying data. She quickly realizes that even though quite a number of customers did in fact use the coupons and increased their shopping baskets, their average number of baskets per month actually went down—so quite a number of people seem to have simply combined smaller shopping trips to be able to benefit from the discount offer. However, for some shoppers, the combined monthly shopping basket value did go up markedly, so there might be value here. Laura wonders how she can discriminate between those customers who simply use the coupons to discount their existing purchases and those who are actually enticed to purchase additional items. She notices that one of the earlier generated customer segments correlates better than others with the group of customers whose revenue went up—this fraction of customers is significantly higher than in the other groups. She considers using this very simple, manually designed predictor for a future campaign but wants to first make sure that she cannot do better with some smarter techniques. She decides that in the end it is not so important if she can actually understand the extracted model but only how well it performs.

To provide good starting points for the modeling technique, she decides to generate a few potentially informative attributes first. Models that rely on thousands of details typically perform poor, so providing how often every product has been bought by the customer in the last month is not an option for her. To get robust models, she wants to aggregate the tiny bits of information, but what kind of aggregation could be helpful? She returns to her cognitive map to review the dependencies. One aspect is the availability of competitors: She reckons that customers may have alternative (possibly specialized) markets nearby but have been attracted by the coupon this time, keeping them away from the competitors. She decides to aggregate the money spent by the customer per month for a number of product types (such as *beverages*, thinking of the chain of liquor stores again). She conjectures that customers that perform well on average, but underperform in a specific segment only, may be enticed by the coupon to buy products for the underperforming segment also. Providing the segment performance before and after Stan's campaign should help a predictor to detect such dependencies if they exist.

The cognitive map brings another idea into her mind: people who appreciate the full assortment but live somewhat further away from the own stores may see the coupon as a kind of travel compensation. So she adds a variable expressing a coarse estimation of the distance between the customer home and the nearest available market (which is only possible for the shopping card owners). She continues to use her cognitive map to address many different aspects and creates attributes that may help to verify her hypotheses. She then investigates the generated attributes visually and also technically by means of feature selection methods.

After selecting the most promising attributes, she trains a classifier to distinguish the groups. She uses part of the data to simulate an independent test scenario and thereby evaluates the expected impact of a campaign—are the costs created by sending coupons to customers who do not purchase additional products offset by customers buying additional items? After some additional model fine tuning, she reaches satisfactory performance. She discusses the results with the marketing&sales team and deploys the prediction system to control the coupon mailings for the next quarter. She keeps monitoring the performance of these coupon campaigns over future quarters and updates her model sporadically.

2.5 Concluding Remarks

In this chapter we have, very briefly and informally, touched upon a number of issues data analysts may encounter while making sense of real-world data. Many other problems can arise, and many more methods for data analysis exist in the academic literature and in real-world data analysis tools. We will attempt at covering the most prominent and most often used examples in the following chapters.

Note that one of the biggest problems data analysts very often have is that the data they get is not suited to answer the questions they are asked. For instance, if we were supposed to use the data in our customer database to find out how to differentiate Asian shopping behavior from European, we would have a very hard time. This data can only be used to distinguish between different types of European shoppers because it contains data from European markets only. Note also that we are (why ever) assuming that we used a nice, representative sample of all different types of European shoppers to generate the data—very often this is not the case, and the data itself is already biased and will bias our analysis results—in this example we could be heavily biased by the type of supermarket chain we used to record the data in the first place. An upscale delicatessen supermarket will have dramatically different shopping patterns than the low-scale discounter. We will be discussing these points later in more depth as well.

Chapter 3
Project Understanding

We are at the beginning of a series of interdependent steps, where the project understanding phase marks the first. In this initial phase of the data analysis project, we have to map a problem onto one or many data analysis tasks. In a nutshell, we conjecture that the nature of the problem at hand can be adequately captured by some data sets (that still have to be identified or constructed), that appropriate modeling techniques can successfully be applied to learn the relationships in the data, and finally that the gained insights or models can be transferred back to the real case and applied successfully. This endeavor relies on a number of assumptions and is threatened by several risks, so the goal of the project understanding phase is to assess the main objective, the potential benefit, as well as the constraints, assumptions, and risks. While the number of data analysis projects is rapidly expanding, the failure rate is still high, so this phase should be carried out seriously to rate the chances of success realistically. The project understanding phase should be carried out with care to keep the project on the right track.

We have already sketched the data analysis process (CRISP-DM in Sect. 1.2). There is a clear order in the steps in the sense that for a later step, all precedent steps must have been executed. However, this does not mean that we can run once through all steps to deterministically achieve the desired results. There are many options and decisions to be made. Most of them will rely on our (subjective and dynamic) understanding of the problem at hand. The line of argument will not always be from an earlier phase to a later one. For instance, if a regression problem has to be solved, the analyst may decide that a certain method seems to be a promising choice for the modeling phase. From the characteristics of this technique he knows that all input data have to be transformed into numerical data, which has to be carried out beforehand (data preparation phase). This requires a careful look at the multivalued ordinally scaled attributes already in the data understanding phase to see how the order of the values is best preserved. If it is not considered in time, it may happen that later, in the evaluation phase, it turns out that the project owner expected to gain insights into the input–output relationship rather than having a black-box model only. If the analyst had considered this requirement beforehand, he might have chosen a different method. Changing this decision at any point later than in this initial

M.R. Berthold et al., *Guide to Intelligent Data Analysis*,
Texts in Computer Science 42,
DOI 10.1007/978-1-84882-260-3_3, © Springer-Verlag London Limited 2010

Table 3.1 Problems faced in data analysis projects, excerpt from [1]

Problem source	Project owner perspective	Analyst perspective
Communication	Project owner does not understand the technical terms of the analyst	Analyst does not understand the terms of the domain of the project owner
Lack of understanding	Project owner was not sure what the analyst could do or achieve	Analyst found it hard to understand how to help the project owner
	Models of analyst were different from what the project owner envisioned	
Organization	Requirements had to be adopted in later stages as problems with the data became evident	Project owner was an unpredictable group (not so concerned with the project)

project understanding phase often renders some (if not most) of the earlier work in data understanding, data preparation, and modeling useless. While the time spent on project and data understanding compared to data preparation and modeling is small (20% : 80%), the importance to success is just the opposite (80% : 20%) [4].

3.1 Determine the Project Objective

As a first step, a *primary* objective (not a long list but one or two) and some success criteria in terms of the project domain have to be determined (who will decide which results are desired and whether the original project goal was achieved or not). This is much easier said than done, especially if the analysis is not carried out by the domain expert himself. In such cases the project owner and the analyst *speak different languages* which may cause misunderstandings and confusion. In the worst case, the communication problems lead to very soft project goals, just vague enough to allow every stakeholder seeing his own perspective somehow accounted for. At the end, all of a sudden, the stakeholders recognize that the results do not fit their expectations. The challenge here is usually not a matter of technical but of communicative competence.

Table 3.1 shows some typical problems occurring in such projects. To overcome language confusion, a glossary of terms, definition, acronyms, and abbreviations is inevitable. Knowing the terms still does not imply an understanding of the project domain, the objectives, constraints, and influencing factors. One interviewing technique that may help to get most out of the expert is to rephrase all of her statements, which often provokes additional relativizing statements. Another technique is to use explorative tools such as mind maps or cognitive maps to sketch beliefs, experiences, and known factors and how they influence each other.

An example of a **cognitive map** in the shopping domain considered in Sect. 2 is given in Fig. 3.1. Each node of this graph represents a property of the considered product or the customer. The variable of interest is placed in the center: how often

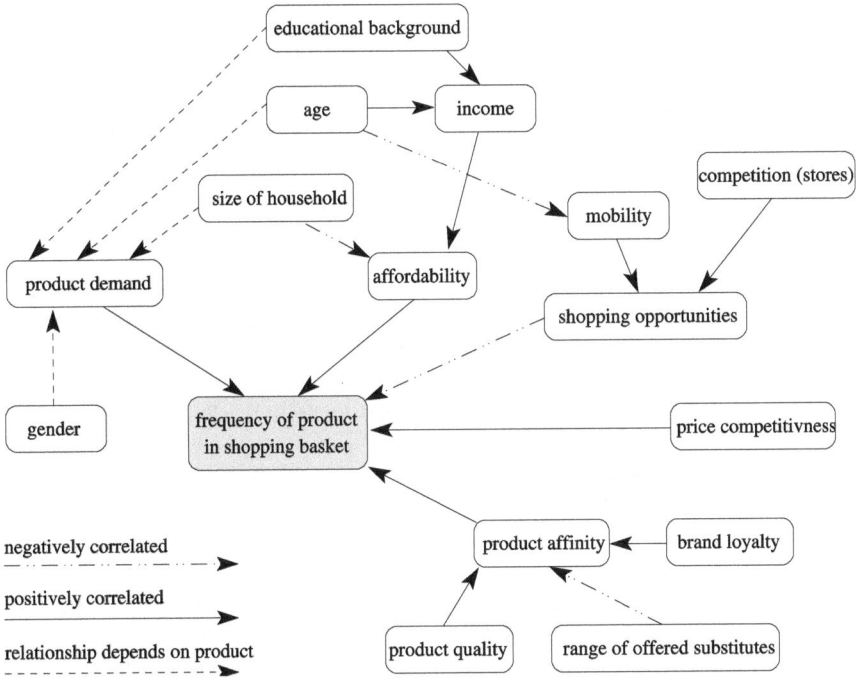

Fig. 3.1 A cognitive map for the shopping domain: How often will a certain product occur in a shopping basket of some customer? The positive correlation between *income* and *affordability* reads like *the higher the income, the higher the affordability*, whereas an example of a negative correlation reads like *the broader the range of offered substitutes, the lower the product affinity*

will a certain product be found in the shopping basket of the customer? This depends on various factors, which are placed around this node. The direction of influence is given by the arrows, and the line style indicates the way how the variables influence each other: The higher the customer's affinity to the product, the more often it will be found in the basket. The author of the cognitive map conjectures that the product affinity itself is positively influenced by a high product quality and the customer's brand loyalty (a loyal customer is less likely to buy substitute products). On the other hand, the broader the range of offered substitutes, the more likely a customer may try out a different product. Other relationships depend on the product itself: The higher the demand of a certain product, the more often it will be found in the shopping basket, but the demand itself may, depending on the product, vary with gender (e.g., razor blades, hairspray), age (e.g., denture cleaner), or family status (e.g., napkins, fruit juice). The development of such a map supports the domain understanding and adjustment of expectations.

While constructing a cognitive map, a few rules should be adhered to: First, to keep the map clear, only direct dependencies should be included in the graph. For instance, the size of the household influences the target variable, but only indirectly via the generated product demand and the affordability, and therefore there is no direct connection from *size of household* to *frequency of product in shopping basket*.

Table 3.2 Clarifying the primary objectives

Objective:	Increase revenues (per campaign and/or per customer) in direct mailing campaigns by personalized offer and individual customer selection
Deliverable:	Software that automatically selects a specified number of customers from the database to whom the mailing shall be sent, runtime max. half-day for database of current size
Success criteria:	Improve order rate by 5% or total revenues by 5%, measured within 4 weeks after mailing was sent, compared to rate of last three mailings

Secondly, the labels of the nodes should be chosen carefully, so that they are easily interpretable when plugged into the relationship templates such as *the higher ...*, *the higher ...*. As an example, the node *size of household* could have been named *family status*, but then it is not quite clear what *the more family status ...* actually means.

Once an understanding of the domain has been achieved, the problem and primary objective have to be identified (see Table 3.2). Again, it is often useful to discuss or model the current solution first, for instance, by using techniques from software engineering (business process modeling, UML use cases, etc.) [3]. When the current solution has been elaborated, its advantages and disadvantages can be explored and discussed. Often, the primary objective is assumed to be known beforehand, probably the project would not have been initiated without having identified a problem first. But as there are many different ways to attack a problem, the objective should be precise about the direction to follow. A general statement about the goal is easily made ("model the profitable customers to increase the sales"), but it is often not precise enough (how do we precisely identify a profitable customer?) and not actionable (how exactly shall this model help to increase the sales?). To render the objective more precise, it is necessary to sketch the target use already at this early stage. Thus it becomes clear what kind of result has to be delivered, which may range from a written technical report with interesting findings to a user-friendly software that uses the final model to automatize decisions.

From the perspective of the project owner some of these elaborate steps may appear unnecessary—they master their domain already, after all. However, these steps must be considered as a preparation of the closely linked data understanding phase (see next section). All the identified factors, situations, and relationships that are assumed to be relevant must be present and recognizable in the data. If they cannot be found in the data, either there is a misconception in the project understanding or (even worse) the data is not substantial or detailed enough to reflect the important relations of the real-world problem. In both cases, it would be fatal to miss this point and proceed unworried.

3.2 Assess the Situation

The next step is to estimate the chances of a successful data analysis project. This requires the review of available resources, requirements, and risks. The most important resources are data and knowledge, that is, databases and experts who can provide background information (about the domain in general and about the databases in particular). Besides a plain listing of databases and personnel, it is important to clarify the access to both: if the data is stored in an operative system, mining the data may paralyze the applications using it. To become independent, it is advisable to provide a database dump. Experts are typically busy and difficult to grasp—but an inaccessible knowledge source is useless. A sufficiently large number of time slots for meetings should be arranged.

Based on the domain exploration (cognitive map, business process model, etc.), a list of explicit and implicit assumptions and risks is created to judge the chances of a successful project and guide the next steps. Data analysis lives on data. This list shall help to convince ourselves that the data is meaningful and relevant to the project. Why should we undertake this effort? We will see whether we can build a model from this data later anyway. Unfortunately, this is only half of the truth. After reviewing a number of reports in a data analysis competition, Charles Elkan noted that *"when something surprising happens, rather than question the expectations, people typically believe that they should have done something slightly different"* [2]. Expecting that the problem can be solved with the given data may lead to continuously changing and "optimizing" the model—rather than taking the possibility into account that the data is not appropriate for this problem. In order to avoid this pitfall, the conjectured relations and expert-proven connections can help us in verifying that the given data satisfy our needs—or to put forward good reasons why the project will probably fail. This is particularly important as in many projects the available data have not been collected to serve the purpose that is intended now. To prevent us from carrying out an expensive project having almost no prospect of success, we have to carefully track all assumptions and verify them as soon as possible. Typical requirements and assumptions include:

- requirements and constraints
    - *model requirements*,
      e.g., model has to be explanatory (because decisions must be justified clearly)
    - *ethical, political, legal issues*,
      e.g., variables such as gender, age, race must not be used
    - *technical constraints*,
      e.g., applying the technical solution must not take more than n seconds
- assumptions
    - *representativeness*:
      If conclusions about a specific target group are to be derived, a sufficiently large number of cases from this group must be contained in the database, and the sample in the database must be representative for the whole population.
    - *informativeness*:
      To cover all aspects by the model, most of the influencing factors (identified in the cognitive map) should be represented by attributes in the database.
    - *good data quality*:
      The relevant data must be of good quality (correct, complete, up-to-date) and unambiguous thanks to the available documentation.

– *presence of external factors*:
  We may assume that the external world does not change constantly—for instance, in a marketing project we may assume that the competitors do not change their current strategy or product portfolio at all.

Every assumption inherently represents a risk (there might be other risks though). If possible, a contingency should be sketched in case the assumption turns out to be invalid, including options such as the acquisition of additional data sources.

3.3 Determine Analysis Goals

Finally, the primary objective must be transformed into a more technical data mining goal. An architecture for the envisaged solution has to be found, composed out of building blocks as discussed in Sect. 1.3 (data analysis tasks). For instance, this architecture might contain a component responsible for grouping the customers according to some readily available attributes first, another component finds interesting deviating subgroups in each of the groups, and a third component predicts some variable of interest based on the customer data and the membership to the respective groups and subgroups. The better this architecture fits the actual situation, the better the chances of finding a model class that will prove successful in practice. To achieve this analogy, the discussions about the project domain are of great help.

Again there is the danger of accepting a reasonable architecture quickly, underestimating or even ignoring the great impact on the overall effort. Suppose that a company wants to increase the sales of some high-end product by direct mailing. One approach is to develop a model that predicts who will buy this product using the company's own customer database. Such a model might be interesting to interpret (useful for a report), but if it is used to decide to whom a mailing should be sent, most of the customers may have the product already (within the same customer database). Applying the model to people not being in the database is impossible as we lack the information about them that is needed by the model. The predictive model may also find out that customers buying the product were loyal customers for many years—but *artificially* increasing the duration of the customer relationship to support the purchase of the product is unfortunately impossible. If a foreseeable result is ignored or a misconception w.r.t. the desired use of the model is not recognized, considerable time may be wasted with building a correct model that turns out to be useless in the end.

For each of the building blocks, we can select a model class and technique to derive a model of this class automatically from data. There is nothing like *the unique best method for predictive tasks*, because they all have their individual weaknesses and strengths and it is impossible to combine all their properties or remove all biases (see Chap. 5). Although the final decision about the modeling technique will be made in the modeling phase, it should be clear already at this point of the analysis which properties the model should have and why. The methods and tools optimize the technical aspects of the model quality (such as accuracy, see also Chap. 5). Other aspects are often difficult to formalize and thus to optimize (such as interestingness or interpretability), so that the choice of the model class has the greatest influence on these properties. Desirable properties may be, for instance:

- **Interpretability**:
  If the goal of the analysis is a report that sketches possible explanations for a certain situation, the ultimate goal is to understand the delivered model. For some *black-box models*, it is hard to comprehend how the final decision is made, and their model lacks interpretability.
- **Reproduceability/stability**:
  If the analysis is carried out more than once, we may achieve similar performance —but not necessarily similar models. This does no harm if the model is used as a black box, but hinders a direct comparison of subsequent models to investigate their differences.
- **Model flexibility/adequacy**:
  A flexible model can adapt to more (complicated) situations than an inflexible model, which typically makes more assumptions about the real world and requires less parameters. If the problem domain is complex, the model learned from data must also be complex to be successful. However, with flexible models the risk of overfitting increases (will be discussed in Chap. 5).
- **Runtime**:
  If restrictive runtime requirements are given (either for building or applying the model), this may exclude some computationally expensive approaches.
- **Interestingness** and **use of expert knowledge**:
  The more an expert already knows, the more challenging it is to "surprise" him or her with new findings. Some techniques looking for associations (see Sect. 7.6) are known for their large number of findings, many of them redundant and thus uninteresting. So if there is a possibility of including any kind of previous knowledge, this may ease the search for the best model considerably on the one hand and may prevent us from rediscovering too many well-known artifacts.

When discussing the various modeling techniques in Chaps. 7–9, we will give hints which properties they possess. The final choice is then up to the analyst.

3.4 Further Reading

The books by Dorian Pyle [4, 5] offer many suggestions and constructive hints for carrying out the project understanding phase. [5] contains a step-by-step workflow for business understanding and data mining consisting of various *action boxes*. An organizationally grounded framework to formally implement the business understanding phase of data mining projects is presented in [6]. In [1] a template set for educing and documenting project requirements is proposed.

References

1. Britos, P., Dieste, O., García-Martínez, R.: Requirements elicitation in data mining for business intelligence projects. In: Advances in Information Systems Research, Education and Practice, pp. 139–150. IEEE Press, Piscataway (2008)
2. Elkan, C.: Magical thinking in data mining: lessons from coil challenge 2000. In: Proc. 7th Int. Conf. on Knowledge Discovery and Data Mining (KDD), pp. 426–431. ACM Press, New York (2001)
3. Marban, O., Segovia, J., Menasalvas, E., Fernandez-Baizan, C.: Towards data mining engineering: a software engineering approach. Inf. Syst. **34**, 87–107 (2009)
4. Pyle, D.: Data Preparation for Data Mining. Morgan Kaufmann, San Mateo (1999)
5. Pyle, D.: Business Modeling and Data Mining. Morgan Kaufmann, San Mateo (2003)
6. Sharma, S., Osei-Bryson, K.-M.: Framework for formal implementation of the business understanding phase of data mining projects. Expert Syst. Appl. **36**, 4114–4124 (2009)

Chapter 4
Data Understanding

The main goal of data understanding is to gain general insights about the data that will potentially be helpful for the further steps in the data analysis process, but data understanding should not be driven exclusively by the goals and methods to be applied in later steps. Although these requirements should be kept in mind during data understanding, one should approach the data from a neutral point of view. Never trust any data as long as you have not carried out some simple plausibility checks. Methods for such plausibility checks will be discussed in this chapter. At the end of the data understanding phase, we know much better whether the assumptions we made during the project understanding phase concerning representativeness, informativeness, data quality, and the presence or absence of external factors are justified.

We first take a general look at single attributes in Sect. 4.1 and ask questions like: What kind of attributes do we have, and what do their domains look like? What is the precision of numerical values? Is the domain of an attribute stable over time, or does it change? We also need to assess the data quality. Methods and criteria for this purpose are introduced in Sect. 4.2.

Data understanding requires taking a closer look at the data. However, this does not mean that we must browse through seemingly endless columns of numbers and other values. In this way we would probably overlook most of the important facts. Looking at the data refers to visualization techniques (Sect. 4.3) that can be used to get a quick overview on basic characteristics of the data and enable us to check the plausibility of the data to a certain extent. Visualization techniques are suitable for the analysis of single attributes and of attributes in combination. Apart from the pure visualization, it is also recommended to compute simple statistical measures for correlation between attributes as described in Sect. 4.4.

Outliers, values, or records that are very different from all others should be identified with methods described in Sect. 4.5. They might cause difficulties for some of the methods applied in later steps, or they might be wrong values due to data quality problems. Missing values (see Sect. 4.6) can lead to similar problems as outliers, and by simply ignoring missing values we might obtain wrong data analysis results, so we must be aware of whether we have to deal with missing values and, if we have to, of what kind the missing values are.

M.R. Berthold et al., *Guide to Intelligent Data Analysis,*
Texts in Computer Science 42,
DOI 10.1007/978-1-84882-260-3_4, © Springer-Verlag London Limited 2010

Data understanding is also a step that is required for data preparation. For example, data understanding will help us to identify and to characterize outliers and missing values. However, how to treat them—whether to leave them as they are, to exclude them from further analysis steps, or to replace them by more plausible values—is a task for data preparation.

Throughout this chapter we will use the Iris data set [2, 7]: a set of 150 data points describing three different types of iris flowers (Iris setosa, Iris virginica, and Iris versicolor) using four different attributes measuring the length and width of the sepal and the petal leaves. This is a classic data set with a few very simple and obvious properties which lends itself naturally to demonstrate how the several different data analysis methods work. However, readers should not let themselves be fooled into believing that any real-world data sets will ever display just nice and well pronounced features. Quite the opposite: in real-world data the odd and undesired effects often far outweigh the interesting ones.

4.1 Attribute Understanding

In most cases, we assume that the data can be described in terms of a table or data matrix whose rows contain the **instances**, records, or **data objects** and whose columns represent the **attributes**, **features**, or **variables**. The data might not be stored directly in one table but in different tables from which the attributes of interest need to be extracted and joined into a single table. For instance, when we are interested in the correlation between the age of a customer and the number of budget products they buy, we would need a table with two columns representing the attributes *age* and *number of budget products bought*. The age of the customers can be extracted directly from the customer table (computing it from the birth date if necessary), whereas the number of budget products a customer has bought is not stored in any specific table but must be counted in the list of items he has bought.

An attribute or its **domain**—the set of possible values for the attribute—can have different principal characteristics. One of the most basic is the **scale type**: an attribute can be *nominal* (or *categorical*), *ordinal*, or *numeric* (details are provided below). Numeric attributes are further subdivided depending on whether they are *discrete* or *continuous* and whether they have an *interval*, *ratio*, or *absolute scale*.

An attribute is called **nominal** or **categorical** if its domain is a finite set. The possible values for a categorical attribute are often considered as classes or categories. For a purely categorical attribute, there is no additional structure on the domain. This means that two values or categories are either equal or different, but not more or less similar. Examples for categorical attributes are the gender with the values *F* (female) and *M* (male) in the customer data set or the species in the Iris data set.

For categorical attributes, there are sometimes different levels of **granularity**, and it is necessary to make a choice on which level the model for the analysis of the data should be built. The different levels of granularities form a hierarchy of more and more refined domains. A typical example for such refined levels of granularity are product categories. The general category might have the values *drinks, food, ...,* a more refined level for the drinks might distinguish between *water, beer, wine, . . . ,* and these categories might even be more refined by incorporating the producer. Even the same producer might offer different brands and variants of water, beer, or wine. This is still not the lowest level of refinement. We might even distinguish identical drinks by the size of the container in which they are sold, for instance, 0.5 l, 1.0 l, and 1.5 l bottles. Of course, the most refined level will always provide the most detailed information. Nevertheless, it is often not very useful to carry out an analysis on this level, since it is impossible to extract general structures or rules when we restrict our analysis on the refined level. A general rule like "Customers tend to buy wine and cheese together" might not be discovered on a refined level of granularity where no combination of a specific wine and specific cheese might be frequent. Therefore, it is crucial to choose an appropriate level of granularity for such attributes taking the aim of the analysis and the size of the data set into account. For smaller data sets, little statistical evidence can be found when we look at a very refined level, since the number of instances per value of the attribute decreases with the level of refinement. Especially, if a number of such attributes with different levels of granularity is considered, the number of possible combinations for the values grows extremely fast for refined levels. However, the choice of a suitable granularity is a task within data preparation. Nevertheless, we must make a decision here as well, on which level we want to understand and take a closer look at the data. We might do this even on different levels of granularity.

Another problem that sometimes comes along with categorical attributes is a dynamic domain. For certain categorical attributes, the domain will never change. The range of possible values for the month of birth is *January, . . . , December*, and it seems quite improbable that it will be changed in the near future. The situation for products is, however, completely different. Certain products might not be offered by a shop anymore or might vanish completely from the market, whereas other new products enter the market, or already existing products are adopted by the shop.

Such a **dynamic domain** of an attribute can cause problems in the analysis later on. When products are analyzed on a long-term basis of, say, several years, then products that have entered the market just recently will not show significant (accumulated) sales numbers compared to products that have been in the market for decades. Therefore, categorical attributes can lead to undesired or even wrong analysis results when such problems like the level of granularity or dynamic domains are not taken into account. We identify such attributes already at this early stage and make sure that we do not forget to handle them later.

A specific type of categorical attributes are **ordinal** attributes with an additional linear ordering imposed on the domain. An attribute for university degrees with the values none, B.Sc., M.Sc., and Ph.D. represents an ordinal attribute. A Ph.D. is a higher degree than an M.Sc., and an M.Sc. is a higher degree than a B.Sc. However, the ordering does not say that the difference between a Ph.D. and an M.Sc. is the same as the difference between an M.Sc. and a B.Sc.

The domain of a **numerical** attribute are numbers. Numerical attributes can be **discrete**, usually taking integer values, or **continuous**, taking arbitrary real values. Discrete numerical attributes often result from counting processes like the number of children or the number of times a customer has ordered from an on-line shop in the last twelve months.

Sometimes, categorical attributes are coded by numerical values. For instance, the three possible values *food, drinks, nonfood* of the attribute *general product category* might be coded by the numbers 1, 2, and 3. However, this does not turn the attribute *general product category* into a (discrete) numerical attribute. We should bear this fact in mind for later steps of the analysis to avoid that the attribute is suddenly interpreted as a numerical attribute: it does not make sense to carry out numerical operations like computing the mean on such coded categorical attributes. For a discrete attribute, though, especially when it represents some counting process, it is meaningful to calculate the mean value, even though the mean value will usually not be an integer number. It is meaningful to say that on average the customers buy products 2.6 times per year in our on-line shop. But it does not make sense that the average *general product category* we sell is 2.6, which we might obtain when we simply compute the mean value of the products we have sold based on the numerical coding of the general product categories.

In contrast to discrete numerical attributes, a continuous attribute can—at least theoretically—assume any real value. However, such numerical values will always be measured and represented with limited precision. It should be taken into account how precise these values are. Drastic round-off errors or truncations can lead to problems in later steps of the analysis. Suppose, for instance, that a cluster analysis is to be carried out later on and that there is one numerical attribute, say X, that is truncated to only one digit right after the decimal point, while all other numerical attributes were measured and stored with a higher precision. When comparing different records, such truncation for the attribute X influences their perceived similarity and might be a dominating factor for the further analysis only for this reason. Truncation errors and measurements with limited precision should be distinguished from values corrupted with noise. The problem of noise will be tackled in the context of data quality in Sect. 4.2 and will also be discussed in more detail in Chap. 5.

Numerical attributes can have an *interval*, a *ratio*, or an *absolute scale*. For an interval scale, the definition of what zero means is more or less arbitrary. The date is a typical example for an attribute measured on an interval scale. There are calendars with different definitions of the time point zero. For instance, the Unix standard time, counted in milliseconds, has its time point zero in the year 1970 of the Gregorian calendar. The same applies to temperature scales like Fahrenheit and Celsius degrees, where zero refers to different temperatures. Certain operations like quotients are not meaningful for interval scales. For example, it does not make sense to say that a temperature of 21°C is three times as warm as 7°C.[1]

In contrast to this, a **ratio scale** has a canonical zero value and thus allows us to compute meaningful ratios. Examples of ratio scales are height, distance, or duration. Distance can be measured in different units like meters, kilometers, or miles.

---

[1] Such a statement may make sense, though, for the Kelvin temperature scale, because on this scale the temperature is directly proportional to the average kinetic energy of the particles—and it is meaningful to compute ratios of energies.

But no matter which unit we choose, a distance of zero will always have the same meaning. Especially ratios, which do not make sense for interval scales, are often useful for ratio scales: the quotient of distances is independent of the measurement unit, so that the distance 20 km is always twice as long as the distance 10 km, even if we change the unit kilometers to meters or miles. Whereas for a ratio scale, only the value zero has a canonical meaning and the meaning of other values depends on the choice of the measurement unit, for an **absolute scale**, there is a unique measurement unit. A typical example for an absolute scale is any kind of counting procedure.

4.2 Data Quality

The saying "garbage in, garbage out" applies to data analysis as to any other area. The results of an analysis cannot be better than the quality of the data, so that we should be concerned about the data quality before we carry out any deeper analysis with the data. **Data quality** refers to how well the data fit to their intended use. There are various data quality dimensions.

Accuracy is defined as the closeness between the value in the data and the true value. For numerical attributes, accuracy means how exact the value in the data set is compared to the true value. Noise or limited precision in measurements can lead to reduced accuracy for numerical attributes. Limited precision is often obvious from the data set. For example, in the Iris data set all numerical values are measured with only one digit right after the decimal point. The magnitude of noise can be estimated when measurements for the same value have been taken repeatedly. Accuracy of numerical values can also be affected by wrong or erroneous measurements or simply by errors like transposition of digits when measurements are recorded manually. For categorical attributes, problems with accuracy can result from misspellings like "fmale" for a value of the attribute *gender*, and also from erroneous entries.

We distinguish between *syntactic* and *semantic accuracy*. **Syntactic accuracy** means that a considered value might not be correct, but it belongs at least to the domain of the corresponding attribute. For a categorical attribute like *gender* for which only the values *female* and *male* are admitted, "fmale" violates syntactic accuracy. For numerical attributes, syntactic accuracy does not only mean that the value must be a number and not a string or text. Also certain numerical values can be out of the range of syntactic accuracy. Attributes like *weight* or *duration* will admit only positive values, and therefore negative values would violate syntactic accuracy. Other numerical attributes have an interval as their range like [0, 100] for the percentage of votes for a candidate. Negative values and values larger than 100 should not occur. For integer-valued attributes like the number of items a customer has bought, floating-point values should be excluded.

Problems with **semantic accuracy** mean that a value might be in the domain of the corresponding attribute, but it is not correct. When the attribute *gender* has the value *female* for the customer John Smith, then this is not a question of syntactic accuracy, since *female* is a possible value of the attribute *gender*. But it is obviously a wrong value for a person named "John".[2]

Discovering problems of syntactic accuracy in a data set is a relatively easy task. Once we know the domains of the attributes, we can easily verify, whether the values lie in the corresponding domains or not. A simple measure for syntactic accuracy is the fraction of values that lies in the domains of their corresponding attribute.

The verification of semantic accuracy is much more difficult or often even impossible. Another source for the same data would enable us to check our data, and differences not caused by problems with syntactic accuracy indicate problems with semantic accuracy. Sometimes also certain "business rules" are known for the data. For instance, if we find a record in our data set with the value *male* for the attribute *gender* and *yes* for the attribute *pregnant*, there must be a problem of semantic accuracy based on the known "business rule" that only women can be pregnant.

Whether or in which detail to check syntactic and semantic accuracy depends very much on how the data were generated. Especially, when data were entered manually, there is a higher chance for accuracy problems. In any case, it is recommended to carry out at least some simple tests to see whether there might be problems with accuracy. However, the usual practice is to keep these tests at a minimum and to find out later on that there are problems with accuracy, namely when the data analysis yields implausible results.

Throughout this book we normally assume that the data are already given, for example, as a database table. This is not the best point in time to cope with data quality problems. Chances of avoiding or reducing data quality problems are highest when the data are entered into the database. For instance, instead of letting a user type in the value of categorical attribute with the danger of misspellings, one could provide a fixed selection of values from which the user can choose.

Another dimension of data quality is **completeness** which can be divided into completeness with respect to attribute values and completeness with respect to records. Completeness with respect to attribute values refers to missing values (which will be discussed in Sect. 4.6). When missing values are explicitly marked as such, then a simple measure for this dimension of data quality is the fraction of missing values among all values. But we will see that missing values are not always directly recognizable, so that the fraction of known missing values might only provide a lower bound for the fraction of actually missing values.

Completeness with respect to records means that the data set contains the necessary information that is required for the analysis. Some records might simply be missing for some technical reasons. Data might have been lost because a few years ago the underlying database system was changed and only those data records were transferred to the new database that were considered to be important at that point in time. In a customer database, customers who had not bought anything for a longer time might not have been transferred (in order to eliminate potential

---

[2]Note, however, that the problem may also reside with the name. Maybe the name of the person was misspelled, and the correct name is "Joan Smith"—then the gender is actually *female*.

zombie customers) to the new database, or older transactions were not stored anymore.

Very often the available dataset itself is biased and not representative. Consider as an example a bank that provides mortgages to private customers. If the aim of the analysis is to predict for future applicants of loans whether they will return the loan, we must take into account that the sample is biased in the sense that we only have information about those customers who have been granted a loan. For those customers who have been denied the loan initially, we have no information whether they would have returned the loan or not. But especially these customers might be the ones for which it is interesting to find a good scheme to predict the risk. For customers with high income and a safe job and no current debt, we need no sophisticated data analysis techniques to predict that there is a good chance that they will return the loan. Of course, it is impossible to obtain a representative sample in the statistical sense in this case. Such a sample would mean that we would have to provide loans to any customer, no matter how bad their financial status is, for a certain period and collect and evaluate these data. Unfortunately, this would be a method entailing almost guaranteed bankruptcy.

The same problems occur in many other areas. For a production plant, we usually have large amounts of data when it is running in a normal mode. For exceptional situations, we will have little or no data. We cannot ask for such data, for instance, by requiring to check what happens if, say, a nuclear plant operates at its limit.

In such cases we might encounter future situations for which we had no corresponding data in our sample. Such possible gaps in the data should be identified. One should be aware that the space of possible values is automatically covered sparsely by the data when we have a larger number of attributes. Consider a set of m numerical attributes, and we want to make sure that we have at least positive and negative values for each attribute in our data set. This does not require a large data set. But if we want to make sure that we have data for all combinations of positive and negative values for the considered attributes, this leads to 2^m possible combinations. If we have $m = 20$ attributes, we have already more than one million possible combinations of positive and negative values. Therefore, if we have a data set with one million records, we have on average one sample for each of these combinations. For a data set with 100,000 records, at least 90% of the combinations will not be covered.

Other problems can be caused by **unbalanced data**. As an example, consider a production line for goods for which an automatic quality control is to be installed. Based on suitable measurements, a classifier is to be constructed that sorts out parts with flaws or faults. The scrap rate in production is usually very small, so that our data might contain far less than 1% examples for parts with flaws or faults.

Timeliness refers to whether the available data are too old to provide up to date information or cannot be considered as representative for predictions of future data. Timeliness is often a problem in dynamically changing domains, where only recently collected data provide relevant information, while older data can be misleading and can indicate trends that have vanished or even reversed.

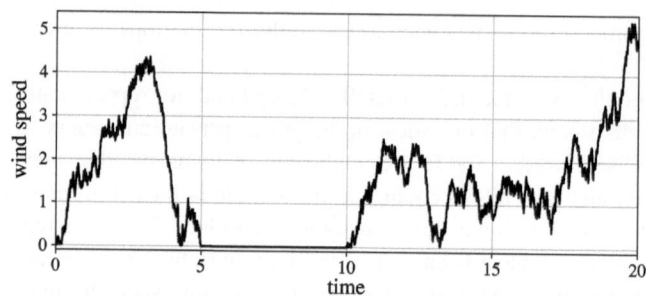

Fig. 4.1 Measured wind speeds with a period of a jammed sensor

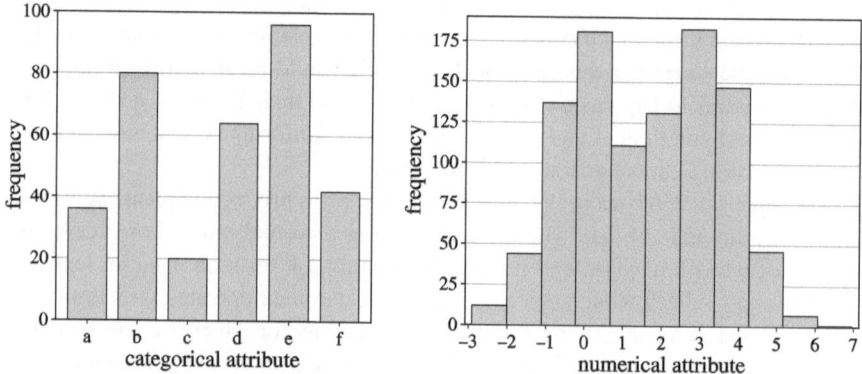

Fig. 4.2 A bar chart (categorical attribute, *left*) and a histogram (numerical attribute, *right*)

4.3 Data Visualization

According to Tukey [24], "there is no excuse for failing to plot and look" when one wants to handle a data analysis problem. The right plots of the data can provide valuable information as the simple time series plot in Fig. 4.1 shows, which enables us to discover zero values that are actually missing values. There are infinitely many ways to plot data, and it is not always easy to find the best ways of plotting a given data set. Nevertheless, there are some very useful standard data visualization techniques that will be discussed in the following.

4.3.1 Methods for One and Two Attributes

A **bar chart** is a simple way to depict the frequencies of the values of a categorical attribute. A simple example for a categorical attribute with six values a, b, c, d, e, and f is shown on the left in Fig. 4.2.

A **histogram** shows the frequency distribution for a numerical attribute. To this end, the range of the numerical attribute is discretized into a fixed number of inter-

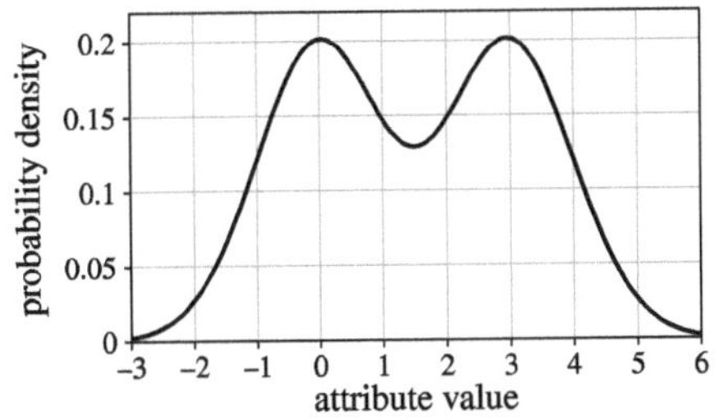

Fig. 4.3 Probability density function of a sample distribution, from which the data for the histogram in Fig. 4.2 was sampled

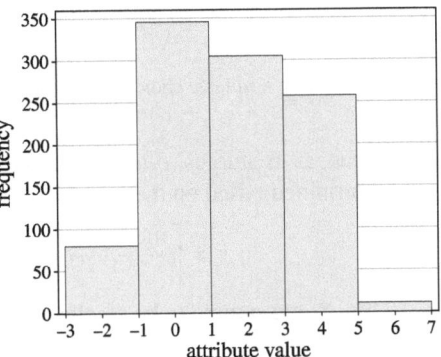

Fig. 4.4 A histogram with too few bins, for the same data set that was depicted in Fig. 4.2 on the right. As a consequence, the distribution looks unimodal but skewed

vals (called *bins*), usually of equal length. For each interval, the (absolute) frequency of values falling into it is indicated by the height of a bar as shown on the right in Fig. 4.2. From this histogram it can be read, for example, that a little bit more than 100 values lie in the vicinity of 1.

The histogram in Fig. 4.2 resulted from a sample of size 1000 from a mixture of two normal distributions with means 0 and 3, respectively, having both a standard deviation of 1. The density of this distribution is shown in Fig. 4.3.

But how should we choose the number of bins, and how much does this choice influence the result? Figure 4.4 shows a histogram with only five bins for the same data set underlying the histogram shown in Fig. 4.2. With only five bins, the two peaks of the original distribution are no longer visible, and one gets the wrong impression that the distribution is unimodal but skewed.

There is no generally best choice for the number of bins, but there are certain recommendations. **Sturges' rule** [22] proposes to choose the number k of bins according to the following formula:

$$k = \lceil \log_2(n) + 1 \rceil, \tag{4.1}$$

where n is the sample size. Although Sturges' rule is still very often used as a default in various statistics software packages, it is tailored to data from normal distributions and data sets of moderate size [21]. The number of bins of the histogram in Fig. 4.2 has been computed based on Sturges' rule. The size of the data set is $n = 1000$.

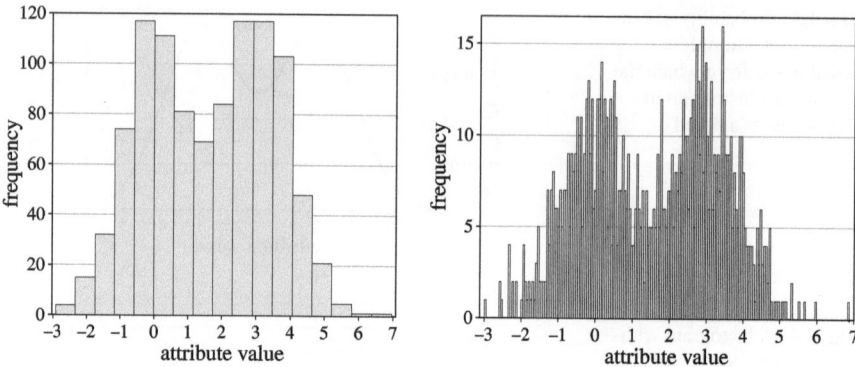

Fig. 4.5 Histograms with a suitable choice for the number of bins (*left*) and too many bins (*right*)

Assuming that, as in Sturges' rule, the bins have equal length, the number of bins can also be determined based on the length h of each bin:

$$k = \left\lceil \frac{\max_i\{x_i\} - \min_i\{x_i\}}{h} \right\rceil, \tag{4.2}$$

where $x_1, \ldots, x_n$ is the sample to be displayed. Reasonable values for h are [20]

$$h = \frac{3.5 \cdot s}{n^{\frac{1}{3}}}, \tag{4.3}$$

where s is the sample standard deviation, and [8]

$$h = \frac{2 \cdot \mathrm{IQR}(x)}{n^{\frac{1}{3}}}, \tag{4.4}$$

where $\mathrm{IQR}(x)$ is the interquartile range of the sample, that is, the length of the interval which covers the middle 50% of the data.

For the data set with the histogram displayed in Fig. 4.2, (4.3) yields $k = 16$, and (4.4) leads to $k = 17$. A histogram for the second choice (that is, for $k = 17$) is shown on the left in Fig. 4.5.

As we have seen in Fig. 4.4, the histogram can be misleading when the number of bins is chosen too small. Choosing the number of bins too high usually leads to a very scattered histogram in which it is difficult to distinguish true peaks from random peaks. An example is shown on the right in Fig. 4.5, where $k = 200$ was chosen for the number of bins for the same data underlying Fig. 4.2.

All of these methods (that is, (4.2), (4.3), and (4.4) for determining the number of bins or the length of the bins) are highly sensitive to outliers, since they divide the range between the smallest and the largest value of the sample into bins of equal size. A single outlier can make this range extremely large, so that for a smaller number of bins, the bins themselves become very large, and for a larger number of bins, most of the bins can be empty. To avoid this problem, one can either leave out extreme values from the sample (for instance, the 3% smallest and the 3% largest values) for calculating and displaying the histogram, or one can deviate from the principle of bins of equal length.

Fig. 4.6 Two boxplots for a sample from a standard normal distribution

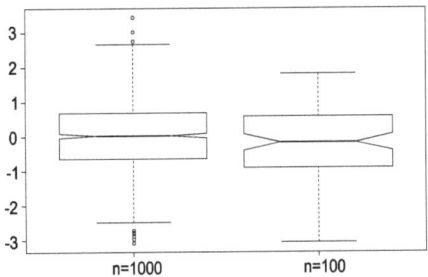

Boxplots are a very compact way to visualize and summarize main characteristics of a sample from a numerical attribute. Figure 4.6 shows two boxplots from samples from a standard normal distribution with mean 0 and variance 1. The left boxplot is based on sample of size $n = 1000$, whereas a sample of size $n = 100$ was used for the right boxplot.

The line in the middle of a boxplot indicates the sample median. The notch in the box is not always shown. It indicates a 95% confidence interval for the median. The box itself corresponds to the interquartile range covering the middle 50% of the data. The whiskers are drawn in the following way. The maximum length of each whisker is 1.5 times the length of the interquartile range. But if there is no data point at the maximum length of a whisker, the corresponding whisker is shortened until it reaches the next data point. Data points lying outside the whiskers are considered as outliers and are indicated in the form of small circles.

Comparing the two boxplots in Fig. 4.6, we can observe the following:

- Although both boxplots come from samples from the same normal distribution, they look different, since they are based on different samples.
- The notch of the left boxplot, representing a 95% confidence interval for the median, is much smaller than the notch of the right boxplot because of the larger sample size for the left boxplot.
- Theoretically, the whiskers for a sample from a symmetric distribution like the normal distribution should have roughly the same length. For the boxplot based on the smaller sample size, we can see that whiskers differ significantly in length, since—by chance—the largest value among the sample of 100 values was not greater than 2, whereas the smallest value was smaller than −3.
- In contrast to the boxplot on the left-hand side, the right boxplot does not contain any outliers. This is again due to the smaller sample size. The theoretical length of the interquartile range for a standard normal distribution is 1.349. Therefore, the probability of a point lying outside the (theoretical) range $[−2.698, 2.698]$ of the whiskers is almost 0.7%. Therefore, for a sample from a normal distribution of size $n = 1000$, we can expect roughly 7 outliers on average in a boxplot and less than one for a sample of size $n = 100$.

The boxplots of asymmetric distributions look completely different. If we sample from an exponential distribution, whose probability density function is shown in Fig. 4.7, we obtain boxplots as they are shown in Fig. 4.8. The boxplots on the left and right represent samples of sizes $n = 1000$ and $n = 100$, respectively.

Fig. 4.7 The probability density function of the exponential distribution with $\lambda = 1$

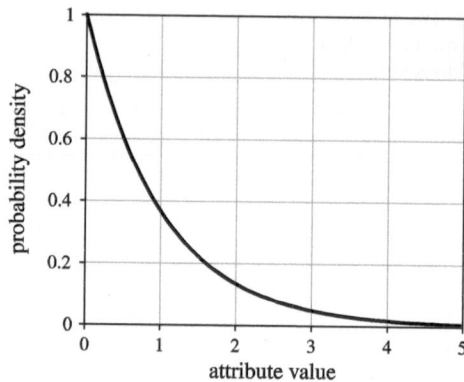

Fig. 4.8 Two boxplots for a sample from an exponential distribution

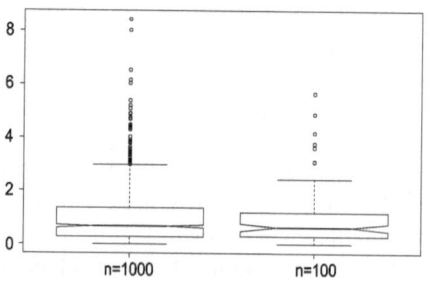

Barcharts, histograms, and boxplots are visualizations for single attributes. In most cases, we have to deal with a number of attributes, and we are not only interested in the characteristics of single attributes but also in the relations and dependencies between the attributes. However, the display for visualizing the data is two-dimensional, and even if we use 3D-techniques from computer graphics, we cannot directly display more than two or three variables at the same time in a simple coordinate system, unless we use additional features such as symbols, color and size. **Scatter plots** refer to displays where two attributes are plotted against each other. The two axes of the coordinate system represent the two considered attributes, and each instance in the data set is represented by a point, a circle, or any other symbol.

Simple scatter plots are not suited for larger data sets. For a data set with one million objects and a window size of 500×500 pixels, we would have on average four data objects per pixel. For larger data sets, many points or symbols in the scatter plot will be plotted at the same position, and we cannot distinguish whether a point in the scatter plot represents one or 100 objects. In the worst case, a scatter plot for a larger data set might simply look like the one in Fig. 4.9, providing only information about the range of the data, but no hint concerning the distribution of the data.

This can be amended by using **density plots** or plots based on **binning**. Using semitransparent points for plotting the data is one way to generate a density plot. Each plotted point is semitransparent, and the more points are plotted at the same place, the less transparent the image will become in this place. Binning was already used to generate histograms, and the principle is used for the scatter plots. The two-

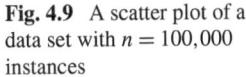

Fig. 4.9 A scatter plot of a data set with $n = 100,000$ instances

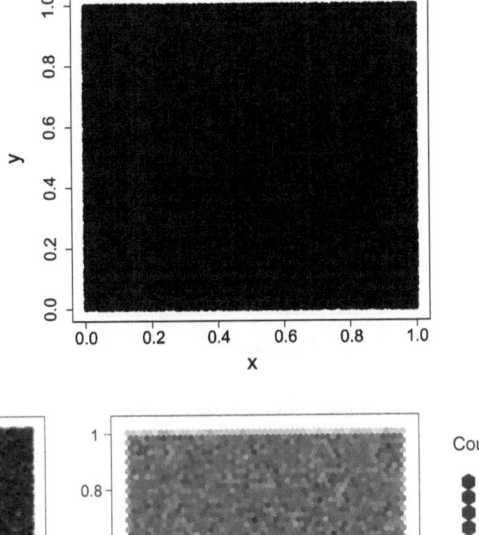

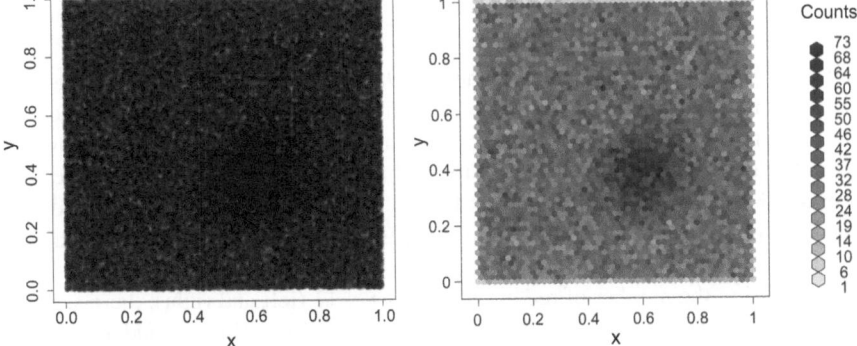

Fig. 4.10 Density plot (*left*) and a plot based on hexagonal binning (*right*) for the same data set as shown in Fig. 4.9

dimensional domain of the data for the scatter plot is partitioned into bins of the same size. Possible forms for the bins are rectangles or hexagons. The intensity of the color for the bin is chosen proportional to the number of data objects falling into the bin. Figure 4.10 shows a density plot on the left and a plot based on a hexagonal binning on the right for the same data set displayed in Fig. 4.9. Both plots indicate a higher density of the data around the point $(0.6, 0.4)$, which cannot be seen in the simple scatter plot in Fig. 4.9.

Scatter plots can be enriched with further information in order to involve more attributes. Different plot symbols or colors can be used for plotting the points in order to include information about a categorical attribute. Color intensity and the size of the symbols are possible means to indicate the value of additional numerical attributes.

Figure 4.11 shows two scatter plots of the Iris data set—one displaying the sepal length versus the sepal width and the other one the petal length versus the petal width—in which different species are displayed by different colors. Both plots show that the red circles, representing the species Iris setosa, can be well distinguished from the other two species Iris versicolor and Iris virginica displayed as triangles

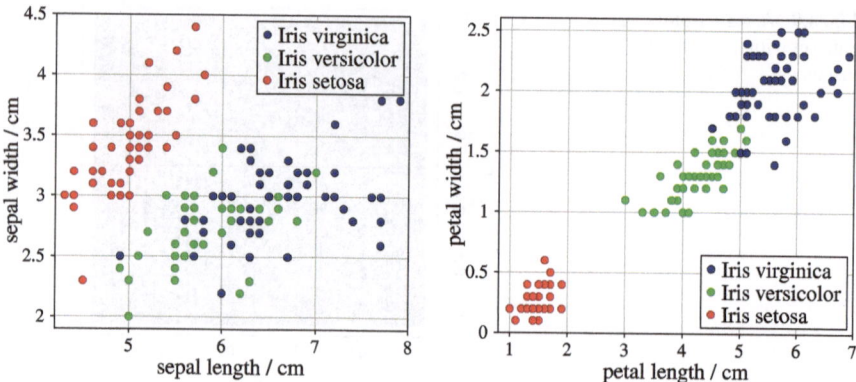

Fig. 4.11 Scatter plots of the iris data set for sepal length vs. sepal width (*left*) and for petal length vs. petal width (*right*). All quantities are measured in centimeters (cm)

and crosses, respectively. However, the left chart in Fig. 4.11 gives the impression that Iris virginica and Iris versicolor are very difficult to distinguish, at least when we only take the sepal length and the sepal width into account. But when we consider the petal length and the petal width (right chart in Fig. 4.11), we can still see the overlap of the corresponding symbols for the species, but there is a clear tendency that Iris virginica tends to larger values than Iris versicolor for the petal length and width.

Comparing the number of red circles in Figs. 4.11 (left and right), there seem to be less red circles on the right. But how can some of the objects suddenly vanish in the scatter plot? When we count the number of red circles, we see that in both scatter plots there are less than 50, although the data set contains 50 instances of Iris setosa that should be displayed by red circles. The circles are not missing in the scatter plots. Some circles are simply plotted at exactly the same position, since their measured sepal length and width or their measured petal length and width coincide. Recall that these values were only measured with a precision of just one digit right after the decimal point. To avoid this impression of seeing less objects than there actually are, one can add **jitter** to the scatter plot. Instead of plotting the symbols exactly at the coordinates specified by the values in the data set, we add a small random value to each original value in the data table. The left chart in Fig. 4.12 shows the resulting scatter plot with jitter where we have added random values from a uniform distribution on the interval $[-0.04, 0.04]$ to the original values. This ensures that a point originally lying left or below another point will always remain left or below the other point, even when the jitter is added.

Jitter is essential when categorical attributes are used for the coordinate axes of a scatter plots, since categorical attributes have only a limited number of possible values, so that plotting of objects at exactly the same position occurs very often when no jitter is added.

From a scatter plot we can already extract important information. Consider again the scatter plot displayed in Fig. 4.12. We can see that the petal length and width are correlated. Objects with larger values for the petal length also tend to have larger

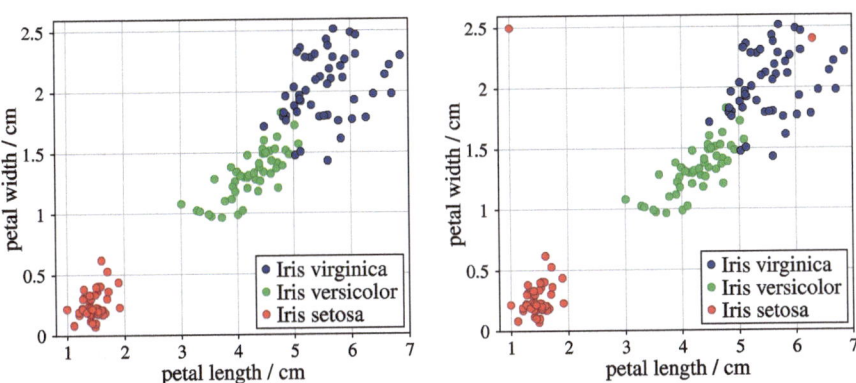

Fig. 4.12 The same scatter plot as in Fig. 4.11 on the right, but with jitter (*left*) and with jitter and two outliers (*right*; the outliers are the *red points* in the top left and top right corners)

values for the petal width. The scatter plot also shows that Iris setosa—the red circles in the scatter plot—can be easily distinguished from the other two species just on the basis of the petal length or width. The scatter plot does not indicate that the other two species cannot be separated clearly. It only shows that, solely based on the petal length and width, it is not possible to distinguish the two species perfectly. Outliers can also be discovered in scatter plots. The left chart in Fig. 4.12 does not have any outliers. In the right chart, however, we have added two artificial outliers. The data point in the upper left corner is a clear outlier with respect to the whole data set. Note that the values for the attributes petal length and width are both in the general range of the corresponding attributes in the data set. But there is no other object in the data set with a similar combination of these attribute values. The second outlier in the right chart of Fig. 4.12—the circle in the upper right corner—is not an outlier with respect to the values for the petal length and width or their combination. However, it is an outlier for the class Iris setosa displayed by red circles. Whenever such outliers are discovered, one should check the data or the data generating process again to ensure that the outliers are not due to erroneous data.

It should be noted that the scatter plots—like all other visualization techniques—are very useful tools to discover simple structures and patterns or peculiar deviations like outliers in a data set. But there is no guarantee that a scatter plot or any visualization technique will automatically show all or even any interesting or deviating pattern in the data set. A scatter plot with no outliers does not mean that there are no outliers in the data set. It only means that there are no outliers with respect to the combination of the attributes displayed in the scatter plot. In this sense, visualization techniques are like test cases for computer programs. Test cases can discover errors in a program. But if the test cases have not indicated any errors, this does not imply that the program does not have any bugs. In the same way, a visualization technique might give hints to certain interesting patterns in the data set. But if one cannot see any interesting patterns in a visualization, it does not mean that there are no patterns in the data set.

4.3.2 Methods for Higher-Dimensional Data

In this section, we restrict our considerations to numerical attributes. Scatter plots are projections of the data set to a two-dimensional plane where the two dimensions of the projection plane correspond to two selected attributes. The data set can be considered as a subset of a space that has as many dimensions as the data set has attributes. How can such a higher-dimensional data set be represented in two or three dimensions? The idea is to preserve as much of the "structure" of the higher-dimensional data in the lower-dimensional representation. But what does it mean to preserve the structure? There is no unique answer to this question, and various approaches have been proposed to solve this problem. A very efficient method is principal component analysis that will be introduced in the next section.

4.3.2.1 Principal Component Analysis

Principal component analysis (PCA), which is also briefly described in the appendix on statistics, is a method from statistics to find a projection to a plane—or more generally, to a linear subspace—which preserves as much as possible of the original variance in the data. In order to restrict the search for the best projection plane to planes through the origin of the coordinate system, the data are first centered around the origin by subtracting the mean value for each attribute from the attribute values. In this way, the projection to the plane can be represented by a matrix M mapping the data points $\mathbf{x} \in \mathbb{R}^m$ to the plane by

$$\mathbf{y} = M \cdot (\mathbf{x} - \bar{\mathbf{x}}), \tag{4.5}$$

where $\bar{\mathbf{x}}$ denotes the (empirical) mean value (or vector of (empirical) mean values)

$$\bar{\mathbf{x}} = \frac{1}{n} \sum_{i=1}^{n} \mathbf{x}_i.$$

M is a $2 \times m$ matrix when PCA is used for visualizing the data on a plane. As we will see later on, PCA can be understood as a more general dimension reduction technique. Therefore, we do not restrict our considerations to projections to a plane of dimension 2 but to any linear subspace of dimension $q \le m$. In this case, M is a $q \times m$ matrix. It should be noted that arbitrary matrices are not admitted for the projection, since only projections, but no scalings, are allowed. Scalings would enable the overall variance of the projection to be changed, which is not desired.

The solution of the optimization problem of finding the projection plane such that the variance of the projected data is maximized leads to an eigenvalue problem. The projection matrix M for PCA is given by

$$M = (\mathbf{v_1}, \dots, \mathbf{v_q}),$$

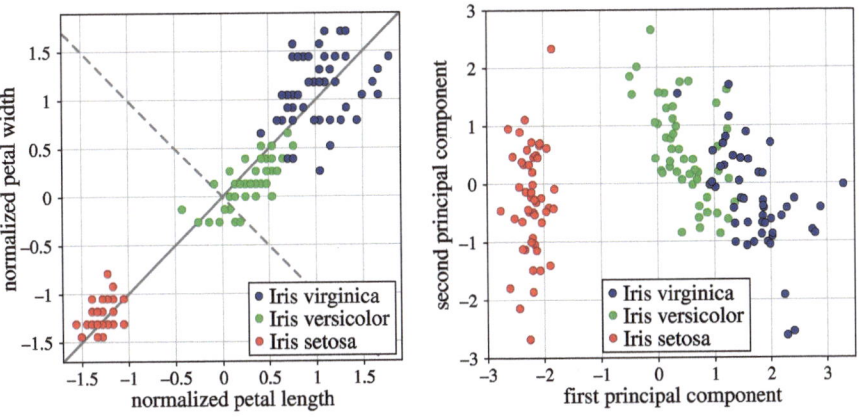

Fig. 4.13 *Left*: the first (*solid line*) and the second principal component (*dashed line*) of an example data set (Iris data). *Right*: the example data set projected to the space that is spanned by the first and second principal components (resulting from a PCA involving all four attributes)

where $\mathbf{v_1}, \ldots, \mathbf{v_q}$ are the normalized eigenvectors[3] of the **covariance matrix** of the data

$$C = \frac{1}{n-1} \sum_{i=1}^{n} (\mathbf{x_i} - \bar{\mathbf{x}})(\mathbf{x_i} - \bar{\mathbf{x}})^{\top}$$

for the q largest eigenvalues $\lambda_1 \geq \cdots \geq \lambda_q$.

The vectors $\mathbf{v_1}, \ldots, \mathbf{v_q}$ are called **principal components**. Figure 4.13 illustrates the concept of principal components on a two-dimensional example. Of course, for such an example, a dimension reduction is not necessary. There are two principal components. The vector corresponding to the largest eigenvalue is the first component, and its direction is indicated by the solid line. The dashed line represents the direction of the second component, corresponding to the second largest eigenvalue, which in this two-dimensional case is the smallest eigenvalue. It is obvious that the projection of the data set to the first principal component preserves a large fraction of the original variance, whereas a projection to the second principal component only would lead to a significant loss of the original variance.

The left chart in Fig. 4.13 is a plot of the petal length and width of the Iris data set. However, apart from the necessary centering of the data around the origin by subtracting the mean, the data have been z-score standardized by the transformation

$$x \mapsto \frac{x - \hat{\mu}_X}{\hat{\sigma}_X}, \tag{4.6}$$

where $\hat{\mu}_X$ and $\hat{\sigma}_X$ are the mean value and empirical standard deviation of attribute X (see also Sect. 6.3.2). Without standardization, the result of PCA would depend on

---

[3]λ is called an eigenvalue of a matrix A if there is a nonzero vector $\mathbf{v}$ such that $A\mathbf{v} = \lambda\mathbf{v}$. The vector $\mathbf{v}$ is called an eigenvector to the eigenvalue λ.

Table 4.1 Preservation of the variance of the Iris data set depending on the number of principal components

	Principal component			
	PC1	PC2	PC3	PC4
Proportion of variance	0.73	0.229	0.0367	0.00518
Cumulative proportion	0.73	0.958	0.9948	1.00000

the scaling of the attributes. If no standardization is carried out, the attribute with the largest variance can easily dominate the first principal component. For the example in Fig. 4.13 with z-score standardization, the first principal component is the vector $(\sqrt{2}/2, \sqrt{2}/2)$. If the petal length is measured in meters instead of centimeters, but the petal width is still measured in centimeters, the first principal component without z-score standardization becomes the vector $(0.0223, 0.9998)$, since the variance of the petal length has been decreased drastically by the scaling factor 0.01 resulting from the change from centimeters to meters, so that more or less only the petal width contributes to the variance in the data.

PCA can be used for visualization purposes by restricting to the first two principal components. More generally, PCA can carry out a dimension reduction to any lower-dimensional space; even more, PCA also provides information about over how many dimensions the data set actually spreads. This information can be extracted from the eigenvalues $\lambda_1 \geq \cdots \geq \lambda_m$ of the covariance matrix. When we project the data to the first q principal components $v_1, \ldots, v_q$ corresponding to the eigenvalues $\lambda_1, \ldots, \lambda_q$, this projection will preserve a fraction of

$$\frac{\lambda_1 + \cdots + \lambda_q}{\lambda_1 + \cdots + \lambda_m} \tag{4.7}$$

of the variance of the original data. Table 4.1 shows the corresponding result of PCA applied to the Iris data set without the categorical attribute for the species. A projection of this four-dimensional data set to the first principal component, i.e., to only one dimension, covers already 73% of the variance of the original data set. A projection to a plane defined by the first two principal components covers already 95.8% of the variance. This means that the four numerical attributes of the Iris data set are not located on a two-dimensional plane in the four-dimensional space but do not deviate too much from the plane defined by the first two principal components. The right chart in Fig. 4.13 shows the projection of the Iris data set to the first two principal components where PCA was carried out after the z-score standardization had been applied.

The importance of the scaling effect carried out by the z-score standardization can also be observed by revisiting the example where we had considered only the petal length and width of the Iris data set. We had applied PCA to the original data and the data where we had changed the measurement of the petal length from centimeters to meters without scaling, resulting in the vector $(0.7071, 0.7071)$ as the first principal component for the original data and the vector $(0.0223, 0.9998)$ for the modified data. The variance preserved by the projection to the first principal component is 98.1% in the first case and 99.996% in the second case. In the latter case, the first principal component corresponds more or less to the petal width,

Fig. 4.14 A data set distributed over a cube in a chessboard-like pattern. The colors are only meant to make the different cubes more easily discernible. They do *not* indicate classes. Note the outlier in the upper left corner

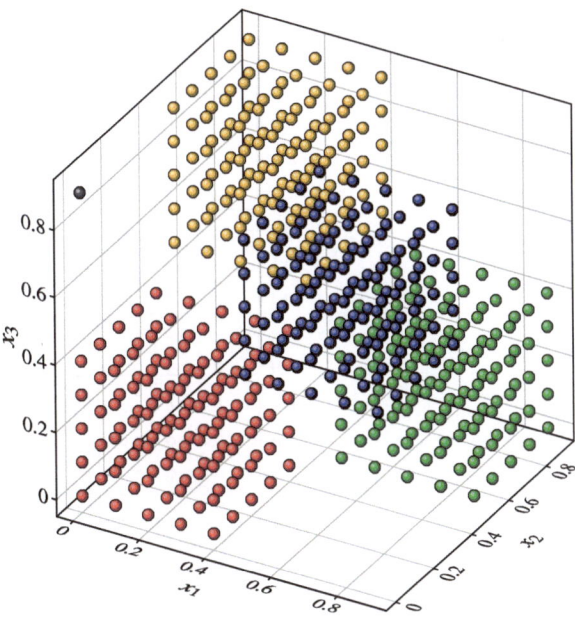

since the variance of the petal length measured in meters is almost negligible, and therefore the projection can preserve close to 100% of the original variance.

In order to illustrate the advantages and limitations of PCA for visualization purposes, we consider an artificial three-dimensional data set illustrated in Fig. 4.14. The data fill the unit cube in chessboard-like manner. When the unit cube is divided into eight subcubes, these subcubes are alternatingly empty and filled with data. There is also one outlier close to the upper left corner of the surface in the front of the cube. The scatter plots resulting from projections to two axes of the coordinate system are shown in Fig. 4.15. All these scatter plots give the wrong impression that the data are uniformly distributed over a grid in the data space. The scatter plots provide neither a hint to the chessboard pattern in the three-dimensional data space nor to the single outlier.

Figure 4.16 shows the projection to the first two principal components of the data set after a z-score standardization has been carried out. The outlier can now be identified easily. From Fig. 4.16 it is also obvious that the data cannot be distributed uniformly over the original three-dimensional space but that there must be some inherent pattern. Of course, it is impossible to recover the three-dimensional chessboard pattern in a two-dimensional projection completely.

4.3.2.2 Projection Pursuit

PCA has the advantage that the best projection with respect to the given criterion—the preservation of the variance—can be computed directly based on the eigenvectors of the covariance matrix. **Projection pursuit** [10] takes a different approach.

Fig. 4.15 The scatter plots
for the data set shown in
Fig. 4.14

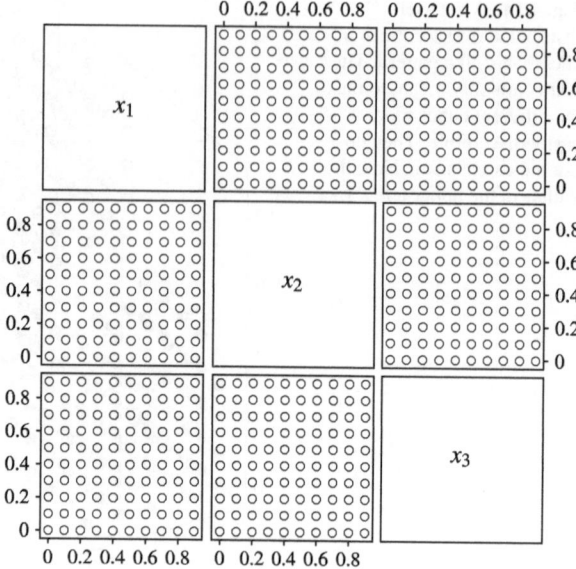

Fig. 4.16 The projection to
the first two principal
components of the data set
shown in Fig. 4.14

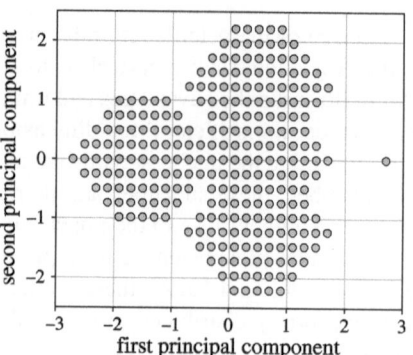

The projection of the data should show interesting aspects of the data. But what does interesting mean? Normally, for projection pursuit, interestingness of a projection is defined as the deviation from a normal distribution, according to the observation that most of the projections of high-dimensional data will resemble a normal distribution [6]. The more the projected data deviate from a normal distribution, the more interesting is the projection. Various criteria [5, 9–12] can be defined to measure how much a projection deviates from a normal distribution.

In contrast to PCA, there is no way to find the most interesting projections with respect to the deviation from a normal distribution. Projection pursuit simply generates random projections and chooses the ones yielding the best values for the measures of interestingness.

4.3.2.3 Multidimensional Scaling

PCA is a method for dimension reduction that is based on two assumptions:

- The representation of the data in a lower-dimensional space should be obtained by a projection to a linear subspace.
- The criterion to evaluate the representation of the data in the lower-dimensional space is the preservation of the variance.

Projection pursuit shares the first assumption of PCA but replaces the second one by the deviation of the projection from the normal distribution. **Multidimensional scaling** (**MDS**) is also a dimension-reduction technique but differs in both assumptions from PCA and projection pursuit.

- MDS is not restricted to mappings in the form of simple projections. In contrast to PCA, MDS does not even construct an explicit mapping from the high-dimensional space to the low-dimensional space. It only positions the data points in the low-dimensional space.
- The representation of the data in the low-dimensional space constructed by MDS aims at preserving the distances between the data points and not the variance in the data set, for which PCA was designed.

For a data set with n data objects, MDS requires a distance matrix $[d_{ij}^{(X)}]_{1 \le i, j \le n}$, where $d_{ij}^{(X)}$ is the distance between data object i and data object j. MDS assumes that the distance is symmetric, i.e., $d_{ij}^{(X)} = d_{ji}^{(X)}$ for all $i, j \in \{1, \ldots, n\}$. The distances $d_{ii}^{(X)}$ must be zero, so that each data object has no distance to itself. MDS does not require the data, but only the distance matrix. Usually, the distance matrix is derived from the data by computing the Euclidean distances between the data objects (see Sect. 7.2 for more details on distance measures).

It is important to note that it is recommended to carry out a normalization of the data first, for instance, by applying a z-score standardization. Without such a normalization, MDS has to cope with the same problems as PCA. The variance and the difference between values of an attribute highly depend on the scaling or the measurement unit of the attribute. In the same way as a single attribute or few attributes might dominate the variance, just because the corresponding measurement units tend to larger values, these attributes will contribute more to the Euclidean distance than attributes with very small values and ranges.

Since MDS is mainly used for visualization purposes, the original high-dimensional data should be represented by points in two or sometimes also in three dimensions. Each data object should be represented by a point in the low-dimensional space, usually in $\mathbb{R}^2$ or $\mathbb{R}^3$. Let us restrict our considerations for the moment to a two-dimensional visualization. Then MDS must define a point $\mathbf{p}_i = (p_x^{(i)}, p_y^{(i)})$ to be plotted in the plane for each data object x_i. The plot of these points should preserve the original distances $d_{ij}^{(X)}$ between the data points in the original high-dimensional space. Of course, it is usually impossible to preserve the distances exactly in the two-dimensional representation. Therefore, we require that the distances $d_{ij}^{(Y)} = \|\mathbf{p_i} - \mathbf{p_j}\|$ between the points in the two-dimensional representation should deviate as little as possible from the original distances $d_{ij}^{(X)}$. One way to measure this deviation is the sum of the squared errors between the original distances and the distances in the two-dimensional representation.

$$E_0 = \sum_{i=1}^{n} \sum_{j=i+1}^{n} \left(d_{ij}^{(Y)} - d_{ij}^{(X)} \right)^2. \tag{4.8}$$

This equation does not refer explicitly to the two-dimensional representation. The distances $d_{ij}^{(Y)}$ in the lower-dimensional space can be derived from points in $\mathbb{R}^2$, but the distances could also be computed for points in $\mathbb{R}^q$ for any $q \in \mathbb{N}$.

The sum of squared errors depends on the number of data objects and on the values for the original distances. For more data points and for larger distances $d_{ij}^{(X)}$, E_0 will tend to become larger as well. In order to obtain an error measure independent of these effects, E_0 is often normalized to

$$E_1 = \frac{1}{\sum_{i=1}^{n} \sum_{j=i+1}^{n} (d_{ij}^{(X)})^2} \sum_{i=1}^{n} \sum_{j=i+1}^{n} \left(d_{ij}^{(Y)} - d_{ij}^{(X)} \right)^2. \tag{4.9}$$

The factor $\frac{1}{\sum_{i=1}^{n} \sum_{j=i+1}^{n} (d_{ij}^{(X)})^2}$ is independent of the distances $d_{ij}^{(Y)}$ and therefore independent of the lower-dimensional representation of the data. It does not affect the result of the minimization of E_0 or E_1.

The relative error

$$E_2 = \sum_{i=1}^{n} \sum_{j=i+1}^{n} \left(\frac{d_{ij}^{(Y)} - d_{ij}^{(X)}}{d_{ij}^{(X)}} \right)^2 \tag{4.10}$$

is an alternative to the absolute error in (4.9). Very often, neither the absolute nor the relative error is considered for MDS, but a compromise between them given by

$$E_3 = \frac{1}{\sum_{i=1}^{n} \sum_{j=i+1}^{n} d_{ij}^{(X)}} \sum_{i=1}^{n} \sum_{j=i+1}^{n} \frac{(d_{ij}^{(Y)} - d_{ij}^{(X)})^2}{d_{ij}^{(X)}}. \tag{4.11}$$

When MDS is carried out based on the error measure E_3, it is also called **Sammon mapping**. The error E_3 for a concrete representation of the data in the lower-dimensional space is called **stress**.

So far, we have only proposed error measures for MDS, but we still need to find a way to minimize these error measures. The minimization requires finding n suitable points in $\mathbb{R}^m$ and therefore involves $m \times n$ variables, where m is the dimension for the MDS representation of the data, and n is the number of data objects. This means that even a two-dimensional MDS representation of a small data set like the Iris data leads to a minimization problem with $2 \times 150 = 300$ parameters to be optimized. Unfortunately, there is no known analytical solution for any of the error measures in (4.8)–(4.11). Therefore, a heuristic optimization strategy is needed. Typically, a **gradient method** is applied. Gradient methods are discussed in more

Table 4.2 Multidimensional scaling

Algorithm MDS($\mathcal{D}$)

input: data set $\mathcal{D} \subset \mathbb{R}^m$ with $|\mathcal{D}| = n$ or distance matrix $[d_{i,j}^{(X)}]_{1 \leq i,j \leq n}$

parameter: dimension q for the representation, stepwidth $\alpha > 0$, stop criterion SC

output: set Y of n points in $\mathbb{R}^q$

1   Initialize $Y = \{\mathbf{y_1}, \ldots, \mathbf{y_n}\} \subset \mathbb{R}^q$ randomly or better with a PCA projection
2   If the input is a data set, compute the distances $d_{ij}^{(X)}$ between the data objects
3   **do**
4   Compute $d_{i,j}^{(Y)} = \| \mathbf{y}_i - \mathbf{y}_j \|$ (for all $i, j = 1, \ldots, n$)
5   Compute $\partial E_1 / \partial \mathbf{y_k}$ according to (4.12) (for all $k = 1, \ldots, n$)
6   update $\mathbf{y_k^{new}} = \mathbf{y_k^{old}} - \alpha \cdot \partial E_1 / \partial \mathbf{y_k}$ (for all $k = 1, \ldots, n$)
7   **while** SC is not satisfied

detail in Sect. 5.3.2. In order to minimize a differentiable function—here any of the error measures (4.8)–(4.11)—one can compute the vector of partial derivatives with respect to the parameters of the object function and then follow the opposite direction of the gradient, since the gradient points in the direction of steepest ascend and our aim is not to maximize but to minimize the error function. The gradient method starts with a random or a suitable initial solution, computes the gradient at the given solution, goes in the opposite direction of the gradient, and takes the resulting solution as the next solution. For this new solution, the gradient is computed again, and the process is repeated until no or little improvements can be achieved.

As an example, we compute the gradient for error measure E_1 in (4.9) with respect to one data point $\mathbf{y}_k$ in the lower-dimensional space:

$$\frac{\partial E_1}{\partial \mathbf{y_k}} = \frac{2}{\sum_{i=1}^n \sum_{j=i+1}^n (d_{ij}^{(X)})^2} \sum_{j \neq k} (d_{kj}^{(Y)} - d_{kj}^{(X)}) \frac{\mathbf{y_k} - \mathbf{y_j}}{d_{kj}^{(Y)}}, \tag{4.12}$$

where we have used

$$\frac{\partial d_{ij}^{(Y)}}{\partial \mathbf{y_k}} = \frac{\partial}{\partial \mathbf{y_k}} \|\mathbf{y_i} - \mathbf{y_j}\| = \begin{cases} \frac{\mathbf{y_k} - \mathbf{y_j}}{d_{kj}^{(Y)}} & \text{if } i = k, \\ 0 & \text{otherwise.} \end{cases} \tag{4.13}$$

The basic structure of the MDS algorithm is described in Table 4.2.

It is recommended to start with a good initialization which can be obtained by first carrying out a principal component analysis and then projecting the data to the first q principal components. The stepwidth α determines how far to go in the opposite direction of the gradient. If α is chosen too large, the algorithm might never find a (local) minimum of the error function. Small values for α lead to slow convergence of the algorithm. It is also possible to use an adaptive stepwidth α that determines a new value for α in each step of the algorithm. The stop criterion can be a fixed number of iterations or a lower bound for the change of the error function, i.e., when E_1 changes less than a given threshold in one step of the algorithm, the iteration scheme is terminated.

The algorithm is identical for the error measures (4.10) and (4.11) except for lines 5 and 6 where the gradient of E_1 has to be replaced by the gradients of the corresponding error measures.

Fig. 4.17 The Sammon
mapping of the Iris data,
scatter plots of which are
shown in Fig. 4.11 on p. 46

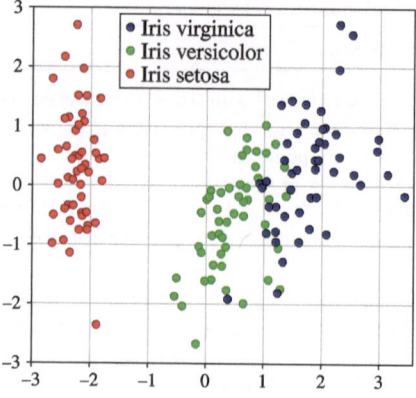

Fig. 4.18 The Sammon
mapping of the data set that is
shown in Fig. 4.14 on p. 51

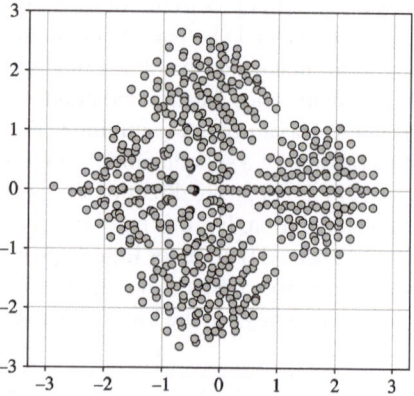

Figure 4.17 shows the result of MDS in the form of the Sammon mapping applied to the Iris data set. The categorical attribute for the species has not been used for MDS. But the corresponding points resulting from the Sammon mapping are marked by colors corresponding to the Iris flower species. Before the distances $d_{ij}^{(X)}$ in the original four-dimensional space defined by the sepal and petal length and width are computed, a z-score standardization is applied to each of the four numerical attributes.

Before the MDS algorithm can be applied, we have to make sure that no distance between two different data objects is zero because these distances occur in the denominator of the gradient, so that a zero distance would lead to a division by zero error. There are two Iris flowers—number 102 and number 143—with exactly the same values for all attributes, leading to a distance of zero between them. Therefore, we removed one of them from the data set before carrying out the computations for MDS.

Figure 4.18 shows the result of the Sammon mapping applied to the "3D chessboard pattern" data set in Fig. 4.14, where we have again carried out z-score standardization in advance. The four cubes filled with data can even be recognized in this two-dimensional representation. However, the outlier—the left-most point—is not as well separated from the other data points as in the case of PCA in Fig. 4.16.

4.3.2.4 Variations of PCA and MDS

MDS differs from PCA in various aspects. MDS is based on the idea of preserving the distances among the original data objects, whereas PCA focuses on the variance.

PCA provides an explicit mapping from the space in which the data objects are located to the lower-dimensional space, whereas MDS provides only an explicit representation of the data objects in the lower-dimensional space. This means that when a new or hypothetical data object is considered, it can be represented immediately in case of PCA simply by projecting the data object to the corresponding principal components. This is not possible for MDS.

The computational complexity of PCA is lower than the complexity of MDS. The computation of the covariance matrix can be carried out in linear time with respect to the size of the data set. Once the covariance matrix has been calculated, the computation of the eigenvalues and eigenvectors depends only on the number of attributes, but not on the number of data anymore. The number of attributes is usually much smaller than the number of data objects. For MDS, it is necessary to consider the pairwise distances between data objects leading to a quadratic complexity in the number of data objects. Although a quadratic complexity is often considered as feasible in computer science, it is unacceptable for larger data sets. Consider a data set with one million data objects. The distance matrix contains 10^{12} entries in this case. Since the distance matrix is symmetric and the entries in the diagonal are all zero, we only need to know $(10^{12} - 10^6)/2 = 4999995 \cdot 10^5$ entries. If we want to store the distances in 4-byte floating-point values, this would require more than 1800 gigabytes! There are various modifications of MDS [3] that try to overcome these complexity problems by sampling [15] or variations of the error measures [16, 17].

All these dimension-reduction methods generate scatter plots with abstract coordinate axes that do not correspond to attributes of the original data. In this way, the scatter plots of projection pursuit can even be extended and evaluated by the measures of interestingness proposed for projection pursuit and also other statistical measures and tests in order to select the most interesting visualizations [23].

4.3.2.5 Parallel Coordinates

So far, the visualization techniques for multidimensional data we have introduced here are based on scatter plots with abstract coordinate axes. Since coordinate axes should be drawn pairwise perpendicular and only two axes can satisfy this condition on a plane and three axes in 3D-space, we cannot use more than three axes for scatter plot-like visualizations. **Parallel coordinates** draw the coordinate axes parallel to each other, so that there is no limitation for the number of axes to be displayed. For a data object, a polyline is drawn connecting the values of the data object for the attributes on the corresponding axes.

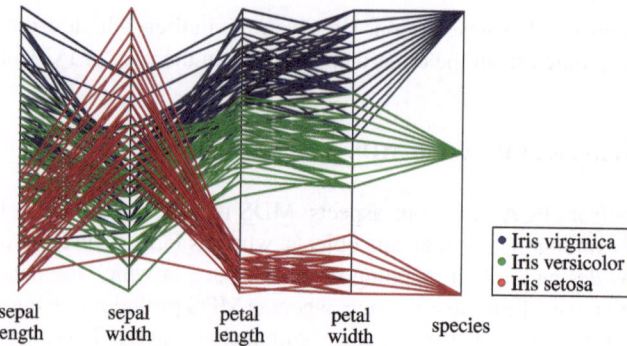

Fig. 4.19 Parallel coordinates plot for the Iris data set

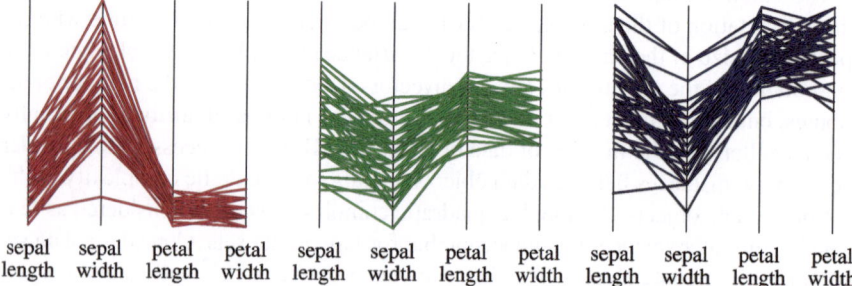

Fig. 4.20 Parallel coordinate plots for Iris setosa, Iris versicolor, and Iris virginica (*left* to *right*)

Figure 4.19 shows the plot of the Iris data set with parallel coordinates. The species is a categorical attribute and can only assume three possible values. The polylines for the different species are displayed with different colors. The plot clearly shows that the setosa has smaller values for the petal length and width than the other two species.

For larger data sets, it becomes more or less impossible to track the lines that correspond to a data object in parallel coordinates plots or even to discover general structures. It can be helpful to generate separate parallel coordinate plots for different subsets of the data. For instance, in the case of the Iris data set, we could generate a separate plot for each of the three species as shown in Fig. 4.20. In order to keep the three plots comparable, we have not rescaled the axes for each of the three plots. Normally, the axes will be scaled in such a way that the minimum and maximum of all values for the corresponding attribute of the displayed data are lowest and highest point of the axes.

Figure 4.21 shows the parallel coordinates plot for the "3D chessboard pattern" data set in Fig. 4.14. This plot is identical to the plot that we would obtain if we had not used the "3D chessboard pattern," but had filled the cube uniformly with data.

Fig. 4.21 Parallel
coordinates plot for the data
set shown in Fig. 4.14

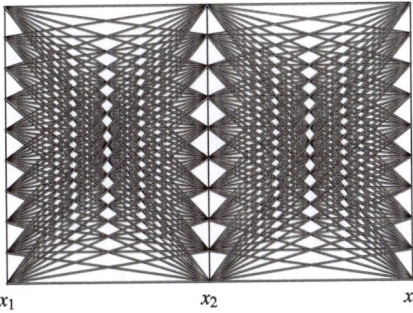

x_1 x_2 x_3

Fig. 4.22 Radar plot for
the numerical attributes of
the Iris data set

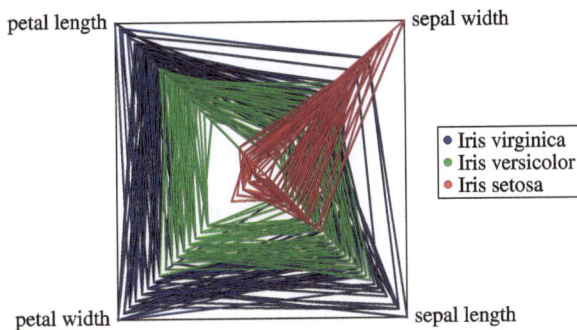

petal length   sepal width

- Iris virginica
- Iris versicolor
- Iris setosa

petal width   sepal length

4.3.2.6 Radar and Star Plots

Radar plots are based on a similar idea as parallel coordinates with the difference
that the coordinate axes are not drawn as parallel lines, but in a star-like fashion
intersecting in one point. Figure 4.22 shows a radar plot for the four numerical
attributes of the Iris data set. Due to the fact that radar plots sometimes resemble
spider webs, they are also called **spider plots**.

Radar plots are only suited for smaller data sets. For such smaller data sets, it is
sometimes better not to draw all data objects in the system of coordinate axes but
to draw each data object separately, which is then called a **star plot**. A star plot for
the numerical attributes of the Iris data set is shown in Fig. 4.23. The first 50 "stars"
correspond to the objects from the species setosa, the next 50 to versicolor, and the
last 50 to virginica. The star plot also shows clearly that setosa differs much from
the two species.

4.4 Correlation Analysis

In the previous section, we have introduced methods to visualize the distribution
of values for one, two, or even more attributes simultaneously. Scatter plots can
uncover dependencies or correlations between two attributes. Instead of a purely
visual approach, it is possible to compute measures of correlation between attributes

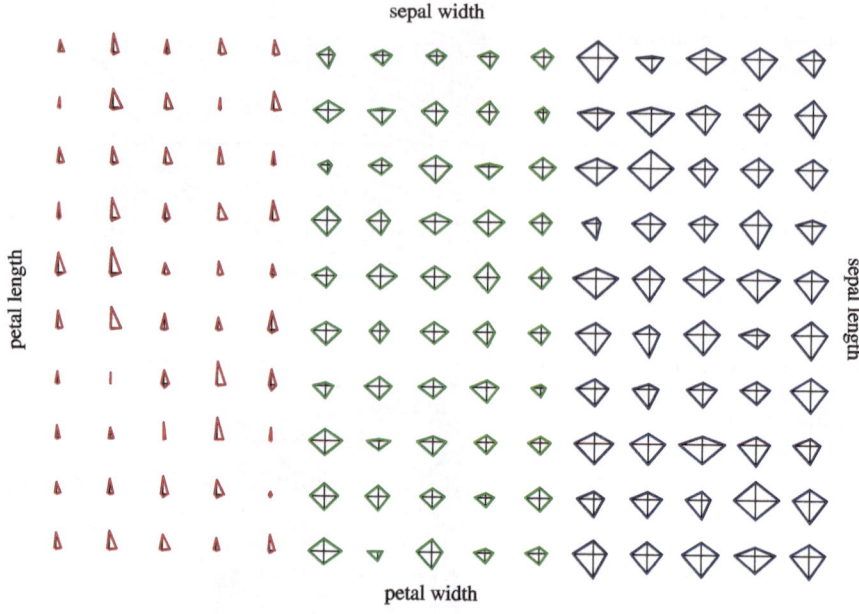

Fig. 4.23 Star plot for the numerical attributes of the Iris data set

Table 4.3 Pearson's correlation coefficients for the numerical attributes of the Iris data set

	Sepal length	Sepal width	Petal length	Petal width
Sepal length	1.000	−0.118	0.872	0.818
Sepal width	−0.118	1.000	−0.428	−0.366
Petal length	0.872	−0.428	1.000	0.963
Petal width	0.818	−0.366	0.963	1.000

to confirm expected dependencies or to discover unexpected correlations between attributes.

The (sample) **Pearson's correlation coefficient** is a measure for a linear relationship between two numerical attributes X and Y and is defined as

$$r_{xy} = \frac{\sum_{i=1}^{n}(x_i - \bar{x})(y_i - \bar{y})}{(n-1)s_x s_y}, \qquad (4.14)$$

where $\bar{x}$ and $\bar{y}$ are the mean values of the attributes X and Y, respectively, and s_x and s_y are the corresponding (sample) standard deviations. Pearson's correlation coefficient yields values between -1 and 1. The larger the absolute value of the Pearson correlation coefficient, the stronger the linear relationship between the two attributes. For $|r_{xy}| = 1$, the values of X and Y lie exactly on a line. For $r_{xy} = 1$, the line has a positive slope, for $r_{xy} = -1$ a negative one. Table 4.3 lists the Pearson correlation coefficients for the numerical attributes of the Iris data set.

Of course, the values in the diagonal must be equal to 1, since an attribute correlates fully with itself. When we plot an attribute against itself in a scatter plot, the points lie on a perfect line, the diagonal, so that Pearson's correlation coefficient must be 1 in this case. The matrix with the coefficients must also be symmetric.

It is also not surprising that the Pearson's correlation coefficient between the length and the petal width is very high. This means more or less that the leaves roughly keep their shape. A short leaf will also not be very broad, and a long leaf will be broader. It seems counterintuitive that there is more or less no correlation between the sepal length and the sepal width or even a very small negative correlation. When we take a look at the scatter plots in Fig. 4.11 on page 46, this can be explained easily. The negative correlation originates from the fact that we have the measurements from different species. Setosa has short but broad leaves compared to the other species. When we compute Pearson's correlation coefficient separately for the species for the sepal length and width, we obtain the values 0.743, 0.526, and 0.457 for Iris setosa, Iris versicolor, and Iris virginica, respectively. The correlation is not as high as for the petal length and width, but at least it is positive and not negative.

Pearson's correlation coefficient measures linear correlation. Even if there is a functional dependency between two attributes, but the function is nonlinear but monotone, Pearson's correlation coefficient will not be -1 or 1. It can even be far away from these values, depending on how much the function describing the functional relationship deviates from a line.

Rank correlation coefficients avoid this problem by ignoring the exact numerical values of the attributes and considering only the ordering of the values. Rank correlation coefficients intend to measure monotonous correlations between attributes where the monotonous function does not have to be linear.

Spearman's rank correlation coefficient or **Spearman's rho** is defined as

$$\rho = 1 - 6 \frac{\sum_{i=1}^{n} (r(x_i) - r(y_i))^2}{n(n^2 - 1)}, \tag{4.15}$$

where $r(x_i)$ is the rank of value x_i when we sort the list $(x_1, \ldots, x_n)$. $r(y_i)$ is defined analogously.

Spearman's rho measures the sum of quadratic distances of ranks and scales this measure to the interval $[-1, 1]$. When the rankings of the x- and y-values are exactly in the same order, Spearman's rho will yield the value 1; if they are in reverse order, we will obtain the value -1.

Spearman's rho assumes that there are no ties, i.e., no two values of one attribute are equal. If two or more values coincide, their rank is not defined. When ties exist, the rank $r(x_i)$ is usually defined as the mean value of all ranks of consecutive coinciding values in the sorted list. So if we have the (already sorted) list of values 0.6, 1.2, 1.4, 1.4, 1.6 for an attribute, the corresponding ranks would be 1, 2, 3.5, 3.5, 4.

Kendall's tau rank correlation coefficient or simply **Kendall's tau** is not, like Spearman's rho, based on ranks, but rather on the comparison of the orders of pairs of values. Assuming that $x_i < x_j$, the two pairs (x_i, x_j) and (y_i, y_j) are called **concordant** if $y_i < y_j$, i.e., when the two pairs are in the same order. They are called **discordant** when they are in reverse order, which means that $y_i > y_j$.

Table 4.4 Spearman's rank correlation coefficients for the numerical attributes of the Iris data set

	Sepal length	Sepal width	Petal length	Petal width
Sepal length	1.000	−0.167	0.882	0.834
Sepal width	−0.167	1.000	−0.289	−0.289
Petal length	0.882	−0.289	1.000	0.938
Petal width	0.834	−0.289	0.938	1.000

Table 4.5 Kendall's tau for the numerical attributes of the Iris data set

	Sepal length	Sepal width	Petal length	Petal width
Sepal length	1.000	−0.077	0.719	0.655
Sepal width	−0.077	1.000	−0.186	−0.157
Petal length	0.719	−0.186	1.000	0.807
Petal width	0.655	−0.157	0.807	1.000

Kendall's tau is computed as

$$\tau_a = \frac{C - D}{\frac{1}{2}n(n - 1)},$$

where C and D denote the numbers of concordant and discordant pairs, respectively:

$$C = |\{(i, j) \mid x_i < x_j \text{ and } y_i < y_j\}|, \tag{4.16}$$

$$D = |\{(i, j) \mid x_i < x_j \text{ and } y_i > y_j\}|. \tag{4.17}$$

Tables 4.4 and 4.5 contain the values for Spearman's rho and Kendal's tau for the numerical attributes of the Iris data set. The values are not identical but very similar to the ones in Table 4.3 for Pearson's correlation coefficient.

Rank correlation coefficients like Spearman's rho and Kendall's tau depend only on the order (ranks) of the values and are therefore more robust against extreme outliers than Pearson's correlation coefficient.

For categorical attributes, these correlation coefficients are not applicable. Instead, one can carry out independence tests like the χ^2 test for independence, as it is described in the appendix in Sect. A.4.3.6. This test can also be used for numerical attributes when they are discretized by some binning strategy.

4.5 Outlier Detection

Since it has an intuitive, though imprecise meaning, we have used the term outlier already before without giving a precise definition. Actually there is no formally precise definition of outliers. An **outlier** is simply a value or data object that is far away or very different from all or most of the other data.

Outliers can be a hint to data quality problems as they will be discussed in the next section. Outliers can correspond to erroneous data coming from wrong measurements or typing mistakes when data are entered manually. Erroneous data should be corrected, or, if this is not possible, they should be excluded from the data set before further analysis steps are carried out. However, outliers can also be correct data that differ from the rest of the data just by chance or for other reasons like special exceptional situations. Even if outliers are correct data, it might be worthwhile to exclude them from the data set for further analysis and to consider them separately. Some data analysis methods are robust against outliers and more or less not influenced by them, whereas other are extremely sensitive to outliers and might produce incoherent or nonsense results, just because of the presence of one or a few outliers. As a simple example, consider the median and the mean value. If one replaces the largest value in a sample by an extremely large value or even infinity, the median will not change. But the mean value will tend to infinity, even though only one value in the sample goes to infinity.

4.5.1 Outlier Detection for Single Attributes

For a categorical attribute, one can consider the finite set of values. An outlier is a value that occurs with a frequency extremely lower than the frequency of all other values. However, in some cases, this might be actually the target of our analysis. If we want to set up an automatic quality control system and want to train a classifier, classifying the parts as correct or with failures based on measurements of the produced parts, we will probably have so many correct parts in comparison the ones with failures that we would consider them as outliers. However, removing these "outliers" from the data set would actually make it impossible to achieve our original goal to derive a classifier from the data set that can identify the parts with failures.

For numerical attributes, outlier detection is more difficult. We have already classified certain data points in a boxplot as outliers. However, the definition of outliers in a boxplot does not take the number of data into account, so that for larger data sets, boxplots will usually contain points marked as outliers. We have seen this already in the boxplots in Fig. 4.6, both showing samples from a standard normal distribution. The left boxplot in Fig. 4.6 for a sample of size $n = 1000$ contains eight outliers corresponding to what we expect theoretically, namely seven points outside the whiskers. As mentioned before, for a normal distribution, we can expect roughly 0.7% points to be marked as outliers in a boxplot.

For asymmetric distributions, boxplots tend to contain more outliers. For example, in boxplots for samples from an exponential distribution with $\lambda = 1$ as they are shown in Fig. 4.8, we would expect roughly 4.8% points marked as outliers. Heavy-tailed distributions tend to show more outliers in a boxplot, whereas for a sample from a uniform distribution, we would expect no outliers at all, no matter how large the sample size is.

Even if we try to adjust the definition of outliers according to the sample size, the above examples show that what we consider as an outlier depends strongly on the underlying distribution from which the data are sampled. Therefore, statistical tests for outliers are usually based on assumptions about the underlying distribution, although we might not know from which distribution the data are sampled.

The standard assumption for outlier tests for continuous attributes is that the underlying distribution is a normal distribution. **Grubb's test** is a test for outliers for normal distributions taking the sample size into account. It is based on the statistics

$$G = \frac{\max\{|x_i - \bar{x}| \mid 1 \le i \le n\}}{s}, \tag{4.18}$$

where $x_1, \ldots, x_n$ is the sample, $\bar{x}$ is its mean value, and s is its empirical standard deviation. For a given significance level α, the null hypothesis that the sample does not contain outliers is rejected if

$$G > \frac{n-1}{\sqrt{n}} \sqrt{\frac{t_{1-\alpha/(2n),n-2}^2}{n - 2 + t_{1-\alpha/(2n),n-2}^2}}, \tag{4.19}$$

where $t_{1-\alpha/(2n),n-2}$ denotes the $(1 - \alpha/(2n))$-quantile of the t-distribution with $(n - 2)$ degrees of freedom.

For the Iris data set, Grubb's test yields the p-values[4] 0.92, 0.13, 1.0, and 1.0 for the sepal length and width and the petal length and width, respectively. Even the lowest p-value does not indicate that there is an outlier. Note that this is the case although the assumption that the attributes follow normal distributions is not realistic at all as can be seen from the scatter plot in Fig. 4.11 on page 46. The attributes for each species separately might follow normal distributions, but not the attributes considered for all species together.

When Grubb's test with its assumption of normal distribution is applied to very skewed and heavy-tailed distributions like the exponential distribution, an outlier will almost always be indicated.

4.5.2 Outlier Detection for Multidimensional Data

Outlier detection in multidimensional data is usually not based on specific assumptions on the distribution of the data and is not carried out in the sense of statistical tests. Visualization techniques provide a simple method for outlier detection in multidimensional data. Scatter plots can be used when only two attributes are considered. In the right scatter plot in Fig. 4.12 on page 47, we had added two outliers. The one in the upper left corner would not have been detected when only single

---

[4]For Grubb's test, the null hypothesis is that there are no outliers. Then the point in the sample with the largest distance to the mean is considered. In the case of Grubb's test, the p-value is the probability that in a sample of size n, such a large or even large deviation from the mean would occur. For a more formal and general definition of p-values, see also Appendix A.

attributes are considered. Instead of using projections to two attributes, one can also use dimension-reduction methods like PCA or multidimensional scaling in order to identify outliers in the corresponding plots as in Figs. 4.16 and 4.18.

There are many approaches for finding outliers in multidimensional data based on clustering the data and defining those data objects as outliers that cannot be assigned reasonably to any cluster [18, 19]. There are also distance-based [13, 14], density-based [4], and projection-based methods [1] for outlier detection.

4.6 Missing Values

In the ideal case, all attributes of all objects in the data table have well-defined values. However, in real data sets, one has to deal with **missing values** quite often. The occurrence of missing values can have different causes. A sensor might be broken leading to a missing value. People might have refused or forgotten to answer a question in a questionnaire, or an attribute might not be applicable for a certain object. For instance, the attribute *pregnant* with the two possible values *yes* and *no* does not make sense for men. Of course, one could always enter the value *no* for the attribute *pregnant*. But this might lead to a grouping of men with nonpregnant women.

Especially in a customer database, missing values will be quite common. Attributes like *age*, *sex*, and *email address* might be available for some customers, but probably not for all customers.

In the best case, missing values are indicated in a unique way. A symbol like "?" or null entry might be used for missing value. Sometimes a special numeric value is used to indicate missing values. For example, the value -1 could indicate a missing value for a counting attribute (like number of products bought by a customer in a certain period, number of children, etc.). Since the value -1 can never occur in a counting attribute, it cannot be misunderstood as meaningful (nonmissing) value. One must, however, take this into account when simple characteristics like the mean value or the median of an attribute are computed. If one simply adds up all values—including the -1's for the missing values—and divide it by the number of values to obtain the mean value, the result might even be negative if there are many missing values. If we know the number of children for only very few of our customers, we might find out in this way that the average number of children of our customers is -0.3, which is, of course, complete nonsense.

Sometimes, missing values are not even marked in a specific way. We have seen already an example in Fig. 2.2 on page 18 where a missing value of the customer's birth date was replaced by a default value that could also be just a usual value.

As another example, consider an anemometer, a sensor for measuring the wind speed. Such an anemometer is equipped with a small wheel that is turned by the wind, and from the speed of the wheel the speed of the wind can be deduced. Sometimes, the wheel might be blocked for some reason. In this case, the sensor will constantly yield the value zero for the wind speed. This value zero is also a possible wind speed although it is very improbable in most geographical regions that

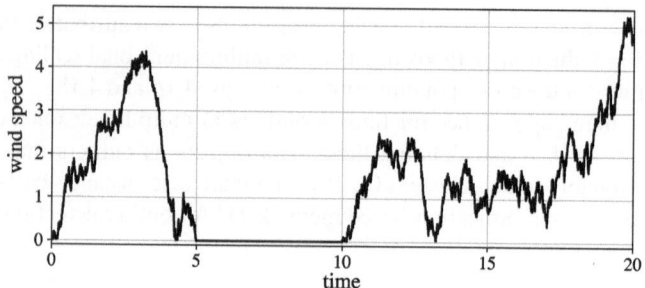

Fig. 4.24 Measured wind speeds with a period of a jammed sensor

the wind speed is absolutely zero over a longer time. When we see a diagram like in Fig. 4.24, it is very probable that during the period between 5 and 10, the wind speed was not zero, but that there are missing values in this period due to a jammed anemometer.

It is very important to identify such hidden missing values. Otherwise, the further analysis of the data can be completely mislead by such erroneous values.

When there are missing values, one should also take into account how missing values enter the data set. The simplest and most common assumption about missing values is that they are **missing completely at random (MCAR)** or are **observed at random (OAR)**. This means that special circumstances or special values in the data lead to higher or lower chances for missing values. One can imagine that one has printed out the data table on a large sheet of paper with no missing values and then someone has dropped accidentally some random spots of ink on the paper so that some values cannot be read anymore.

In a more formal way, the situation missing completely at random can be described in the following way. We consider the random variable X for which we might have some missing entries in the data set. The random variable X itself does not have missing values. The corresponding value is just not available in our data set. The random variable X_{miss} can assume the value 0 for a missing values and 1 for a nonmissing value for the random variable X of interest. This means that for $X_{\text{miss}} = 1$, we see the true value of X in our data set, and for $X_{\text{miss}} = 0$, instead of the true value of X, we see a missing value. We also consider the random vector Y representing all attributes except X in our data set. The situation observed at random means that X_{miss} is independent of X and Y, i.e.,

$$P(X_{\text{miss}}) = P(X_{\text{miss}}|X, Y) \qquad \text{(MCAR)}.$$

Consider a sensor for the outside air temperature X whose battery might run out of energy once in a while, leading to missing values of the missing completely at random. This is the best case for missing values. It can, for instance, be concluded that the unknown missing values follow the same distribution as the known values of X.

The situation **missing at random (MAR)** is more complicated but might still be manageable with appropriate techniques. The probability for a missing value depends on the value of the other attributes Y but is conditionally independent[5] of the true value of X given Y:

$$P(X_{miss}|Y) = P(X_{miss}|X, Y) \qquad \text{(MAR).}$$

To illustrate what missing at random means, consider the above example of a sensor for the outside air temperature with the battery running out of energy once in a while. But assume now that the staff responsible for changing the batteries does not change the batteries when it is raining. Therefore, missing values for the temperature are more probable when it is raining, so that the occurrence of missing values depends on other attributes, in this case the attribute for rain or amount of rain.

Note that the occurrence of a missing value is in this case not independent of the value of the variable X, it is only conditionally independent given the attribute for the rain. Usually, the outside temperatures are lower when it is raining, so that missing values correlate with lower temperatures. In contrast to missing completely at random, here the missing values of X do not follow the same distribution as the measured values of X. If we compute the average temperature based only on the measured values, we would obviously overestimate the average temperature. Therefore, special care has to be taken when missing values occur at random, but not completely at random.

The worst case are **nonignorable missing values**, where the occurrence of missing values directly depends on the true value, and the dependence cannot be resolved by other attributes. Consider a bad sensor for the outside temperature that always fails when there is frost. This would mean that we have no information about temperatures below $0°C$. In the case of missing at random, we might still be able to obtain information from the data set about the missing values. For the nonignorable values, we have no chance to make correct guesses, at least based on the available data.

How can we distinguish between the three different types of missing values? First of all, the knowledge and background information about why missing values occur should be taken into account. It is also possible to try to distinguish the two cases observed at random and missing at random based on the given data. If an attribute X has missing values and one wants to find out of which type the missing values are, one way would be to apply independence tests checking directly whether $P(X_{miss}) = P(X_{miss}|Y)$. Usually, it is not feasible to apply a test that checks the independence of X_{miss} against the multivariate attribute Y standing for all other attributes, but only against single attributes. Applying independence test of X_{miss} against each other attribute Y will not uncover multiple dependencies, but if even these tests indicate a dependency already, then we can be quite sure that the values of X are not missing completely at random.

Another way to approach the problem of distinguishing between missing completely at random and missing at random is the following procedure:

1. Turn the considered attribute X into a binary attribute, replacing all measured values by the values *yes* and all missing values by the value *no*.

---

[5] A formal definition of conditional independence is given in Sect. A.3.3.4 in the appendix.

2. Build a classifier with now binary attribute X as the target attribute and use all other attributes for the prediction of the class values *yes* and *no*.
3. Determine the misclassification rate. The misclassification rate is the proportion of data objects that are not assigned to the correct class by the classifier.

In the case of missing values of the type observed at random, the other attributes should not provide any information, whether X has a missing value or not. Therefore, the misclassification rate of the classifier should not differ significantly from pure guessing, i.e., if there are 10% missing values for the attribute X, the misclassification rate of the classifier should not be much smaller than 10%. If, however, the misclassification rate of the classifier is significantly better than pure guessing, this is an indicator that there is a correlation between missing values for X and the values of the other attributes. Therefore, the missing values for X might not be of the type observed at random but of the type missing at random or, even worse, non-ignorable. Note that it is in general not possible to distinguish the case nonignorable from the other two cases based on the data only.

4.7 A Checklist for Data Understanding

In this chapter, we have provided a selection of techniques for data understanding. It is not necessary to go through all of them in the phase of data understanding, but rather to keep all of them in mind.

- First of all, one should keep in mind that there are general and specific goals for data understanding. One important part of data understanding is to get an idea of the data quality. There are standard data quality problems like syntactic accuracy which are easy to check.
- Outliers are another problem, and there are various methods to support the iden- tification of outliers. Apart from methods exclusively designed for outlier detec- tion as they have been discussed in Sect. 4.5, there are especially visualization techniques like boxplots, histograms, scatter plots, projections based on PCA and MDS that can help to find outliers but are also useful for other purposes.
- Missing values are another concern of data quality. When there are explicit miss- ing values, i.e., entries that are directly marked as missing, then one should still try to find out of which type—OAR, MAR, or nonignorable—they are. This can sometimes be derived from domain knowledge, but also based on classification methods as described in the previous section. We should also be aware of the possibility of hidden missing values that are not explicitly marked as missing. The simplest case might be hidden missing values that have a default value. His- tograms might help to identify candidates for such hidden missing values when there are unusual peaks as in the example in Fig. 2.2 on page 18. However, there is no standard test or technique to identify possible hidden missing values. There- fore, whenever we see something unexpected in the data, hidden missing values of a specific type might be one explanation.

- Apart from these data quality issues, data understanding should also help to discover new or confirm expected dependencies or correlations between attributes. Techniques like the ones mentioned in Sect. 4.4 are one way to solve this task. Apart from this, scatter plots can show correlations between pairs of attributes.
- Specific application dependent assumptions—for instance, the assumption that a specific attribute follows a normal distribution—should also be checked during data understanding.
- Representativeness of the data cannot always be checked just based on the data, but we have to compare the statistics with our expectations. If we suspect that there is a change in a numerical attribute over time, we can compare histograms or boxplots for different time periods. We can do the same with bar charts for categorical attributes.

No matter what the specific application dependent purposes of data understanding are, here is a checklist what we consider as a must for the data understanding phase:

- Check the distributions for each attribute whether there are unusual or unexpected properties like outliers. Are the domains or ranges correct? Do the medians of numerical attributes look correct? This should be done based on
  - histograms and boxplots for continuous attributes and
  - bar charts for categorical attributes.
- Check correlations or dependencies between pairs of attributes with scatter plots which should be density-based for larger data sets. For small numbers of attributes, inspect scatter plots for all pairs of attributes. For higher numbers of attributes, do not generate scatter plots for all pairs, but only for those ones where independence or a specific dependency is expected. Generate in addition scatter plots for some randomly chosen pairs.

To cite Tukey [24] again, if problems in later phases of the data analysis process occur that could have been discovered or even avoided in an early plotting and looking at the data, there is absolutely no excuse for this failure.

Of course, one should also exploit the other methods described in this section for these and other more specific purposes.

4.8 Data Understanding in Practice

A key issue in data understanding (and actually the entire data analysis process) often causes considerable headaches: the loading of the data into the analysis tool of choice. One of the strengths of KNIME is its versatility in terms of powerful file importing nodes and database connectors. R, on the other hand offers the entire breadth of analysis and visualizations, although they are often not all that intuitive to use.

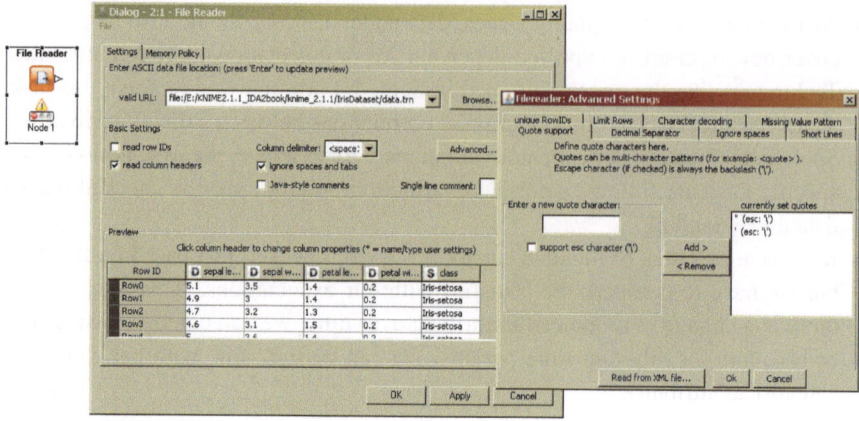

Fig. 4.25 The dialog of the file reader node offers manifold options to control the reading of diverse file formats

4.8.1 Data Understanding in KNIME

Data Loading Data understanding starts with loading the data: KNIME offers a "File Reader" node which hides an entire file import wizard in its dialog. Manifold options allow one to choose the underlying character encoding, column types, and separators and escape characters, to name just a few. The ability to quickly check in the preview tab if the selected options fit the given file make it possible to smoothly find suitable settings also for complex files. Figure 4.25 shows the dialog of this node together with the "expert tab" which hides another set of options.

KNIME also allows one to read data from specialized file formats, such as the Weka ARFF format or the compressed KNIME table format. For these formats, specialized nodes are available in the "IO" category as well.

Reading data from databases is often neglected in stand-alone analytic tools. KNIME offers flexible connectivity to access databases of various types by specifying the corresponding JDBC driver (Java Database Connectivity) in addition to the table location and name. For basic data filtering, specialized nodes are also available, which allow one to do, e.g., column and row filtering also within the database, without loading the data into KNIME explicitly. This saves time for larger databases if only a much smaller subset is to be analyzed. Figure 4.26 shows a part of a workflow reading data from a database and filtering some rows and columns before reading the data into KNIME itself.

Data Types KNIME supports all basic data types, such as string, integers, and numbers, but also date and time types and nominal values. In addition, various extensions add the ability to process images, molecular structures, sequences, and textual data. The repository of type extension is constantly growing as well. For a first glance at the data read it into KNIME and in order to check domain and nominal value range, add the "Statistics" and "Value Counter" node. The first one computes

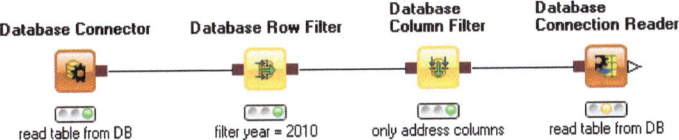

Fig. 4.26 Specialized nodes for database access allow one to filter rows and columns using database routines before loading the data into KNIME

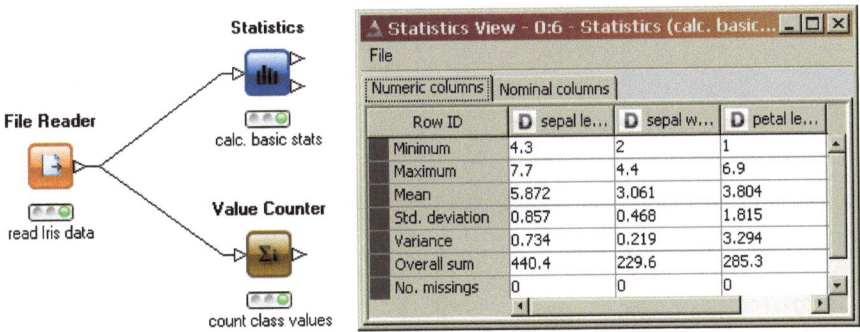

Fig. 4.27 Looking at basic information of the data's attributes often helps one to spot outliers or other errors in the data

basic statistical measures (minimum, maximum, mean, standard deviation, variance, overall sum), counts the number of missing values for numerical attributes, and lists the most and least frequent values for nominal values. The "Value Counter" node operates on one nominal attribute and lists the occurrence frequencies of each value. Figure 4.27 shows an example flow and the view of the statistics node for the iris data.

Visualization Checking basic descriptors of your data is important, but, as pointed out earlier in this chapter, looking at visualizations often gives dramatically better insights into your data. KNIME offers all of the visualizations described in Sect. 4.3: histograms, boxplots (also conditional), scatter and scatter matrix plots, and parallel coordinates. Principal Component Analysis and Multidimensional Scaling methods are also available and can be used to calculate new coordinates for, e.g., scatter plot views. All of these views make use of color, shape, and size information whenever possible. Such additional **visual dimensions** are added by using "Color/Shape/Size Manager" nodes as the additional advantage that all subsequent nodes are using the same color/shape/size information. Figure 4.28 shows an example of a KNIME workflow with a number of views connected to a workflow which assigns color and shape based on selected columns.

KNIME has been developed to also support explorative data analysis and therefore most of the views allow for interactivity, or **visual brushing**. This means that points (or other visual objects) representing records in the underlying database can be marked and will then be marked in other, related views as well. KNIME allows

Fig. 4.28 Views in KNIME
are simply connected to the
data pipeline as well. Adding
nodes which set additional
visual properties allows to
control color, shape and size
in the subsequent views

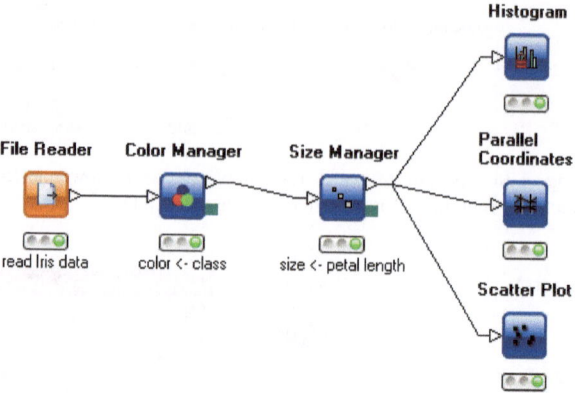

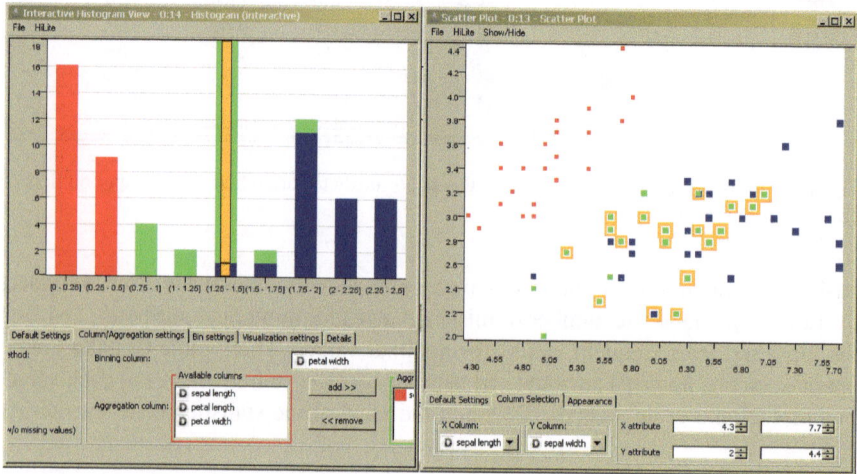

Fig. 4.29 Interactive views in KNIME allow one to propagate selections to other views within the
same workflow. The points falling into the selected bar in the histogram are automatically selected
by KNIME's Hiliting mechanism

for this selection process (called *hiliting*) to propagate along the data pipeline as
long as a meaningful translation between records (or rows in the table) is possible. For instance, selecting a rule will hilite all points contained within that rule in
all other views. Visual brushing is a powerful tool to allow interactive exploration
of data since it allows one to quickly select interesting elements in one view and
see details about the selected elements in other views, for instance, the underlying
customer data, the images covered by a rule, or the molecular structures that seem
to look like outliers in a scatter plot. Figure 4.29 illustrates the hiliting mechanism
in KNIME. We see an interactive histogram of one attribute and a scatter plot de-
picting two other attributes. One of the bars in the histogram was selected, and the
corresponding points in the scatter plot are now marked as well.

4.8.2 Data Understanding in R

4.8.2.1 Histograms

Histograms are generated by the function hist. The simplest way to create a histogram is to just use the corresponding attribute as an argument of the function hist, and R will automatically determine the number of bins for the histogram based on Sturge's rule. In order to generate the histogram for the petal length of the Iris data set, the following command is sufficient:

```
> hist(iris$Petal.Length)
```

The partition into bins can also be specified directly. One of the parameters of hist is breaks. If the bins should cover the intervals $[a_0, a_1), [a_1, a_2), \ldots, [a_{k-1}, a_k]$, then one can simply create a vector in R containing the values a_i and assign it to breaks. Note that a_0 and a_k should be the minimum and maximum values of the corresponding attribute. If we want the boundaries for the bins at 1.0, 3.0, 4.5, 4.0, 6.1, then we would use

```
> hist(iris$Petal.Length,breaks=c(1.0,3.0,4.5,4.0,6.9))
```

to generate the histogram. Note that in the case of bins with different length, the heights of the boxes in the histogram do not show the relative frequencies. The areas of the boxes are chosen in such a way that they are proportional to the relative frequencies.

4.8.2.2 Boxplots

A boxplot for a single attribute is generated by

```
> boxplot(iris$Petal.Length)
```

yielding the boxplot for the petal length of the Iris data set. Instead of a single attribute, we can hand over more than one attribute

```
> boxplot(iris$Petal.Length,iris$Petal.Width)
```

to show the boxplots in the same plot. We can even use the whole data set as an argument to see the boxplots of all attributes in one plot:

```
> boxplot(iris)
```

In this case, categorical attributes will be turned into numerical attributes by coding the values of the categorical attribute as $1, 2, \ldots$, so that these boxplots are also shown but do not really make sense.

In order to include the notches in the boxplots, we need to set the parameter notch to true:

```
> boxplot(iris,notch=TRUE)
```

If one is interested in the precise values of the boxplot like the median, etc., one can use the `print`-command:

```
> print(boxplot(iris$Sepal.Width))
$stats
       [,1]
[1,]   2.2
[2,]   2.8
[3,]   3.0
[4,]   3.3
[5,]   4.0

$n
[1] 150

$conf
          [,1]
[1,] 2.935497
[2,] 3.064503

$out
[1] 4.4 4.1 4.2 2.0

$group
[1] 1 1 1 1

$names
[1] "1"
```

The first five values are the minimum, the first quartile, the median, the third quartile, and the maximum value of the attribute, respectively. n is the number of data. Then come the boundaries for the confidence interval for the notch, followed by the list of outliers. The last values `group` and `names` only make sense when more than one boxplot is included in the same plot. Then `group` is needed to identify to which attribute the outliers in the list of outliers belong. `names` just lists the names of the attributes.

4.8.2.3 Scatter Plots

A scatter plot of the petal width against petal length of the Iris data is obtained by

```
> plot(iris$Petal.Width,iris$Petal.Length)
```

All scatter plots of each attribute against each other in one diagram are created with

```
> plot(iris)
```

If symbols representing the values for some categorical attribute should be included in a scatter plot, this can be achieved by

```
> plot(iris$Petal.Width,iris$Petal.Length,
        pch=as.numeric(iris$Species))
```

where in this example the three types of Iris are plotted with different symbols.

If there are some interesting or suspicious points in a scatter plot and one wants to find out which data records these are, one can do this by

```
> plot(iris$Petal.Width,iris$Petal.Length)
> identify(iris$Petal.Width,iris$Petal.Length)
```

and then clicking on the points. The index of the corresponding records will be added to the scatter plot. To finish selecting points, press the ESCAPE-key.

Jitter can be added to a scatter plot in the following way:

```
> plot(jitter(iris$Petal.Width),
        jitter(iris$Petal.Length))
```

Intensity plots and density plots with hexagonal binning, as they are shown Fig. 4.9, can be generated by

```
> plot(iris$Petal.Width,iris$Petal.Length,
        col=rgb(0,0,0,50,maxColorValue=255),
        pch=16)
```

and

```
> library(hexbin)
> bin<-hexbin(iris$Petal.Width,
               iris$Petal.Length,
               xbins=50)
> plot(bin)
```

respectively, where the library hexbin does not come along with the standard version of R and needs to be installed as described in the appendix on R. Note that such plots are not very useful for such a small data sets like the Iris data set.

For three-dimensional scatter plots, the library scatterplots3d is needed and has to be installed first:

```
> library(scatterplot3d)
> scatterplot3d(iris$Sepal.Length,
                 iris$Sepal.Width,
                 iris$Petal.Length)
```

4.8.2.4 Principal Component Analysis

PCA can be carried out with R in the following way:

```
> species <- which(colnames(iris)=="Species")
> iris.pca <- prcomp(iris[,-species],
                center=T,scale=T)
> print(iris.pca)
Standard deviations:
[1] 1.7083611 0.9560494 0.3830886 0.1439265

Rotation:
                    PC1          PC2
Sepal.Length   0.5210659 -0.37741762
Sepal.Width   -0.2693474 -0.92329566
Petal.Length   0.5804131 -0.02449161
Petal.Width    0.5648565 -0.06694199
                    PC3          PC4
               0.7195664   0.2612863
              -0.2443818  -0.1235096
              -0.1421264  -0.8014492
              -0.6342727   0.5235971

> summary(iris.pca)
Importance of components:
                           PC1    PC2    PC3     PC4
Standard deviation        1.71  0.956 0.3831 0.14393
Proportion of Variance    0.73  0.229 0.0367 0.00518
Cumulative Proportion     0.73  0.958 0.9948 1.00000

> plot(predict(iris.pca))
```

For the Iris data set, it is necessary to exclude the categorical attribute *Species* from PCA. This is achieved by the first line of the code and calling prcomp with iris[,-species] instead of iris.

The parameter settings center=T, scale=T, where T is just a short form of TRUE, mean that z-score standardization is carried out for each attribute before applying PCA.

The function predict can be applied in the above-described way to obtain the transformed data from which the PCA was computed. If the computed PCA transformation should be applied to another data set x, this can be achieved by

```
> predict(iris.pca,newdata=x)
```

where x must have the same number of columns as the data set from which the PCA has been computed. In this case, x must have four columns which must be numerical. predict will compute the full transformation, so that the above command will also yield transformed data with four columns.

4.8.2.5 Multidimensional Scaling

MDS requires the library MASS which is not included in the standard version of R and needs installing. First, a distance matrix is needed for MDS. Identical objects leading to zero distances are not admitted. Therefore, if there are identical objects in a data set, all copies of the same object except one must be removed. In the Iris data set, there is only one pair of identical objects, so that one of them needs to be removed. The Species is not a numerical attribute and will be ignored for the distance.

```
> library(MASS)
> x <- iris[-102,]
> species <- which(colnames(x)=="Species")
> x.dist <- dist(x[,-species])
> x.sammon <- sammon(x.dist,k=2)
> plot(x.sammon$points)
```

$k = 2$ means that MDS should reduce the original data set to two dimensions.

Note that in the above example code no normalization or z-score standardization is carried out.

4.8.2.6 Parallel Coordinates, Radar, and Star Plots

Parallel coordinates need the library MASS. All attributes must be numerical. If the attribute *Species* should be included in the parallel coordinates, one can achieve this in the following way:

```
> library(MASS)
> x <- iris
> x$Species <- as.numeric(iris$Species)
> parcoord(x)
```

Star and radar plots are obtained by the following two commands:

```
> stars(iris)
> stars(iris,locations=c(0,0))
```

4.8.2.7 Correlation Coefficients

Pearson's, Spearman's, and Kendall's correlation coefficients are obtained by the following three commands:

```
> cor(iris$Sepal.Length,iris$Sepal.Width)
> cor.test(iris$Sepal.Length,iris$Sepal.Width,
          method="spearman")
> cor.test(iris$Sepal.Length,iris$Sepal.Width,
          method="kendall")
```

4.8.2.8 Grubb's Test for Outlier Detection

Grubb's test for outlier detection needs the installation of the library `outliers`:

```
> library(outliers)
> grubbs.test(iris$Petal.Width)
```

References

1. Aggarwal, C., Yu, P.: Outlier detection for high dimensional data. In: Proc. ACM SIGMOD Int. Conf. on Management of Data (SIGMOD 2001, Santa Barbara, CA), pp. 37–46. ACM Press, New York (2001)
2. Anderson, E.: The irises of the Gaspe Penisula. Bull. Am. Iris Soc. **59**, 2–5 (1935)
3. Borg, I., Groenen, P.: Modern Multidimensional Scaling: Theory and Applications. Springer, Berlin (1997)
4. Breunig, M., Kriegel, H.-P., Ng, R., Sander, J.: LOF: identifying density-based local outliers. In: Proc. ACM SIGMOD Int. Conf. on Management of Data (SIGMOD 2000, Dallas, TX), pp. 93–104. ACM Press, New York (2000)
5. Cook, D., Buja, A., Cabrera, J.: Projection pursuit indices based on orthonormal function expansion. J. Comput. Graph. Stat. **2**, 225–250 (1993)
6. Diaconis, P., Freedman, D.: Asymptotics of graphical projection pursuit. Ann. Stat. **17**, 793–815 (1989)
7. Fisher, R.A.: The use of multiple measurements in taxonomic problems. Ann. Eugen. **7**(2), 179–188 (1936)
8. Freedman, D., Diaconis, P.: On the histogram as a density estimator: L_2 theory. Z. Wahrschein-lichkeitstheor. Verw. Geb. **57**, 453–476 (1981)
9. Friedman, J.: Exploraory projection pursuit. J. Am. Stat. Assoc. **82**, 249–266 (1987)
10. Friedman, J., Tukey, J.: A projection pursuit algorithm for exploratory data analysis. IEEE Trans. Comput. **C-23**, 881–890 (1974)
11. Hall, P.: On polynomial-based projection indices for exploratory projection pursuit. Ann. Stat. **17**, 589–605 (1989)
12. Huber, P.: Projection pursuit. Ann. Stat. **13**, 435–475 (1985)
13. Knorr, E., Ng, R.: Algorithms for mining distance-based outliers in large datasets. In: Proc. 24th Int. Conf. on Very Large Data Bases (VLDB 1998, New York, NY), pp. 392–403. Morgan Kaufmann, San Mateo (1998)
14. Knorr, E., Ng, R., Tucakov, V.: Distance-based outliers: algorithms and applications. Very Large Data Bases **8**, 237–253 (2000)
15. Morrison, A., Ross, G., Chalmers, M.: Fast multidimensional scaling through sampling. Inf. Vis. **2**, 68–77 (2003)
16. Pekalska, E., Ridder, D., Duin, R., Kraaijveld, M.: A new method of generalizing Sammon mapping with application to algorithm speed-up. In: Proc. 5th Annual Conf. Advanced School for Computing and Imaging, pp. 221–228, Delft, Netherlands (1999)
17. Rehm, F., Klawonn, F., Kruse, R.: Mds$_{polar}$: a new approach for dimension reduction to visualize high dimensional data. In: Advances in Intelligent Data Analysis, vol. VI, pp. 316–327. Springer, Berlin (2005)
18. Rehm, F., Klawonn, F., Kruse, R.: A novel approach to noise clustering for outlier detection. Soft. Comput. **11**, 489–494 (2007)
19. Santos-Pereira, C., Pires, A.: Detection of outliers in multivariate data: a method based on clustering and robust estimators. In: Proc. 5th Annual Conference of the Advanced School for Computing and Imaging, pp. 291–296. Physica, Berlin (2002)
20. Scott, S.: On optimal and data-based histograms. Biometrika **66**, 605–610 (1979)

21. Scott, D.: Sturges' rule. In: Wiley Interdisciplinary Reviews: Computational Statistics, vol. 1, pp. 303–306. Wiley, Chichester (2009)
22. Sturges, H.: The choice of a class interval. J. Am. Stat. Assoc. **21**, 65–66 (1926)
23. Tschumitschew, K., Klawonn, F.: Veda: statistical tests for finding interesting visualisations. In: Knowledge-Based and Intelligent Information and Engineering Systems 2009, Part II, pp. 236–243. Springer, Berlin (2009)
24. Tukey, J.W.: Exploratory Data Analysis. Addison-Wesley, Reading (1977)

Chapter 5
Principles of Modeling

After we have gone through the phases of project and data understanding, we are either confident that modeling will be successful or return to the project understanding phase to revise objectives (or to stop the project). In the former case, we have to prepare the dataset for subsequent modeling. However, as some of the data preparation steps are motivated by modeling itself, we first discuss the principles of modeling. Many modeling methods will be introduced in the following chapters, but this chapter is devoted to problems and aspects that are inherent in and common to all the methods for analyzing the data.

All we need for modeling, it might seem, is a collection of methods from which we have to choose the most suitable one for our purpose. By now, the project understanding has already ruled out a number of methods. For example, when we have to solve a regression problem, we do not consider clustering methods. But even within the class of regression problems, there are various methods designed for this task. Which one would be the best for our problem, and how do we find out about this? In order to solve this task, we need a better understanding of the underlying principles of our specific data analysis methods. Most of the data analysis methods can be viewed within the following four-step procedure:

- **Select the Model Class**.
  First of all, we must specify the general structure of the analysis result. We call this the "architecture" or "model class." In a regression problem, one could decide to consider only linear functions; or instead, quadratic functions could be an alternative; or we could even admit polynomials of arbitrary degree. This, however, defines only the structure of the "model." Even for a simple linear function, we still would have to determine the coefficients (Sect. 5.1).

- **Select the Score Function**.
  We need a score function that evaluates the possible "models," and we aim to find the best model with respect to our goals—which is formalized by the score function. In the case of the simple linear regression function, our score function will tell us which specific choice of the coefficients is better when we compare different linear functions (Sect. 5.2).

M.R. Berthold et al., *Guide to Intelligent Data Analysis*,
Texts in Computer Science 42,
DOI 10.1007/978-1-84882-260-3_5, © Springer-Verlag London Limited 2010

- **Apply the Algorithm**.

  The score function enables us to compare models, but it does not tell us how to *find* the model that obtains the best evaluation from the score function. Therefore, an algorithm is required that solves this optimization problem (Sect. 5.3).

- **Validate the Results**.

  Even if our optimization algorithm is so cleverly designed that it will find the best model in the model class we consider, we still do not know, whether this is the best model among very good models for our data or just the best one among even worse choices. Therefore, we still need other means than the score function to validate our model (Sect. 5.5).

Unfortunately, any of these steps can cause serious problems and can completely spoil the result of the data analysis process. If we are not satisfied with the result of the process, the reason might be an improper choice in one of these four steps. But it might also be the case that our data are too noisy or do not contain enough useful information to yield a satisfactory analysis result. Therefore, in order to better understand what can lead to unsatisfactory data analysis results, we need to take a closer look at what types of errors we have to deal with in the four-step procedure and the data (see Sect. 5.4).

We will not distinguish between the terms "model class" and "model" too pedantically, because usually the model class is characterized by writing down a model in which, say, the coefficients are replaced by formal parameters. As soon as we have decided which values these parameters will get, we obtain a concrete model. That is, by looking at a parameterized model we usually know the model class, and therefore both terms are used interchangeably.

5.1 Model Classes

In order to extract information from data, it is necessary to specify which general form the analysis result should have. We call the form or structure of the analysis result the **model class** or **architecture** to be derived from the data. This very general concept of a model should not be understood in a narrow sense that it must reflect a structure inherent in the data or represent the process of data generation.

(a) Examples for models in a more strict sense are linear models. The simplest case of a linear model is a regression line

$$y = ax + b, \tag{5.1}$$

describing the idealized relationship between attributes x and y. The parameters a and b still need to be determined. They should be chosen in such a way that the regression line fits best to the given data. A perfect fit, i.e., the data points lie exactly on the regression line, cannot be expected. In order to fit the regression line to the data, a criterion or error measure is required, defining how well a line with given parameters a and b fits the data. Such criteria or error measures are discussed in more detail in Sect. 5.2. The error measure itself does not tell us how to find the regression line with the best fit. Depending on the error measure, an algorithm is needed that finds the best fit or at least a good fit.

(b) Another example of an even simpler model is a constant value as a representative or prototype of a numerical attribute (or an attribute vector) either for the full dataset or for various subsets of it. The mean value or median are typical choices for such prototypes. Although the mean value and median can be viewed as purely descriptive concepts, they can also be derived as optimal fits in terms of suitable error measures, as we will see in Sect. 5.3.

(c) Multidimensional scaling as it was introduced in Sect. 4.3.2.3 can also be interpreted as a model. Here, the model is a set of points in the plane or 3D-space, representing the given data set. Again, a criterion or error measure is needed to evaluate whether a given set of points is a good or bad representation of the data set under consideration.

(d) All examples so far were numerical in nature, but there are, of course, also models for nominal data. A simple model class is that of propositional rules like *"If temperature = cold and precipitation = rain, then action = read book"* or *"If a customer buys product A, then he also buys product B with a probability of 30%"* can also be viewed as models. The latter rules are called association rules, because they *associate* features of arbitrary variables.

The four problems or models described in (a) the regression line, (b) the constant value, (c) multidimensional scaling, and (d) rules will serve as simple examples throughout this chapter. In most cases the models are parameterized as in the above first three examples. The number of parameters depends on the model. The constant value and the regression line have only one and two parameters, respectively, whereas $2n$ parameters are needed for multidimensional scaling for the representation of a data set with n records in the plane. The freedom of choice in the rule models are the variables and values that build the conditions in the rule's antecedent.

When we are looking for a regression function $f : \mathbb{R} \to \mathbb{R}$, we might decide that we want to restrict f to lines of the form $f(x) = a_1 x + a_0$, so that we have fixed set of two parameters. But we might also not know, whether a line, a quadratic function or even a polynomial of higher degree $f(x) = a_k x^k + \cdots + a_1 x + a_0$ might be the best choice. When we do not fix the degree of the polynomial for the regression function, we have not a fixed but a variable set of parameters. A propositional rule (example (d)) or sets of rules also belong to this case, as the number of conditions and rules is typically not fixed in advance. The same holds for decision trees (see Chap. 8), which can also not be represented in terms of a simple fixed set of parameters.

Another distinction can be made between, say, linear regression models and a propositional rule, by their applicability: a linear regression model can (in principle) be applied to all possible values of x and yields a resulting y. As they can be applied to all data from the data space, they are often called **global model**. This is different with rules: the consequent of the rule applies only in those cases where all conditions of the antecedent hold—there is no information returned otherwise. As such models can be applied to a somewhat limited fraction of the whole dataspace, they are called **local models**. We will use the term local model and **pattern** synonymously.

In any case, we need to define the class of possible models that we consider as possible candidates for solving our data analysis task. Project understanding aims at identifying the goals of the data analysis process and will therefore already restrict the choice of possible model classes. But even a well-defined goal will in most cases not imply a unique class of possible models. For instance, in the case of a regression function to be learned from the data, it is often not obvious what type of regression function should be chosen. Unless there is an underlying model—for instance, for a physical process—that provides the functional dependencies between the attributes, a polynomial of arbitrary degree or even other functions might be good candidates for a model.

Finding the "best" model for a given data set is not a trivial task at all, especially since the question what a good or the best model means is not easy to answer. How well the model fits the data is one criterion for a good model which can often be expressed in a more or less obvious manner, for instance, in the case of regression as the mean square or the mean absolute error of the values predicted by the model and the values provided in the data set. Such fitting criteria will be discussed in more detail in the next section.

Another important aspect is the simplicity of the model. Simpler models will be preferred for various reasons:

- They are usually easier to understand and to be interpreted.
- Their computational complexity is normally lower.
- A model can also be seen as a summary of the data. A complex model might not be a summary but, in the worst case, just more or less a one-to-one representation of the data. General structures in the data can only be discovered by a model, when it summarizes the data to a certain extent. The problem that too complex models often fail to reveal the general relations and structures in the data is called overfitting and will be discussed in more detail in Sect. 5.4.

Interpretability is often another requirement for the model derived from the data. In some cases, when only a classifier or regression function with a low error is the target, **black-box models** like neural networks might be accepted. But when the model should deliver an understandable description of the data or when the way how a classifier selects the predicted class needs an explanation, black-box models are not a proper choice. It is impossible to define a general measure for interpretability, because interpretability depends very much on the context. For instance, a quadratic regression function might be perfectly interpretable in a mechanical process where acceleration, velocity, and distance are involved, whereas in other cases a quadratic function would be nothing else than a black-box model, although a very simple one.

Computational aspect also plays a role in the choice of the model class. Finding the model from a model class that satisfies the desired criteria—data fitting and low model complexity and interpretability—best or at least reasonably well, is also a question of computational effort. In some cases, there might be simple ways to derive a model based on the given data. For example, for the regression line example (a), it might be the goal to minimize sum of squared errors. In this case, an explicit analytical solution for the best choice of the parameters a and b can be

Fig. 5.1 A small dataset to which a line is to be fitted

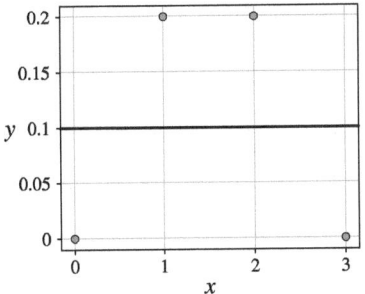

provided as we will see in the next section. Although the search space in the association rule example (d) is finite in contrast to the search space $\mathbb{R}^2$ for the regression line, it is so large that highly efficient strategies are needed to explore all potentially interesting rules.

Although all these aspects, the fitting criterion or score function, the model complexity, interpretability, and the required computational complexity for finding the model, are important, the focus is very often put on the fitting criterion that can be usually defined in a more or less obvious way.

5.2 Fitting Criteria and Score Functions

The choice of a model class determines only the general structure of the model, very often in terms of a set of parameters. In order to find the best or at least a good model for the given data, a fitting criterion is needed, usually in the form of an **objective function**

$$f : \mathcal{M} \to \mathbb{R}, \tag{5.2}$$

where $\mathcal{M}$ is the set of considered models.

In the case of example (a), a regression line of the form (5.1), a model is characterized by two real parameters, so that the objective function simplifies to

$$f : \mathbb{R}^2 \to \mathbb{R}. \tag{5.3}$$

Figure 5.1 shows a small dataset to which a line $y = ax + b$ should be fitted. The most common error measure is the mean squared error,[1] i.e.,

$$E(a, b) = \frac{1}{n} \sum_{i=1}^{n} (ax_i + b - y_i)^2. \tag{5.4}$$

The goal is to choose the parameters a and b in such a way that (5.4) is minimized. The error function (5.4) does not tell us directly how to find the best choice

---

[1] Note that minimizing the mean squared error or the sum of squared errors leads to the same solution, since the parameters a and b do not depend on the constant factor $\frac{1}{n}$.

Fig. 5.2 The mean squared
error for the dataset in
Fig. 5.1

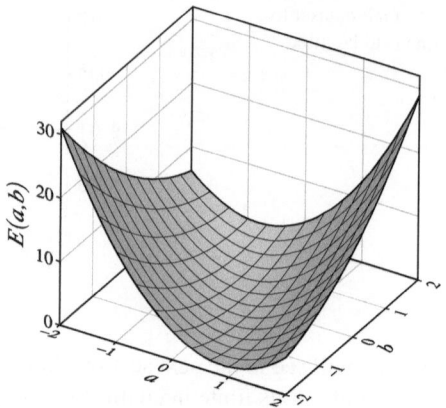

for the parameters a and b. It only provides a criterion telling us whether parameter combination (a_1, b_1) is considered better than parameter combination (a_2, b_2).
This is the case where $E(a_1, b_1) < E(a_2, b_2)$. When we plot the error function
(5.4), which is a function of the parameters a and b, we obtain the graph shown
in Fig. 5.2.

The mean squared error is not the only reasonable choice to measure how well a
model fits the data. Other obvious examples for alternative measures are the mean
absolute error

$$E(a, b) = \frac{1}{n} \sum_{i=1}^{n} |ax_i + b - y_i| \qquad (5.5)$$

or the mean Euclidean distance of the data points to the regression line. Instead
of the mean or the sum of the errors, the maximum of the errors could also be
considered.

All these error measures have in common that they only yield the value zero
when the regression line fits perfectly to the data and that they increase with larger
distance of the data points to the regression line. Properties of such error measures
and their advantages and disadvantages are discussed in more detail in Sect. 8.3.3.3.

Error functions for the even simpler model (b) (on page 83) of a single value m
representing a sample can be defined in a similar way. It can be shown that the
minimization of mean squared errors

$$E(m) = \frac{1}{n} \sum_{i=1}^{n} (x_i - m)^2 \qquad (5.6)$$

leads to the mean value

$$m = \bar{x} = \frac{1}{n} \sum_{i=1}^{n} x_i, \qquad (5.7)$$

Table 5.1 A cost matrix for
the tea cup classifier

True class	Predicted class	
	OK	broken
OK	0	c_1
broken	c_2	0

whereas the median minimizes the mean absolute error

$$E(m) = \frac{1}{n} \sum_{i=1}^{n} |x_i - m|.$$

(5.8)

For model (c), multidimensional scaling, three different error functions were already introduced in Sect. 4.3.2.3.

5.2.1 Error Functions for Classification Problems

A propositional rule may be regarded as a very simple classifier: if all conditions in the antecedent hold, it predicts the value of some target variable. If the rule does not apply, there is no prediction (which is usually wrong, of course). For classifiers, the most common error function is the **misclassification rate**, i.e., the proportion of records that are wrongly classified. However, the misclassification rate can be considered as a special form of a **cost function** or **cost matrix**. Such a cost matrix takes into account that misclassifications for one class can have stronger consequences or higher costs than another class. Consider the example of a tea cup producer. An automatic quality control device checks the produced cups before they are delivered to the shops. The quality control device classifies the tea cups into the two classes *broken* and *ok*. The device might not be absolutely correct, so that some broken cups are classified as *ok* and some intact cups are wrongly recognized as broken. Table 5.1 shows the four possible cases that can occur when a tea cup is classified. The diagonal of the table represents the situations where the tea cup was classified correctly. In these cases, the classifier does not cause any extra costs because of an erroneous classification.

When the device classifies an intact cup as broken, the cup will be wasted, and it must be produced again. Therefore, the costs c_1 caused by this error are equal to the production costs for one cup. The costs c_2 for a broken cup that was considered as intact are more difficult to estimate. The cup must be produced again, after the buyer's complaint. But these costs for reproduction are not caused by the erroneous classification, so that they are not included in c_2. The costs c_2 must cover the mailing costs for the reproduced cup and also the costs for the loss of reputation of the company for delivering broken cups. The costs for misclassifications differ significantly in certain cases. If instead of cups, safety critical parts like brake pads are produced, the costs for classifying a broken item as intact might be enormous, since they might lead to serious accidents.

Table 5.2 A general cost matrix

True	Predicted class		
class	c_1	c_2	... c_m
c_1	0	$c_{1,2}$	... $c_{1,m}$
c_1	$c_{2,1}$	0	... $c_{2,m}$
⋮	⋮	⋮	⋱ ⋮
c_m	$c_{m,1}$	$c_{m,2}$	... 0

When a cost matrix for a classification problem with m classes is provided as in Table 5.2, instead of the misclassification rate, the **expected loss** given by

$$\text{loss}(c_i|E) = \sum_{j=1}^{m} P(c_j|E)c_{ji} \tag{5.9}$$

should be minimized. E is the evidence, i.e., the observed values of the predictor attributes used for the classification, and $P(c_j|E)$ is the predicted probability that the true class is c_j given observation E.

Sometimes, classification can be considered as a special form of regression. A classification problem where each instance belongs to one of two classes can be reformulated as a regression problem by assigning the values 0 and 1 to the two classes. This means that the regression function must be learned from data where the values y_i are either 0 or 1. Since the regression function will only approximate the data, it will usually not yield the exact values 0 and 1. In order to interpret such a regression function as a classifier, it is necessary to assign arbitrary values to the classes 0 and 1. The obvious way to do this is to choose 0.5 as a threshold and consider values lower than 0.5 as class 0 and values greater than 0.5 as class 1.

When a classifier is learned as a regression function based on error measures as they are used for regression problems, this can lead to undesired results. The aim of regression is to minimize the approximation error which is not the same as the misclassification rate. In order to explain this effect, consider a classification problem with 1000 instances, half of them belonging to class 0 and the other half to class 1. Assume that there are two regression functions f and g as possible candidates to be used as a classifier.

- Regression function f yields 0.1 for all data from class 0 and 0.9 for all data from class 1.
- Regression function g always yields the exact and correct values 0 and 1, except for 9 data objects where it yields 1 instead of 0 and vice versa.

Although f does not yield the exact values 0 and 1 for the classes, it classifies all instances correctly when the threshold value 0.5 is used to make the classification decision. As a regression function, the mean squared error of f is 0.01. The regression function g has a smaller mean squared error of 0.009, but classifies 9 instances incorrectly. From the viewpoint of regression, g is better than f, and from the viewpoint of the misclassification rate, f should be preferred.

5.2.2 Measures of Interestingness

When we search for patterns, for instance, for single classification rules or association rules in the form *if A = a then B = b*, statistical measures of interest are typically used to evaluate the rule which is then considered as the "model." Assume that $A = a$ for n_a records, $B = b$ for n_b records, and $A = a$ and $B = b$ at the same time holds for n_{ab} records, where we have n records altogether. If $\frac{n_b}{n} \approx \frac{n_{ab}}{n_a}$, the rule has no significance at all, since replacing the records with $A = a$ by a random sample of size n_A, we would expect roughly the same fraction of records with $B = b$ in the random sample. The rule can only be considered as relevant if $\frac{n_{ab}}{n_a} \gg \frac{n_b}{n}$. In order to measure the relevance of the rule, we can compute the probability that a random sample of size n_a contains at least n_{ab} records with $B = b$. This probability can be derived from a hypergeometric distribution:

$$\sum_{i=n_{ab}}^{\min\{n_a,n_b\}} \frac{\binom{n_b}{i} \cdot \binom{n-n_b}{n_a-i}}{\binom{n}{n_a}}. \tag{5.10}$$

This can be interpreted as the *p*-value for the statistical test[2] with the null hypothesis that the rule applies just by chance. The lower this *p*-value, the more relevant the rule can be considered.

A very simple measure of interestingness that is often used in the context of association rules and frequent item set mining is the support or frequency of the rule in the data set. There, a rule is considered "interesting" if the support exceeds a given lower bound. As we have seen above, the support alone is not a good advisor when looking for unexpected, interesting rules. The focus on support in the context of association rules/frequent patterns has merely technical/algorithmic reasons.

These are two examples for statistical measures of interest. There are many others depending on the type of pattern we are searching [5]. We will encounter some of them in the subsequent chapters.

5.3 Algorithms for Model Fitting

The objective functions, mostly error functions, as they were explained in the previous section do not tell us directly how to find the best or at least a good model with respect to the fitting criterion. For this purpose, methods from optimization are needed.

5.3.1 Closed Form Solutions

In the best case, a closed-form solution for the optimization problem can be obtained directly. This is, however, not possible for most of the objective functions we

---

[2]The test here is Fisher's exact test.

consider. A positive example, for which we can find a closed form solution, is our case (a) (see page 82), the linear regression function (see also Sect. 8.3 for more details). For a minimum of the error function of linear regression, it is necessary that the partial derivatives with respect to the parameters a and b of the error function (5.4) vanish. This leads to the system of two linear equations

$$\frac{\partial E}{\partial a} = \frac{2}{n} \sum_{i=1}^{n} (ax_i + b - y_i)x_i = 0,$$

$$\frac{\partial E}{\partial b} = \frac{2}{n} \sum_{i=1}^{n} (ax_i + b - y_i) = 0.$$

The solution of this system of equations is

$$a = \frac{n \sum_{i=1}^{n} x_i y_i - (\sum_{i=1}^{n} x_i)(\sum_{i=1}^{n} y_i)}{n \sum_{i=1}^{n} x_i^2 - (\sum_{i=1}^{n} x_i)^2},$$

$$b = \bar{y} - a\bar{x}.$$

For the least squares error, such an analytical solution can be derived as long as the regression function is a linear function in the model parameters (for details, see Sect. 8.3).

Sometimes, constraints are defined on the parameters to be optimized, for instance, that the parameters must be positive values. For continuous optimization problems with such constraints, so-called Lagrange functions or Kuhn–Tucker conditions can be used to incorporate the constraints in the optimization.

However, in many cases the simple technique of computing partial derivatives of the objective function—with or without constraints—and solving for the parameters is not applicable, since the resulting system of equations is nonlinear or the derivatives might not exist, for instance, when the absolute error instead of the squared error is considered.

5.3.2 Gradient Method

If the model to be learned from the data is described by k real-valued parameters, i.e., the objective function is of the form $f : \mathbb{R}^k \to \mathbb{R}$, the problem of optimizing the objective function might be as simple as in the case of least squares linear regression as illustrated in Fig. 5.2, but for other models and other objective functions, it could look more like in Fig. 5.3 with a number of local minima and maxima. Even worse, the landscape in Fig. 5.3 is three-dimensional because the model is assumed to be described by two parameters only. But when the model is described by k parameters, where k is greater than two, the objective function corresponds to a landscape in $(k + 1)$ dimensions, so that is not even possible to plot it adequately. For instance, in the case of multidimensional scaling, the landscape defined by the objective function represents a $2n$-dimensional surface in a $(2n + 1)$-dimensional space, where n is

Fig. 5.3 A landscape with many local minima and maxima

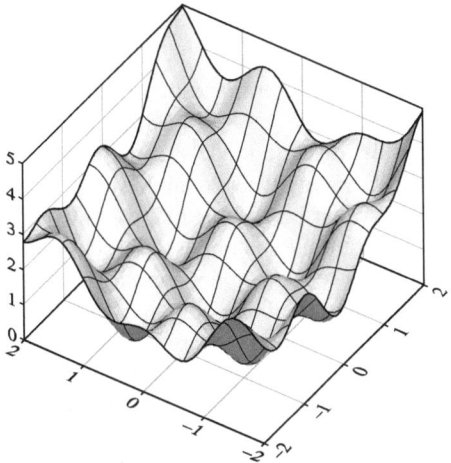

the number of instances in the data set displayed in the plane by multidimensional scaling.

When the objective function is differentiable, a **gradient method** can be applied. The gradient, i.e., the vector of partial derivatives with respect to the model parameters, points in the direction of steepest ascend. The idea of optimization based on a gradient method is to start at a random point—an arbitrary choice of the parameters to be optimized—and then to go a certain step in the direction of the gradient, when the objective function should be maximized, and in the opposite direction of the gradient, when the objective function should be minimized, leading to a new point in the parameter space. If this point yields a better value for the objective function, the gradient in this point is computed, and the next point in the direction or, respectively, in the opposite direction of the new gradient is chosen. This procedure is continued until no more improvements can be achieved or a fixed number of gradient steps has been carried out.

The stepwidth can be chosen constant. However, the problem with constant stepwidth is that, with a large stepwidth, one might "jump" over or oscillate around a local optimum. On the other hand, a very small stepwidth can lead to extreme slow convergence or even to starving, which means that the algorithm converges before a local optimum is reached. Therefore, an adaptive stepwidth is usually preferred, however, for the price of higher computational costs.

Applying a gradient method to minimize an objective function, it can only find the local minimum in the same "valley" of the landscape where the starting point is located.[3] Therefore, it is recommended to run a gradient method repeatedly, starting with different initial points in order to increase the chance to find the global or at least a good local optimum.

---

[3]For maximization, the same holds, except that the gradient method will just climb the "mountain" or even small "hill" on which the starting point is located.

5.3.3 Combinatorial Optimization

When the model is determined by k real-valued parameters, the domain $\mathcal{M}$ of the objective function (5.2) is $\mathbb{R}^k$ or a subset of it. It is also very common in data analysis to consider discrete models, usually leading to finite domains of the objective function. Rule mining is an example where the domain of the objective function is finite, at least when we assume that we only consider categorical attributes. In such cases, taking derivatives is impossible, so that neither a least squares technique as in the case of linear regression nor a gradient method is applicable. In principle, an **exhaustive search** of the finite domain $\mathcal{M}$ is possible; however, in most cases it is not feasible, since $\mathcal{M}$ is much too large. As an example, consider the problem of finding the best possible association rules with an underlying set of 1000 items (products). Every combination of items, i.e., every nonempty subset is a possible candidate set from which several rules may be constructed. The number of subsets alone contains $2^{1000} - 1 > 10^{300}$ elements, making it impossible to apply an exhaustive search.

Fortunately, in certain cases there are algorithms that exploit the specific structure of the finite search space and avoid a complete search by suitable branch and bound techniques. Examples for such techniques will be provided in Sect. 7.6.

5.3.4 Random Search, Greedy Strategies, and Other Heuristics

No matter, whether the domain of the objective function is a subspace $\mathbb{R}^k$, a finite set, or even a space composed of real and discrete model parameters, in many cases it is impossible to find the optimum by an analytical method. Therefore, it is very common to apply heuristic search strategies that cannot guarantee to find the global optimum but will—hopefully—find at least some good solutions.

In most general terms, we want to find the minimum (or maximum) of an objective function as in (5.2) where $\mathcal{M}$ is an arbitrary search space.

Random search is a brute force method that simply generates many elements of the search space $\mathcal{M}$ and chooses the best one among them. As long as we do not impose any assumptions on the search space and the objective function, random search or systematic search are the only ways to explore the search space. Usually, there is a neighborhood structure on the search space, and the objective function will often yield similar values for neighboring elements (models) in the search space. If the search space consists of real-valued parameters, then a very small change of the parameters will usually not lead to an extreme difference in the values of the objective function.

With such an assumption, it is normally possible to formulate a **greedy strategy** in order to find a good solution faster than by pure random search. **Hillclimbing** is a very simple example of a greedy strategy. Instead of generating arbitrary random solutions, hillclimbing starts from a random initial solution and then explores only the neighborhood of the so far best solution for better solutions. This means, hillclimbing restricts the random search to the neighborhood of the best solution that was found so far.

Gradient methods are also greedy strategies. In contrast to hillclimbing, they do not explore the neighborhood of the so far best solution randomly, but in a more systematic manner, namely in the direction of the gradient. Other greedy strategies do not try to greedily improve a given solution but construct a solution step by step in a greedy manner. Decision trees are usually built in this way (see Sect. 8.1). In each step, the attribute is chosen as a node that seems to lead to the best split for the data.

Greedy strategies have the advantage that they have the ability to quickly improve a given solution. But they will only find a local optimum of the objective function. Good optimization heuristics need a greedy component in order to be efficient and to find good solutions quickly. But the greedy component alone is too limited since it will get lost in local optima. Therefore, a random component is also required to provide a chance to escape from local optima. A simple mixture of random search and a greedy strategy is to apply the greedy strategy repeatedly starting from different initial random solutions.

Better optimization heuristic will mix the greedy and the random component in a more intelligent way by focusing on those parts of the search space which seem to have a higher chance to contain good solutions. **Genetic algorithms** and **evolution strategies** or, more generally, **evolutionary algorithms** are examples for such methods that combine random with greedy components, using a population of solutions in order to explore the search space in parallel and efficiently.

There are many other problem-dependent heuristics that are applied to model fitting. For instance, when the set of parameters can be split into disjoint subsets in such a way that for each subset, an analytical solution for the optimum can be provided, given that the parameters in the other subsets are considered to be fixed, **alternating optimization** is a useful technique. Alternating optimization computes the analytical solution for the parameter subsets alternatingly and iterates this scheme until convergence.

No matter which kind of heuristic optimization strategy is applied to fit a model to data, one should be aware that, except when an analytical solution is known or exhaustive search is feasible, the result may only represent a local optimum that may have a much lower quality than the global optimum.

5.4 Types of Errors

When fitting a model to data based on an error measure, a perfect fit will be seldom possible. A perfect fit with zero error is suspicious in most cases. It is often an indication that the model fitted to the data may be too flexible and is able to represent the data exactly, but does not show the actual structures or dependencies in the data. Consider a simple regression problem as it is illustrated in Fig. 5.4. The data are approximated by a regression line which fits quite well the data with a small error. The figure also shows a polynomial—the blue line—that interpolates the data points with zero error. If we just consider the error, the polynomial is a better choice than the simple regression line.

Fig. 5.4 An example for overfitting

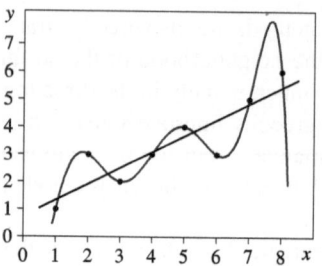

However, if we assume that the data points are corrupted with noise, the polynomial may mainly fit the noise rather than capture the underlying relationship. When we obtain further data points, the regression line is likely even a better approximation for these points than the polynomial. Therefore, a model with a smaller error is not necessarily a better fit for the data. In general, more complex models can fit the data better but have a higher tendency to show this bad effect called **overfitting**.

Once we have fitted a model to given data, the fitting error can be composed into four components.

- The pure or experimental error,
- the sample error,
- the lack of fit or model error, and
- the algorithmic error.

5.4.1 Experimental Error

The **pure error** or **experimental error** is inherent in the data and is due to noise, random variations, imprecise measurements, or the influence of hidden variables that cannot be observed. It is impossible to overcome this error by the choice of a suitable model. This error is inherent in the data. Therefore, it is also called **intrinsic error**. In the context of classification problems, it is also called **Bayes error**.

5.4.1.1 Bayes Error

Classification problems are often illustrated by very simplified examples as shown on the left in Fig. 5.5, where samples from two classes are shown. The samples from one class are depicted as red circles, and samples from the other class as blue circles. In this case, the two classes can be separated easily as indicated by the grey lines. The upper left and the lower right region correspond to the blue class, whereas the upper right and the lower left region correspond to the red class.

However, in real-world problems, the situation is usually more complicated. Classes tend to overlap. An example of such a case is shown in the middle of Fig. 5.5. Here the samples from both classes tend to scatter a little bit more, so that

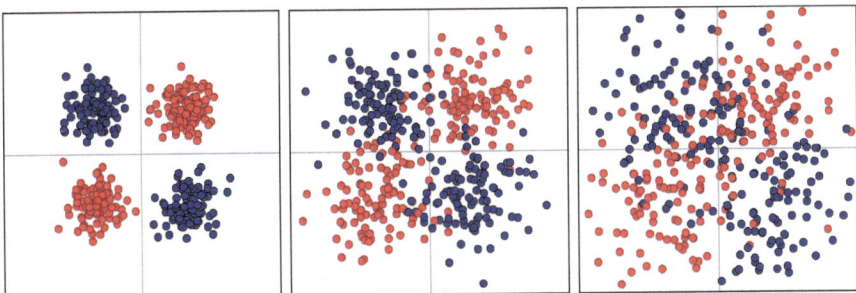

Fig. 5.5 A simple classification problem with perfect separation (*left*) and more difficult classification problems with slightly (*middle*) and strongly overlapping classes (*right*)

it is impossible to classify all objects correctly by a simple partition of the product space defined by the predictor variables. Nevertheless, the number of incorrectly classified samples will still be small, and the corresponding samples can be considered as exceptions. On the right in Fig. 5.5, however, we see an even more extreme example with a high overlap of the classes.

In all three cases, the upper left and the lower right regions are dominated by the blue class, and the upper right and the lower left regions by the red class. In the left chart of Fig. 5.5, the regions are exclusively occupied by the corresponding class, and in the middle chart, the regions are not exclusively occupied by one class, but at least highly dominated. Finally, in the right chart, although the domination of regions by classes is still there, incorrectly classified samples can no longer be considered as exceptions. Although there are two prototypical examples for each class—the centers of the four regions—the elements of the classes tend to deviate strongly from the ideal prototypes leading to the high overlap of the classes and making it impossible to achieve a high rate of correctly classified samples.

Of course, the data sets shown in Fig. 5.5 are artificial examples for illustration purposes. The samples from each class were drawn from two bivariate normal distributions, so that the data come from four normal distributions altogether. In all cases, the means of the distributions are the same, and only the variances differ, leading to a higher overlap of the classes for higher variance. Despite the overlap, the prototypical examples, i.e., the means of the normal distributions, differ significantly. However, there are applications where even the prototypical examples from different classes may be more or less identical. Consider, for instance, the example of three darts players who try to hit the middle of the dartboard (the so-called em bull's eye). The three dart players correspond to the classes, and the goal is to predict who has thrown the dart, given the hitting point of the dart. Assume that one of the players is professional, one is a hobby player, and one is a complete beginner.

The results from a number of trials of the three players may look like as shown in Fig. 5.6. The red dots correspond to the results of the professional player, the blue dots come from the hobby player, and the green dots were caused by the complete beginner. Figure 5.7 shows the results of the three player separately.

In all three cases, the results scatter around the center of the dartboard. The difference between the three players comes from *how much* their results scatter around

Fig. 5.6 Results of three dart players, one of which is a beginner (*green*), one a hobby player (*blue*), and one a professional (*red*)

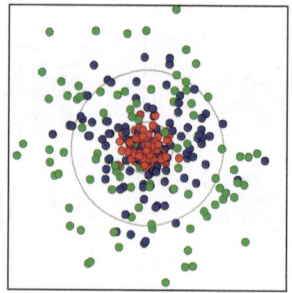

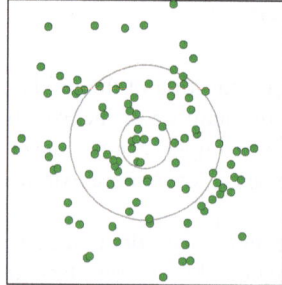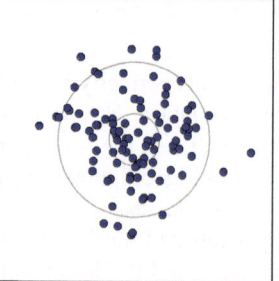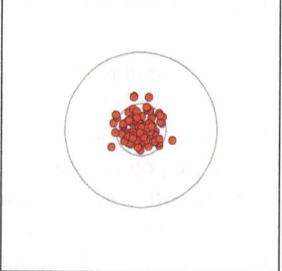

Fig. 5.7 Darts results for a beginner (*left*), a hobby player (*middle*), and a professional (*right*)

Fig. 5.8 Three normal distributions with identical means but different standard deviations (*green*: $\sigma = 5$, *blue*: $\sigma = 3$, *red*: $\sigma = 1$)

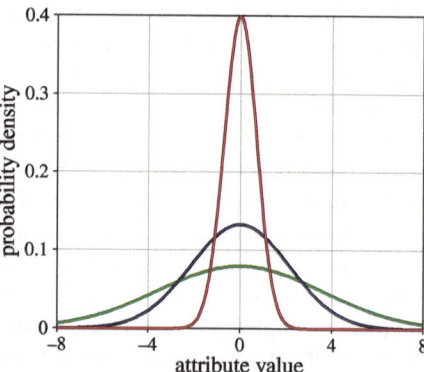

the center of the dartboard. The hits of the professional player will all be close to the center, whereas the complete beginner sometimes almost hits the center but has the highest deviation.

In terms of a classification problems, all three classes have the same prototype or center, and only the deviation differs. In such cases, a classifier can still be constructed, however, with a high misclassification rate.

We take an abstract simplified look at the dart player example. Assume that we draw samples from three univariate normal distributions, representing three different classes, with the same mean but with different variances, as shown in Fig. 5.8. This corresponds more or less to the classification problem of the dart players when we consider only the horizontal deviation to the center of the dartboard. Samples from the normal distribution with the smallest variance correspond to hits of the professional player, whereas the normal distribution with the largest variance represents to the complete beginner. In this theoretical example, it is obvious how to find the classification decision. Assuming that the three classes have the same frequency, an object, i.e., a simple value in this case, should be assigned to the normal distribution (class) with the highest likelihood, in other words, to the normal distribution with the highest value of the corresponding probability density function at the position of the given value. In this way, the region in the middle would be assigned to the normal distribution with smallest variance, the left and right outer region would be assigned to the normal distribution with highest variance, and the region in between would be assigned to the remaining normal distribution.

In this sense, it is obvious how to make best guesses for the classes, although these best guesses will still lead to a high misclassification rate. For the best guesses, there are clear boundaries for the classification decision. However, one should not mix up the classification boundaries with class boundaries. Classification boundaries refer to the boundaries drawn by a classifier by assigning objects to classes. These boundaries will always exist. But these classification boundaries do not necessarily correspond to class boundaries that separate the classes. In most cases, class boundaries do not even exist, since classes tend to overlap in real applications and cannot be clearly separated as in Fig. 5.5. This is due to the Bayes or the pure error.

5.4.1.2 ROC Curves and Confusion Matrices

For many classification problems, there are only two classes which the classifier is supposed to distinguish. Let us call the two classes *plus* and *minus*. The classifier can make two different kinds of mistakes. Objects from the class *minus* can be wrongly assigned to the class *plus*. These objects are called **false positives**. And vice versa, objects from the class *plus* can be wrongly classified as *minus*. Such objects are called **false negatives**. The objects that are classified correctly are called **true positives** and **true negatives**, respectively.

There is always a trade-off between false positives and false negatives. One can easily ensure that there are no false positives by simply classifying all objects as *minus*. However, this means that all objects from the class *plus* become false negatives. The other extreme is to classify all objects as *plus*, in this way avoiding false negatives but accepting that all objects from the class *minus* are false positives. A classifier must find a compromise between these two extremes, trying to minimize both the number of false positives and false negatives. A classifier biased to the class *plus* will have fewer false negatives but more false positives, whereas a classifier biased to the class *minus* will have fewer false positives but more false negatives. Cost functions, as they were explained in Sect. 5.2.1, are one way to introduce such biases to false positives or false negatives.

Some classifiers also provide for each object a probability, whether it belongs to a class or not. The usual decision is then to assign the object to the class with

Fig. 5.9 Examples of ROC
curves

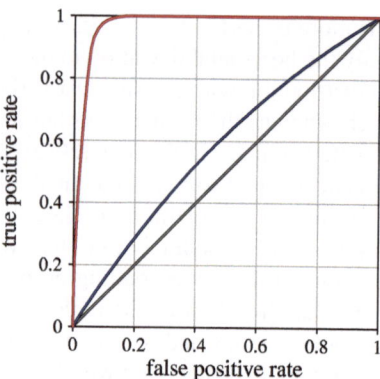

the highest probability. So in the case of the two classes *plus* and *minus*, we would
assign an object to the class *plus* if and only if the probability for this class is greater
than 0.5. But we could also decide to be more careful and to assign objects to the
class *plus* only when the corresponding probability is higher than $\tau = 0.8$, leading
to fewer false positives but more false negatives. If we choose a threshold τ for
assigning an object to the class *plus* lower than 0.5, we will reduce the number of
false negatives for the price of having false positives.

This trade-off between false positives and false negatives is illustrated by the
receiver operating characteristic or **ROC curve** showing the false positive rate
versus the true positive rate (in percent). Figure 5.9 shows examples for possible
ROC curves. For various choices of τ, a new point is drawn at the respective coordinates of false positive rate and true positive rate. These dots are connected to a
curve. The ROC curves in Fig. 5.9 are idealized. Normally, the ROC curves based
on sampled data look less smooth and more ragged.

The best case for a ROC curve would be to jump immediately from 0% to 100%,
so that we could have a classifier with 100% true positives and no false positives.
The red line shows a very good, but not perfect, ROC curve. We can have a very high
true positive rate together with a low false positive rate. The diagonal, shown as a
gray line, corresponds to pure random guessing, so that such a classifier has actually
learned nothing from the data, or there is no connection between the classes and the
attributes used for prediction. Therefore, the diagonal line is the worst case for a
ROC curve.

The **area under curve** (**AUC**), i.e., the area under the ROC curve, is an indicator how well the classifier solves the problem. The larger the area, the better the
solution for the classification problem. The area is measured relative to the area of
the square $[0, 100] \times [0, 100]$ in which the ROC curve is drawn. The lowest value
for AUC is 0.5, corresponding to random guessing, and the highest value is 1 for a
perfect classifier with no misclassifications. The blue line in Fig. 5.9 is a ROC curve
with a lower performance. The reason for the low performance may be due to the
Bayes error, but also because of other errors that will be discussed in the following
sections.

True class	Predicted class		
	Iris setosa	Iris versicolor	Iris virginica
Iris setosa	50	0	0
Iris versicolor	0	47	3
Iris virginica	0	2	48

Table 5.3 A possible confusion matrix for the Iris data set

When there are more than two classes, it is not possible to draw a ROC curve as described above. One can only draw ROC curves with respect to one class against all others.

The **confusion matrix** is another way to describe the classification errors. A confusion matrix is a table where the rows represent the true classes and the columns the predicted classes. Each entry specifies how many objects from a given class are classified into the class of the corresponding column. An ideal classifier with no misclassifications would have only entries different from zero in the diagonal.

Table 5.3 shows a possible confusion matrix for the Iris data set. From the confusion matrix we can see that all objects from the class setosa are classified correctly and no object from another class is wrongly classified as setosa. A few objects from the other classes—three from versicolor and two from virginica—are wrongly classified.

5.4.2 Sample Error

The **sample error** is caused by the fact that the data is only an imperfect representation of the underlying distribution of the data.

A finite sample, especially when its size is quite small, will seldom exactly reflect the true distribution of the probability distribution generating the data. According to the laws of large numbers, the sample distribution converges with probability one to the true distribution when the sample size approaches infinity. However, a finite sample can deviate significantly from the true distribution, although the probability for such a deviation might be small. The bar chart in Fig. 5.10 shows the result for throwing a fair die 60 times. In the ideal case, one would expect each of the numbers $1, \ldots, 6$ to occur 10 times. But for this sample, the sample distribution does not look uniform. Another source for sample errors are measurements with limited precision and round-off errors.

Sometimes the sample is also biased. Consider a bank that supplies loans to customers. Based on the data available on the customers who have obtained loans, the bank wants to estimate the probability for paying back a loan for new customers. However, the collected data will be biased in the direction of better customers because customers with a more problematic financial status have not been granted loans, and therefore, no information is available for such customers whether they

Fig. 5.10 A sample of
60 trials from a fair die

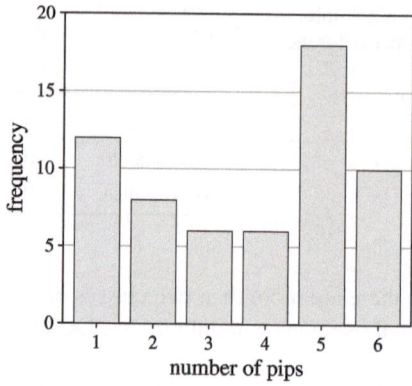

Fig. 5.11 A large error due
to lack of fit

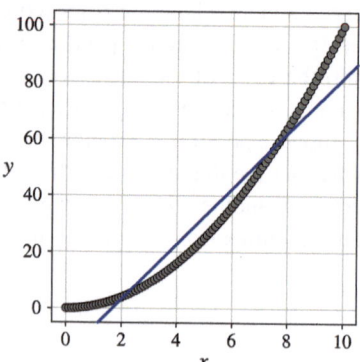

might have paid back the loan nevertheless. From the perspective of a representative sample of all applicants, we deal with a sample error when using the bank's database.

5.4.3 Model Error

A large error may be caused by a high pure error, but it may also be due to a **lack of fit**. When the set of considered models is too simple for the structure inherent in the data, no model will yield a small error. Such an error is also called **model error**. Figure 5.11 shows how a regression line is fitted to data with no pure error. But the data points originate from a quadratic and not from a linear function.

The line shown in the figure is the one with the smaller mean squared error. Such a line can always be computed, no matter from which true function the data come. But the line that fits best such data does not reflect the structure inherent in the data.

Unfortunately, it is often difficult or even impossible to distinguish between the pure error and the error due to the lack of fit. Simple models tend to have a large error due to the lack of fit, whereas more complex models lead to small errors for the given data but tend to overfitting and may lead to very large errors for new data.

A discussion of the problem of finding a compromise between too simple models with large and too complex models with overfitting is discussed in Sect. 5.5.

5.4.4 Algorithmic Error

There is the **algorithmic error** caused by a method that is used to fit the model or the model parameters. In the ideal case, when an analytical solution for the optimum of the objective function exists, the algorithmic error is zero or is only caused by numerical problems. But as we have seen in Sect. 5.3, in many cases an analytical solution cannot be provided, and heuristic strategies are needed to fit the model to the data.

Even if a model exists with a very good fit—the global optimum of the objective function—the heuristic optimization strategy might only be able to find a local optimum with a much larger error that is caused neither by the pure error nor by the error due to the lack of fit.

Most of the time, the algorithmic error will not be considered, and it is assumed that the heuristic optimization strategy is chosen well enough to find an optimum that is at least close to the global optimum.

5.4.5 Machine Learning Bias and Variance

The types of errors mentioned in the previous four subsections can be grouped into two categories. The algorithmic and the model errors can be controlled to a certain extend, since we are free to choose a suitable model and algorithm. These errors are also called **machine learning bias**. We have no influence on the pure or intrinsic error. The same applies to the sample error when the data to be analyzed have already been collected. The error caused by the intrinsic and the sample error sometimes is also called **variance**.

It is also well known from statistics that the mean squared error (MSE) of an estimator θ^* for an unknown parameter θ can be decomposed in terms of the variance of the estimator and its bias:

$$MSE = \text{Var}(\theta^*) + (\text{Bias}(\theta^*))^2. \tag{5.11}$$

Note that this decomposition deviates from the classification into model bias and variance above as it is popular in machine learning. The variance in (5.11) depends on the intrinsic error, i.e., on the variance of the random variable from which the sample is generated, and also on the choice of the estimator θ^* which is considered as part of the model bias in machine learning.

A more detailed discussion on the different meanings and usages of the terms variance and bias can be found in [3].

5.4.6 Learning Without Bias?

The different types of errors or biases discussed in the previous section have an interesting additional impact on the ability to find a suitable model for a given data set: if we have no model or learning bias, we will not be able to generalize. Essentially this means that we need to constrain either the types of models that are available or the way we are searching for a suitable model (or both). Tom Mitchell demonstrates this very convincingly in his hypothesis learning model [8]—in this toy world he can actually prove that in the unrestricted case of a boolean classification problem, the fitting models we can possibly find predict "false" in exactly half of the cases and "true" for the other half. This means that without any constraint we always leave all choices open. The learner or model bias is essential to put some sort of a priori knowledge into the model learning process: we either limit what we can express, or we limit how we search for it.

5.5 Model Validation

In the previous section, algorithms were discussed that fit a model from a predefined model class to a given data set. Complex models can satisfy a simple fitting criterion better than simple models. However, the problem of overfitting increases with the complexity of the model. Especially, when the model is built for prediction purposes, the error of the model based on the data set from which it was computed is usually smaller than for data that have not been used for determining the model. For instance, in Fig. 5.4 on page 94, the polynomial fits the data perfectly. The error for the given data set is zero. Nevertheless, the simple line might be a better description of the data, at least when we assume that the data are corrupted by noise. Under this assumption, the polynomial would lead to larger errors for new data, especially in those regions where it tends to oscillate. How do we find out, which model is actually suited best to our problem?

5.5.1 Training and Test Data

The most common principle to estimate a *realistic performance* of the model for unknown or future data is separating the data set for training and testing purposes. In the simplest case, the dataset is split into two disjoint sets, the **training data** which are used for fitting the model and the **test data** which only serve for evaluating the trained model but not for fitting the model. Usually, the training set is chosen larger than the test data set, for instance, 2/3 of the data are used for training, and 1/3 for testing.

One way to split the data into a training and a test set is a random assignment of the data objects to these two sets. This means that in average the distributions of the values in the original data set and in the training and the test data set should be roughly the same. However, by chance it can happen that the distributions may differ significantly. When a classification problem is considered, it is usually recommended to draw stratified samples for the training and the test set. **Stratification** means that the random assignments of the data to the test and the training set are carried out per class and not simply for the whole data set. In this way, it is ensured that the relative frequency in the original data set, the training, and the test set are the same.

Sometimes, it is not advisable to carry out a (stratified) random assignment of the data to the training and test set. Consider again the example of the producer of tea cups from Sect. 5.2.1 who wants to classify the cups automatically into *ok* and *broken*. Assume that six different types of cups are produced at the moment and in the future new types of cups might be introduced. Dividing the data set randomly into training and test data would not reflect the classification problem to be encountered in the future. If the producer had no intention to change the types of the cups, it would be correct to draw a random sample for testing from all data. In this way, the classifier will be trained and tested with examples from all six types of cups. But since new models of cups might be introduced in the future, this would yield an over-optimistic estimation of the classification error for future cups. A better way would be to use the data from four types of cups for training the classifier and to test them on the remaining two types of cups. In this way, we can get an idea of how the classifier can cope with cups it has never seen before.

This example shows how important it is in prediction tasks to consider whether the given data are representative for future data. When predictions are made for future data for which given data are not representative, extrapolation is carried out with a higher risk of wrong predictions. In the case of high-dimensional data, it cannot be avoided to have scarce or no data in certain regions of the space of possible values as we have discussed already in Sect. 4.2, so that we always have to be aware of this problem.

Sometimes, the data set is split into three parts: In addition to the training and the test data set, a **validation set** is also used. If, for instance, a classifier should be learned from data, but we do not know which kind of model is the most appropriate one for the classifier, we could make use of a validation set. All classifier models are generated based on the training data only. Then the classifier with the best performance on the validation set is chosen. The prediction error of this classifier is then estimated based on the test set.

5.5.2 Cross-Validation

The estimation of the fitting error for new data based on a test data set that has not been used for learning the model depends on the splitting of the original data set into training and test data. By chance, we might just be lucky that the test set contains more easy examples leading to an over-optimistic evaluation of the model.

Or we might be unlucky when the test set contains more difficult examples and the performance of the model is underestimated. **Cross-validation** does not rely on only one estimation of the model error, but rather on a number of estimations. For k-fold cross-validation, the data set is partitioned into k subsets of approximately equal size. Then the first of the k subsets is used as a test set, and the other $(k-1)$ sets are used as training data for the model. In this way, we get the first estimation for the model error. Then this procedure is repeated by using each of the other k subsets as test data and the remaining $(k-1)$ subsets as training data. Altogether, we obtain k estimations for the model error. The average of these values is taken as the estimation for the model error. Typically, $k = 10$ is chosen.

Small data sets might not contain enough examples for training when 10% are left out for testing. In this case, the **leave-one-out method**, also known as the **jackknife method**, can be applied which is simply n-fold cross-validation for a data set with n data objects, so that each time only one data object is used for evaluating the model error.

5.5.3 Bootstrapping

Bootstrapping is a resampling technique from statistics that does not directly evaluate the model error but aims at estimating the variance of the estimated model parameters. Therefore, bootstrapping is suitable for models with real-valued parameters. Like in cross-validation, the model is computed not only once but multiple times. For this purpose, k bootstrap samples, each of size n, are drawn randomly *with replacement* from the original data set with n records. The model is fitted to each of these bootstrap samples, so that we obtain k estimates for the model parameters. Based on these k estimates, the empirical standard deviation can be computed for each parameter to provide information how reliable the estimation of the parameter is.

Figure 5.12 shows a data set with $n = 20$ data points from which $k = 10$ bootstrap samples were drawn. For each of the bootstrap samples, the corresponding regression line is shown in the figure. The resulting parameter estimates for the intercept

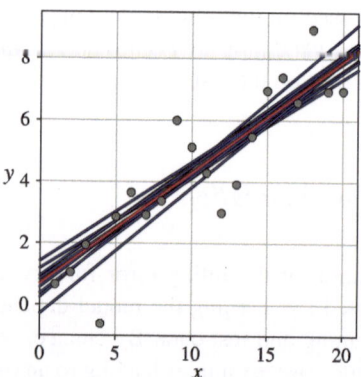

Fig. 5.12 10 lines fitted to a data set based on 10 bootstrap samples (*blue*) and their mean (*red*)

	Bootstrap sample	Intercept	Slope
Table 5.4 The estimators for the 10 bootstrap samples in Fig. 5.12	1	0.3801791	0.3749113
	2	0.5705601	0.3763055
	3	−0.2840765	0.4078726
	4	0.9466432	0.3532497
	5	1.4240513	0.3201722
	6	0.9386061	0.3596913
	7	0.6992394	0.3417433
	8	0.8300100	0.3385122
	9	1.1859194	0.3075218
	10	0.2496341	0.4213876
	Mean	0.6940766	0.3601367
	Standard deviation	0.4927206	0.0361004

and the slope of the regression line are listed in Table 5.4. The standard deviation for the slope is much lower than for the intercept, so that the estimation for the slope is more reliable. It is also possible to compute confidence intervals for the parameters based on bootstrapping [2].

The results from bootstrapping can be used to improve predictions as well by applying **bagging** (bootstrap aggregation). In the case of regression, one would use the average of the predicted values that the k models generated from the bootstrap samples yield. In the example in Fig. 5.12, for a given value x, all 10 lines would provide a prediction for y, and we would use the average of these predictions. For classification, one would generate k classifiers from the bootstrap samples, calculate the predicted class for all k classifiers, and use the most frequent class among the k predicted classes as the final prediction. Bagging will be introduced in more detail in Sect. 9.4.

5.5.4 Measures for Model Complexity

Complex models are more flexible and can usually yield a better fit for the training data. But how well a model fits the training data does not tell much about how well a model represents the inherent structure in the data. Complex models tend to overfitting as we have already seen in Fig. 5.4 on page 94.

Model selection—the choice of a suitable model—requires a trade-off between simplicity and fitting. Based on the principle of **Occam's razor**, one should choose the simplest model that "explains" the data. If a linear function fits the data well enough, one should prefer the linear function and not a quadratic or cubic function. However, it is not clear what is meant by "fitting the data well enough." There is a need for a trade-off between model simplicity and model fit. But the problem is that it is more or less impossible to measure these two aspects in the same unit. In order to combine the two aspects, **regularization** techniques are applied. Regularization is a general mathematical concept that introduces additional information in order to solve an otherwise ill-posed problem. A penalty term for more complex models can be incorporated into to the pure measure for model fit as a regularization technique for the avoidance of overfitting.

5.5.4.1 The Minimum Description Length Principle

The **minimum description length principle** (MDL) is one promising way to join measures for model fit and model complexity into one measure. The basic idea behind MDL is to understand modeling as a technique for data compression. The aim of data compression is to minimize the memory—measured in bits—needed to store the information contained in the data in a file. In order to recover the original data, the compressed data and the decompression rule is needed. Therefore, the overall size of the compressed data file is the sum of the bits needed for the compressed data plus the bits needed to encode the decompression rule. In principle, any file can be compressed to the size of one bit, say the bit with value 1, by defining the decompression rule "if the first bit is 1, then the decompressed file is the original data." This implies that the decompression rule contains the original data set and is therefore as large as the original data, so that no compression is achieved at all when we consider the compressed data and the decompression rule together. The same applies when we do not compress the original data at all. Then we need no space at all for the decompression rule but have not saved memory space at all, since we have not carried out any compression. The optimum lies somewhere in between by using a simple decompression rule allowing a reasonable compression.

The application of these ideas to model selection requires an interpretation of the model as a compression or decompression scheme, the (binary) coding of the model, and the compressed (binary) coding of the data. We illustrate the minimum description length principle by two simplified examples.

The first example is a classification task based on the data set shown in Table 5.5. The attribute C with the two possible values $+$ and $-$ shall be predicted with a decision tree based on the two binary attributes A and B with domains $\{a_1, a_2\}$ and $\{b_1, b_2\}$.

We consider the two decision trees shown in Fig. 5.13 to solve this classification task. The decision tree on the left-hand side is simpler but leads to two misclassifications. The last two records in Table 5.5 are classified wrongly by this decision. The

Table 5.5 A classification task

ID	1	2	3	4	5	6	7	8	9	10	11	12	13	14	15	16	17	18	19	20
A	a_1	a_1	a_1	a_1	a_1	a_1	a_1	a_1	a_1	a_1	a_2	a_2	a_2	a_2	a_2	a_2	a_2	a_1	a_2	a_2
B	b_1	b_2	b_2	b_2	b_1	b_1	b_1	b_2	b_1	b_2	b_1	b_1	b_1	b_1	b_1	b_1	b_1	b_1	b_2	b_2
C	$+$	$+$	$+$	$+$	$+$	$+$	$+$	$+$	$+$	$+$	$-$	$-$	$-$	$-$	$-$	$-$	$-$	$-$	$+$	$+$

Fig. 5.13 Two decision trees
for the data set in Table 5.5

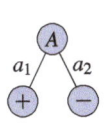

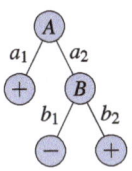

slightly larger decision tree on the right-hand side classifies all records in Table 5.5 correctly. If we were just concerned with misclassification rate for the training data, we would prefer the decision tree on the right-hand side. But the node with the attribute B is only required to correct two misclassifications—10% of the data set. This might just be an artificial improvement of the misclassification rate only for the training data set, and it might lead to overfitting.

In order to decide whether we should prefer the smaller or the larger decision tree, we interpret the two decision trees as compression schemes for the data set in the following way. Based on the corresponding decision tree, we can predict the attribute C, so that we do not have to store the value of the attribute C when we know the decision tree and the values of the attributes A and B for each record. However, this is only true for the larger decision tree. For the smaller decision tree, we have to correct the value of the attribute C for the two records that are misclassified by the smaller decision tree. The length of the compressed file for the data set based on any of the two decision trees is the sum of the lengths needed for coding

- the corresponding decision tree,
- the values of the records for the attributes A and B, and
- the corrected values for the attribute C only for the misclassified records.

The coding of the values of the records for the attributes A and B is needed for any decision tree and can be considered as a constant. The length needed for coding the decision tree depends on the size of the tree. The larger the tree, the more bits are needed to store the tree. However, with a larger tree, the misclassification rate can be reduced, and we need less bits for coding the corrections for attribute C for the misclassified records. When we want to minimize the overall length needed to store the compressed data, we need a compromise between a smaller decision tree with a higher misclassification rate and larger decision with a lower misclassification rate. The smaller decision tree will need less space for its own coding but more for the corrections of the misclassified records, whereas the larger decision tree needs more space for its own coding but can save space due to a lower number of misclassifications. According to the minimum description length principle, we should choose the decision tree with the minimum number of bits needed to code the tree itself and the corrected values for the attribute C.

Of course, the number of bits needed for the coding depends on the binary coding scheme we use for encoding the decision tree and the corrections for the values of the attribute C. If we have a highly efficient coding scheme for decision trees but an inefficient one for the corrections for the values of the attribute C, larger trees would be preferred. The naive MDL approach will ignore this problem and simply try to find the most efficient binary coding for both parts. However, there are more general

Fig. 5.14 A constant (*grey*), a line (*blue*), and a quadratic curve (*red*) fitted to a data set

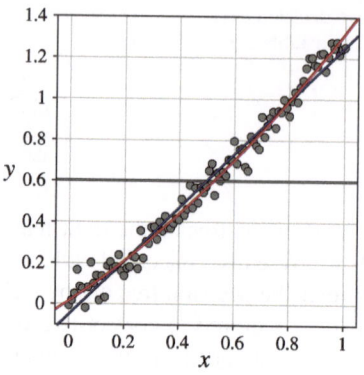

Table 5.6 A regression problem

ID	1	2	3	4	5	6	7	8	9
X	0.00	1.00	2.00	3.00	4.00	5.00	6.00	7.00	8.00
Y	1.19	1.33	2.03	1.21	1.80	2.33	1.63	2.88	2.84

concepts of universal codes and universal models freeing the minimum description length principle from the dependency on the specific coding scheme. For a detailed introduction to universal codes and universal models that are out of the scope of this book, we refer to [4].

As a second example for the application of the minimum description length principle, we consider a regression problem. Figure 5.14 shows a simple data set to which line and a quadratic curve is fitted. Which of these two models should we prefer? Of course, the quadratic curve will lead to a smaller error for the training data set than the simple line. We could also think of a polynomial of higher degree that would reduce the error even more. But then we have to face the problem of overfitting in the same as we have to take this for larger decision trees into account.

For illustration purposes, we consider an even simpler regression problem provided by the data set in Table 5.6.

What would be the best model for this data set? Should we describe the relation between X and Y by a simple constant function, by a linear, or even a quadratic function? Of course, the quadratic function would yield the smallest, and the constant function the largest error. But we want to avoid overfitting by applying the minimum description length principle. As in the example of the classification problem, it will suffice to consider the naive approach and not to bother about universal codings and universal models. Instead of the decision tree, our models are now functions with one, two, and three real-valued parameters. The errors are now also real numbers and not just binary values as in the case of the decision trees. If we insist on exact numbers, the coding of a single error value could need infinite memory, since a real number can have infinitely many digits. To avoid this problem, we restrict our precision to two digits right after the decimal point. For reasons of simplicity, we do not consider a binary coding of the numbers, but a decimal coding. A real number is

Fig. 5.15 A constant (*gray*),
a straight line (*blue*), and a
quadratic curve (*red*) fitted to
the data set of Table 5.6

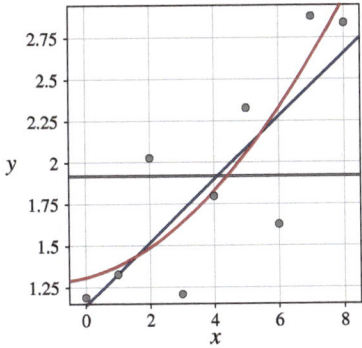

Table 5.7 The errors for the three regression functions

ID	1	2	3	4	5	6	7	8	9
$y = 1.92$	0.73	0.59	−0.11	0.71	0.12	−0.41	0.29	−0.96	−0.92
$y = 1.14 + 0.19x$	−0.05	0.00	−0.51	0.50	0.10	−0.24	0.65	−0.41	−0.18
$y = 1.31 + 0.05x + 0.02x^2$	0.12	0.05	−0.54	0.43	0.03	−0.27	0.70	−0.24	0.15

coded backwards starting from the lowest digit, in our case the second digit after the decimal point. Therefore, the numbers 1.23, 2.05, 0.06, and 0.89 would be coded as 321, 502, 6, and 98, respectively. Note that smaller numbers require less memory because we do not code leading zero digits. We also would have to take the sign of each number into account for the coding. But we will neglect this single bit here in order not to mix the decimal coding for the numbers with a binary coding for the sign. If we use a binary coding for the numbers, the sign integrates naturally to the coding as an additional bit. Figure 5.15 shows a plot of the data set and least squares fit of constant, linear, and quadratic functions.

Table 5.7 lists the errors for the constant, linear, and quadratic functions that have been fitted to the data. How many decimal digits do we need for coding of the data set when we use a constant function? We need three digits for the constant 1.92 representing the function. We also need to encode the errors in order to recover the original values of the attribute Y from our constant function. For the errors of this constant function, we always need to encode to digits. Only the digit before the decimal point is always correct. Altogether, the coding of the data set with the constant function requires $3 + 9 \cdot 2 = 21$ decimal digits.

What about the linear function? The function itself requires the coding of the two coefficients 1.14 and 0.19 for which we need 5 decimal digits. The coding errors require two decimal digits each time, except for the data points with the ID 1 and 2, for which we need only one and zero decimal digits. This means that we have altogether $5 + 7 \cdot 2 + 1 + 0 = 19$ decimal digits. Similar considerations for the quadratic curve lead to $5 + 7 \cdot 2 + 2 \cdot 1 = 20$ decimal digits.

This means that, in terms of our extremely simplified MDL approach, the linear function leads to the most efficient coding for the data set, and we would therefore prefer the linear regression function over the other ones.

It should be emphasized again that there is a more rigorous theory for the minimum description length principle that avoids the problem of finding the most efficient coding and the restriction to a fixed precision for the representation of real numbers. But the naive approach we have described here often suffices to give an idea how complex the chosen model should be.

5.5.4.2 Akaike's and the Bayesian Information Criterion

Akaike's information criterion [1] is based on entropy considerations in terms of how much additional information a more complex model provides with respect to the (unknown) stochastic model generating the data. For large sample sizes, Akaike could show that under these assumptions one should choose the model that minimizes Akaike's information criterion **AIC**, originally called "an information criterion"

$$AIC = 2k - 2\ln(L), \tag{5.12}$$

where k is the number of parameters of the model, and L is the likelihood of the corresponding model. The aim is to minimize the value in (5.12), which is achieved by a compromise between low model complexity—demanding a small value for k— and a good fit, i.e., a high likelihood which is achieved easier for complex models.

For a regression problem with the assumption that the underlying measurements of the "true" function are corrupted with noise coming from a normal distribution, the minimization problem in (5.12) can be simplified to minimizing

$$AIC_{Gauss} = 2k + n\ln(MSE), \tag{5.13}$$

where MSE is the mean squared error

$$MSE = \frac{1}{n}\sum_{i=1}^{n}(f(\mathbf{x}_i) - y_i)^2 \tag{5.14}$$

with regression function f and n data points $(\mathbf{x}_i, y_i)$.

Equation (5.13) applied to our simple regression example described in Table 5.6 yields the values -6.678, -12.130, -9.933 for the constant, the linear, and the quadratic regression functions obtained from the least squares method. Since small values for Akaike's information criterion are better than larger, we would prefer the linear function that was also the best choice in the context of MDL.

The **Bayesian information criterion** (**BIC**) [9] is derived in the context of Bayesian inference for large sample sizes and under certain assumptions on the type of probability distribution that generates the data:

$$BIC = k\ln(n) - 2\ln(L). \tag{5.15}$$

With the same assumptions as for (5.13), BIC becomes in the context of regression

$$BIC_{Gauss} = k \ln(n) + n(MSE)/\sigma^2, \tag{5.16}$$

where σ^2 is the (estimated) variance for the underlying normal distribution that causes the noise.

5.6 Model Errors and Validation in Practice

Although we will introduce realizations of model classes and different optimization algorithms later in this book, we will already demonstrate how to measure error and estimate model accuracy using KNIME and R in this section.

5.6.1 Errors and Validation in KNIME

KNIME offers a number of modules to estimate errors. Most prominently, the *Scorer* node computes a confusion matrix given two columns with the actual and predicted class label. There is also a node to plot a ROC curve and an entropy scorer, which allows one to compute the class–class purities between two columns. So the standard error metrics are available as individual nodes. Figure 5.16 shows the use of the scorer node in practice. The trained Naive Bayes classifier is applied to a second data set, and the output is fed into the scorer node which compares the target with the predicted class. The output of this scorer is a confusion matrix (which is also available as node view) and a second matrix listing some well-known error measures.

More interestingly, however, are methods to run cross validation or other validation techniques. KNIME offers those in the form of so-called meta nodes which encapsulate a series of other nodes. Figure 5.17 shows the inside of such a node. Besides the node to train a model (a neural network in this case) and apply the network to unseen data, there are two special nodes: the begin of the cross validation look which takes care of the repeated partitions of the data and the end node (*X-Aggregator*) which collects the information from all runs.

The special type of "loop nodes" are also available individually in KNIME, and the user can then assemble much more complex looping constructs, but for convenience, frequently used setups, such as the cross validation shown here, are available in preconfigured meta nodes.

5.6.2 Validation in R

In order to apply the idea of using separate parts of a data set for training and testing, one needs to select random subsets of the data set. As a very simple example, we

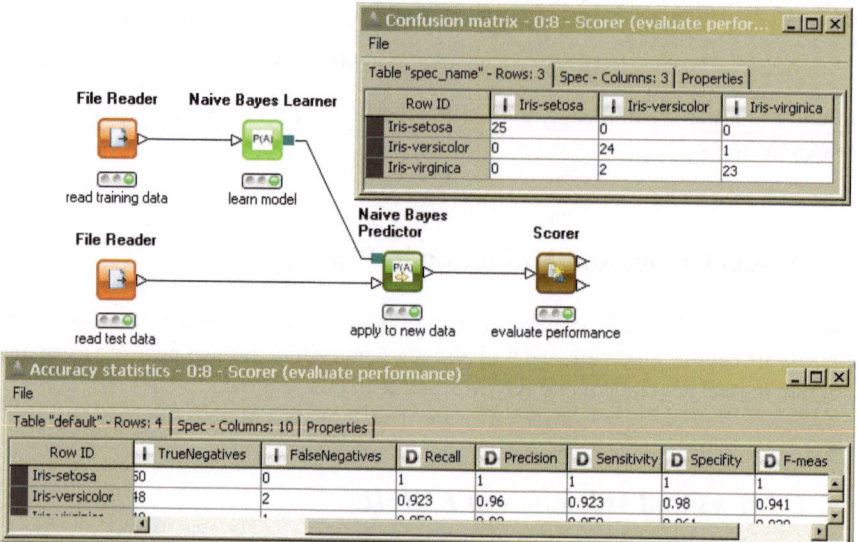

Fig. 5.16 The use of the scorer node together with the two tables it produces on its outports. One table holds the confusion matrix, the second output holds some well-known error measurer

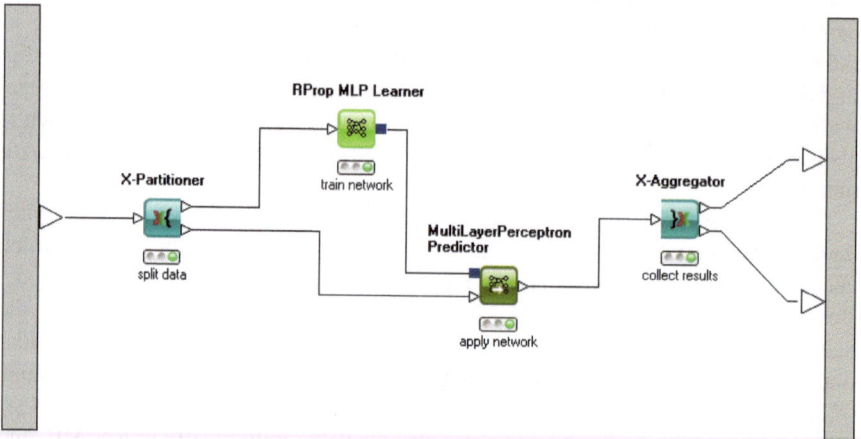

Fig. 5.17 Preconfigured meta nodes allow one to run cross validation in KNIME. In this workflow a neural network is repeatedly applied to different partitions of the incoming data

consider the Iris data set that we want to split into training and test sets. The size of the training set should contain 2/3 of the original data, and the test set 1/3. It would not be a good idea to take the first 100 records in the Iris data set for training purposes and the remaining 50 as a test set, since the records in the Iris data set are ordered with respect to the species. With such a split, all examples of Iris setosa and Iris versicolor would end up in the training set, but none of Iris versicolor, which would form the test set. Therefore, we need random sample from the Iris data set. If the records in the Iris data set were not systematically orderer, but in a random order, we could just take the first 100 records for training purposes and the remaining 50 as a test set.

Sampling and orderings in R provide a simple way to shuffle a data set, i.e., to generate a random order of the records.

First, we need to know the number n of records in our data set. Then we generate a permutation of the numbers $1, \ldots, n$ by sampling from the vector containing the numbers $1, \ldots, n$, generated by the R-command c(1:n). We sample n numbers without replacement from this vector:

```
> n <- length(iris$Species)
> permut <- sample(c(1:n),n,replace=F)
```

Then we define this permutation as an ordering in which the records of our data set should be ordered and store the shuffled data set in the object iris.shuffled:

```
> ord <- order(permut)
> iris.shuffled <- iris[ord,]
```

Now define how large the fraction for the training set should be—here 2/3—and take the first two thirds of the data set as a training set and the last third as a test set:

```
> prop.train <- 2/3
> k <- round(prop.train*n)
> iris.training <- iris.shuffled[1:k,]
> iris.test <- iris.shuffled[(k+1):n,]
```

The R-command sample can also be used to generate bootstrap samples by setting the parameter replace to TRUE instead of F (FALSE).

5.7 Further Reading

The book [6] introduces data mining approaches in a similar view as in this chapter.

A discussion of the problems of bias and errors from the viewpoint of machine learning and statistical learning theory can be found in [8] and [7], respectively.

References

1. Akaike, H.: A new look at the statistical model identification. IEEE Trans. Autom. Control **19**, 716–723 (1974)
2. Chernick, M.: Bootstrap Methods: A Practitioner's Guide. Wiley, New York (1999)
3. Dietterich, T., Kong, E.: Machine learning bias, statistical bias, and statistical variance of decision tree algorithms. Technical report, Oregon State University, USA (1995)

4. Grünwald, P.: The Minimum Description Length Principle. MIT Press, Cambridge (2007)
5. Hamilton, H.J., Hilderman, R.J.: Knowledge Discovery and Measures of Interest. Springer, New York (2001)
6. Hand, D., Mannila, H., Smyth, P.: Principles of Data Mining. MIT Press, Cambridge (2001)
7. Hastie, T., Tibshirani, R., Friedman, J.: The Elements of Statistical Learning: Data Mining, Inference, and Prediction, 2nd edn. Springer, New York (2009)
8. Mitchell, T.: Machine Learning. McGraw-Hill, New York (1997)
9. Schwarz, G.: Estimating the dimension of a model. Ann. Stat. **6**, 461–464 (1978)

Chapter 6
Data Preparation

In the data understanding phase we have explored all available data and carefully checked if they satisfy our assumptions and correspond to our expectations. We intend to apply various modeling techniques to extract models from the data. Although we have not yet discussed any modeling technique in greater detail (see Chaps. 7ff), we have already glimpsed at some fundamental techniques and potential pitfalls in the previous chapter. Before we start modeling, we have to prepare our data set appropriately, that is, we are going to modify our dataset so that the modeling techniques are best supported but least biased.

The data preparation phase can be subdivided into at least four steps. The first step is data selection and will be discussed in Sect. 6.1. If multiple datasets are available, based on the results of the data understanding phase, we may select a subset of them as a compromise between accessibility and data quality. Within a selected dataset, we may concentrate on a subset of records (data rows) and attributes (data columns). We support the subsequent modeling steps best if we remove all useless information, such as irrelevant or redundant data. The second step involves the correction of individual fields, which are conjectured to be noisy, apparently wrong or missing (Sect. 6.2). If something is known, new attributes may be constructed as hints for the modeling techniques, which then do not have to *rediscover* the usefulness of such transformations themselves. For some modeling techniques, it may even be necessary to construct new features from existing data to get them running. Such data transformations will be discussed in Sect. 6.3. Finally, most available implementations assume that the data is given in a single table, so if data from multiple tables have to be analyzed jointly, some integration work has to be done (Sect. 6.5).

6.1 Select Data

Sometimes there is a lot of data available, which sounds good in the first place, but only if the data is actually *relevant* for the given problem. By adding new columns

M.R. Berthold et al., *Guide to Intelligent Data Analysis*,
Texts in Computer Science 42,
DOI 10.1007/978-1-84882-260-3_6, © Springer-Verlag London Limited 2010

with random values to a table of, say, customer transactions, the table may get an impressive volume but does not carry any useful information. Even worse, these additional attributes are considered harmful as it is often difficult to identify them as being irrelevant (we will illustrate this point in a minute). This is the main reason why it is important to restrict the analysis process to potentially useful variables, but other reasons exist such as efficiency: the computational effort often depends greatly on the number of variables and the number of data records. The results of the data understanding phase greatly help to identify useless datasets and attributes, which have to be withheld from the analysis. But for other attributes, things may not be that clear, or the number of attributes that are classified as potentially useful is still quite large. In the following, we investigate how to select the most promising variables (Sect. 6.1.1) and, alternatively, how to construct a few variables that nevertheless carry most of the information contained in all given attributes (Sect. 6.1.2). Having selected the data columns, we finally discuss which data rows should enter our analysis in Sect. 6.1.3.

6.1.1 Feature Selection

Suppose that we have some target or response variable R that we want to predict. For the sake of simplicity, let us assume that we have no data that is relevant for the final outcome of R. What happens if we nevertheless try to build a model that explains or predicts the value of R given some other irrelevant variables? How will it perform? We are going to answer this question following a thought experiment.

Suppose that the probability for a positive response is p. We pick some binary variable A with $P(A = yes) = q$. In this experiment, A has nothing to do with the response R, but usually we do not know this in advance. As R and A are (statistically) independent, we expect the joint probability of observing, say, $A = yes$ and $R = yes$ to be $P(A = yes \land P(R = yes) = P(A = yes) \cdot P(R = yes) = p \cdot q$, as shown in Table 6.1a. Ignoring the variable A and assuming $p = 0.4$, having no other information at hand, we would predict the more likely outcome $R = no$. In the long run, this prediction would be correct in 60% ($= p$) of the cases and wrong in 40% ($= 1 - p$).

Next we take A into account with $q - 0.7$. How would this additional knowledge of A influence our prediction of R? When looking at the expected number of records in Table 6.1b, the knowledge of A does not suggest a different decision: in both rows of the table ($A = yes$ and $A = no$) we have $P(R = no) > P(R = yes)$, so the chances of getting $R = no$ are higher, and this would be our prediction. Clearly, the irrelevant information did not help to improve the prediction, but at least it did no harm—at the first glance. There is, however, one problem: The *true probabilities p* and *q* are not known in advance but have to be estimated from the data (introducing a sample error, see Sect. 5.4). Suppose that our data set consists of 50 records as shown in Table 6.1c. Note that, by pure chance, the sample from the whole population slightly differs in our database from the expected situation: only 2 records out of 50

Table 6.1 (a) Probability distribution of independent variables. (b) Expected number of cases for a total number of $n = 50$ records. (c) Observed number of cases

(a)					(b)				(c)		
		R			$n = 50$	R			$n = 50$	R	
		yes	no			yes	no			yes	no
A	yes	$p \cdot q$	$(1-p) \cdot q$	q	A yes	14	21	35	A yes	14	21	35
	no	$p \cdot (1-q)$	$(1-p) \cdot (1-q)$	$1-q$	no	6	9	15	no	8	7	15
		p	$1-p$			20	30			22	28	

are different (in $A = no$: $R = yes$ instead of $R = no$). Now there is a slight majority for $R = yes$ given $A = no$. If, based on this sample, the prediction of R is changed to yes given $A = no$, the rate of accurate predictions goes down from 60% to 52%. Obviously, such situations occur easily if the numbers in the table are relatively small. If the dataset consists of many more records, say 10,000, the estimates of the joint probabilities are much more robust—as long as the number of cells in our table remains constant. The number of cells, however, increases exponentially in the number of variables; therefore, the number of cases per cell decreases much faster than we can gather new records: To keep the number of cases per cell constant, we would have to double the number of records if we add a single binary variable.

Many methods we will discuss in the following chapters come along with their own weapons against the danger of overfitting, which generally grows with the number of (irrelevant) attributes. By removing irrelevant (and redundant) attributes in advance, the variance of the estimated model is reduced, because we have less parameters to estimate, and the chances of overfitting decrease. In some cases, the application of feature selection methods seems to be more important than the choice of a modeling technique [16].

The goal of **feature selection** is to select an *optimal* subset of the full set of available attributes $\mathcal{A}$ of size n. The more attributes there are, the wider is the range of possible subsets: the number of subsets increases exponentially in the size of $\mathcal{A}$. Feature selection typically implies two tasks: (1) the selection of some evaluation function that enables us to compare two subsets to decide which will perform better and (2) a strategy (often heuristic in nature) to select (some of) the possible feature subsets that will be compared against each other via this measure.

Selecting the k Top-Ranked Features   If we assume that an *attribute subset* is only as good as the *individual attributes*, a straightforward approach is to use an *attribute evaluation* function to rank the attributes only rather than the full subsets. The selection strategy (step 2) could then be as simple as "*take the k top-ranked features.*" This leads to an extremely fast selection process, because the evaluation function has to be evaluated only once per attribute (rather than evaluating all possible subsets once).

If a target variable is given (searching for explanations or predictions), the evaluation function may investigate into the correlation of the considered attribute and

Table 6.2 The full dataset consists of 9 repetitions of the four records on the left plus the four records in the middle, which differ only in the last record. To the right, there are contingency tables for all four variables vs. the target variable

A B C D Target	A B C D Target	A	Target	B	Target	C	Target	D	Target
			no yes		no yes		no yes		no yes
9× $\begin{matrix} + + + - \ no \\ + - + - \ yes \\ - + + - \ yes \\ - - - + \ no \end{matrix}$, 1× $\begin{matrix} + + + - \ no \\ + - + - \ yes \\ - + + - \ yes \\ - - + + \ no \end{matrix}$		+	10 10	+	10 10	+	11 20	+	10 0
		−	10 10	−	10 10	−	9 0	−	10 20

the target attribute. For categorical data, Pearson's χ^2-test (see Sect. A.4.3.4) indicates the goodness of fit for the observed contingency table (of target and considered variable) versus the table expected from the marginal observations. A large deviation from the expected distribution points towards a dependency among both attributes, which is desirable for predictive purposes. Another evaluation function is the information gain or its normalized variants *symmetric uncertainty* or *gain ratio*, which will be discussed in Sect. 8.1. The information gain itself is known to prefer attributes with a large number of values, which is somewhat compensated by the normalized variants (see also Sect. 6.3.2).

As an example, consider the dataset shown in Table 6.2. As we can see from the contingency tables, the values $+/-$ of the attributes A and B are equally distributed among the target values *no/yes* and thus appear useless at first glance. When observing $C = -$, the target value distribution *no* : *yes* is $9 : 0$, which indicates a very clear preference for *no* and makes C valuable. When observing $D = +$, the target value distribution is $10 : 0$, which is even slightly better because it covers one more case. All the mentioned evaluation measures provide the same ranking for this set of variables: $D - C - A, B$ (with A and B having the same rank).

The interesting observation is that in this artificial example the attributes A and B together are sufficient to perfectly predict the target value, which is impossible with attributes C and D, each of them individually ranked higher than A and B. Both, C and D, are quite good in predicting the target value *no*, but they do not complement one another. By our initial assumption the evaluation functions considered so far do not take the interaction among features into account but look at the performance of individual variables only. They do not recognize that C and D are almost redundant, nor do they realize that A and B jointly perform much better than individually. This holds for all evaluation functions that analyze the contingency table of individual attributes only.

However, we may arrive at different conclusions if we account for the values of other attributes, say B, C, D, while evaluating a variable, say A, as it is done by the Relief family of measures. We discuss only one member (Relief, see Table 6.3) and refer to the literature for the more robust extensions ReliefF and RReliefF [15]. An advantage of the Relief family is that it is easily implemented for numerical and categorical attributes (whereas many of the aforementioned evaluation measures are only available for categorical data in most data analysis toolboxes).

Table 6.3 Evaluation of attributes—Relief algorithm. The nearest hit or miss is selected upon the same diff function (sum over all attributes). Rather than applying the algorithm to all data, it may be applied to a sample only to reduce the high computational costs of handling big data sets (caused by the neighborhood search)

Algorithm Relief($\mathcal{D}, \mathcal{A}, C$) $\rightarrow w[\cdot]$	where
input: data set $\mathcal{D}$, $|\mathcal{D}| = n$, attribute set $\mathcal{A}$, target variable $C \in \mathcal{A}$ output: attribute weights $w[A]$, $A \in \mathcal{A}$	$\mathrm{diff}(A, \mathbf{x}, \mathbf{y}) = \begin{cases} 0 & \text{if } \mathbf{x}_A = \mathbf{y}_A \\ 1 & \text{otherwise} \end{cases}$ for categorical attributes A and
1 set all weights $w[A] = 0$ 2 **for all** records $\mathbf{x} \in \mathcal{D}$ 3 find nearest hit $\mathbf{h}$ (same class label $\mathbf{x}_C = \mathbf{h}_C$) 4 find nearest miss $\mathbf{m}$ (different class label) 5 **for all** $A \in \mathcal{A}$: 6 $w[A] = w[A] - \frac{\mathrm{diff}(A,\mathbf{x},\mathbf{h})}{n} + \frac{\mathrm{diff}(A,\mathbf{x},\mathbf{m})}{n}$	$\mathrm{diff}(A, \mathbf{x}, \mathbf{y}) = \frac{|\mathbf{x}_A - \mathbf{y}_A|}{\max(A) - \min(A)}$ for numerical attributes A.

Intuitively, the Relief measure estimates how much an attribute A may help to distinguish between the target classes by means of a weight $w[A]$: if $w[A]$ is large, the attribute is useful. The weights are determined incrementally and respect the interaction between the attributes by the concept of "nearest" or "most similar" records, which takes all attributes (rather than just A) into account. Given some record $\mathbf{x}$, if there are very similar *records* $\mathbf{h}$ and $\mathbf{m}$ with the same or different target label, the features present in this record are apparently not of much help in predicting the target value (and the weights $w[A]$ for all attributes A remain unchanged because the positive and negative diff-terms cancel out in line 6). On the other hand, if the most similar record with the same target label is close (first diff-term is small), but the most similar record with a different target label is far away (second diff-term is large), then the weight will increase overall.

In our particular example, since we have duplicated the records in our dataset, the nearest hit is almost always identical to the selected record $\mathbf{x}$ itself; therefore the contribution of the first diff-term is zero. If we consider the first two records in Table 6.2, they carry different labels but differ otherwise only in the value of B. This makes B attractive, because it is the only attribute that helps to distinguish between both cases. Overall, the ReliefF ranking delivers $A, B - D - C$.

Selecting the Top-Ranked Subset   A computationally more expensive approach is to consider an evaluation function that ranks *attribute subsets* rather than single attributes. This helps to recognize the power of complementing attributes rather than their individual contribution only. The question is how such a function may evaluate the attribute subset. Possible choices include:

Cross product: As a naive solution, we may construct a new attribute as the cross product of all attributes in the subset and evaluate it by the functions discussed earlier (information gain, gain ratio, etc.). Continuing our example, if the subset contains A and B, we construct an attribute AB with $\Omega_{AB} = \{++, +-, -+, --\}$, which would lead to a perfect prediction of the target value in our example. This naive approach is only feasible as long as the number of cases per newly constructed feature remains relatively high, otherwise we risk overfitting.

Wrapper: Once we have methods available to learn models automatically from data (see Chaps. 7ff), we may evaluate the subset by the performance of the derived model itself: For each subset, we build a new model, evaluate it, and consider the result as an evaluation of the attribute subset. This is known as the so-called **wrapper approach** to feature selection. In order to avoid overfitting of the training data, we must carry out the evaluation on hold out validation data.

New measures: As we have already mentioned, we may remove both, irrelevant attributes (carrying no information) and redundant attributes (information is already carried by other attributes). So far we have focussed on the former by investigating how much information a given attribute contains wrt. the target class (which should be close to zero in case of irrelevant attributes). To identify redundant attributes, we may apply the same evaluation function on attribute pairs: if one attribute tells us everything about another, they are obviously redundant, and one of them may be skipped. This is the idea of the **correlation-based filter** (CFS) [7], where a set of features is evaluated by the following heuristic:

$$\frac{k\bar{r}_{ci}}{\sqrt{k+k(k-1)\bar{r}_{ii}}},$$

where $\bar{r}_{ci}$ is the average correlation of the target attribute with all other attributes in the set, whereas $\bar{r}_{ii}$ is the average attribute–attribute correlation between different attributes (except the target) and k denotes the number of attributes. The evaluation function becomes the larger, the better the attributes correlate with the target attribute and the less they correlate with each other.

Once a subset evaluation function has been found, a strategy to select subsets from the full set of attributes has to be found. Two very simple standard procedures are **forward selection** and **backward elimination**. In the former, one starts with a subset that does not use any attributes (and a model must always predict the majority class) and then adds attributes in a greedy manner: in each step the attribute that most improves the subset performance is added. In backward elimination the process is reversed: one starts with a subset that uses all available attributes and then removes them in a greedy manner: in each step the attribute whose removal most improves the subset performance is eliminated. The process stops if there is no attribute that can be added (forward selection) or removed (backward elimination) without actually worsening the model performance. An exhaustive search over all possible attribute sets is usually prohibitive, but at least possible, if only a limited number of attributes is available. Even a random set generation is possible (Las Vegas Wrapper).

The data from Table 6.2 evaluated by a forward selection wrapper with a standard decision tree learner selects the attribute D only. Exhaustive search of all subsets identifies the subset A, B, D as the best. In both cases attribute D is included because of the preference for attributes that correlate directly with the target variable, which is built in with most decision tree learners. On the one hand, having identified the optimal subset does not help if the learner is not capable of constructing the right model out of it (because of representational limitations); therefore it is advantageous to use the method that builds the model to evaluate the attribute subsets directly. On the other hand, if we intend to test different techniques later, the optimal subset of one algorithm is not necessarily the smallest or optimal subset for another. Using exhaustive search together with the subset evaluation function that considers the cross product, the *minimal optimal subset A, B* is identified.

6.1.2 Dimensionality Reduction

Principal component analysis (PCA) has already been introduced in Sect. 4.3.2.1 in the context of data understanding. PCA was used for generating scatter plots from higher-dimensional data and to get an idea of how many intrinsic dimensions there are in the data by looking at how much of the variance can be preserved by a projection to a lower dimension. Therefore, PCA can also be used as a dimension-reduction method for data preparation. In contrast to feature selection, PCA does not choose a subset of features or attributes, but rather a set of linear combinations of features. Although this can be very efficient in certain cases, the extracted features in the form of linear combinations of the attributes are often difficult to interpret, so that the later steps will automatically result in a black-box model.

PCA belongs to a more general class of techniques called **factor analysis**. Factor analysis aims at explaining observed attributes as a linear combination of unobserved attributes. In PCA, the (first) principal components are the corresponding factors. **Independent component analysis (ICA)** is another method to identify such unobserved variables that can explain the observed attributes. In contrast to factor analysis, ICA drops the restriction to linear combinations of the unobserved variables.

Dimension-reduction techniques that construct an explicit, but not necessarily linear mapping, from the high-dimensional space to the low-dimensional space are often called **nonlinear PCA**, even though they might not aim at preserving the variance like PCA, but rather the distances like MDS. Examples for such approaches can be found in [8, 9, 13, 14].

6.1.3 Record Selection

Except for very rare occasions, the available data is already a sample of the whole population. If we had started our analysis a few weeks earlier, the sample would have looked different (smaller, in particular) but probably still representative for the whole population. Likewise we may create a subsample from our data and use it for the analysis instead of the full sample, e.g., for the sake of faster computation or the need for a withhold test dataset (see Sect. 5.5.1). Other reasons for using only a subsample include:

- Timeliness: If we want to draw conclusions to take actions in the future, it is advisable in a changing world to primarily use recent data rather than data that is already several years old. Explicit or implicit time stamps in databases are of great help in selecting the most recent data.
- Representativeness: Sometimes the population of interest is different from the population available in our database (e.g., applicants that are rejected early are not recorded in the database). By choosing underrepresented cases more frequently than overrepresented cases this imbalance may be compensated and lead to a better representativeness in the sample. On the other hand, if the data is already representative and the subsample is relatively small, chances rise that the probability distribution of some attribute of interest becomes quite different in the subsample compared to the full dataset. Stratified sampling takes care of this problem (see Sect. 5.5.1).
- Rare events: If we want to predict some event that occurs only rarely, a remarkably good performance is achieved by simply predicting that the event will not occur (which is true most of the time but obviously does not help very much). As many methods are sensitive to this kind of problem, the cost function has to be chosen appropriately, e.g., by adapting the misclassification costs (see Sect. 5.2.1). Alternatively, the number of rare cases may be increased artificially by sampling them with a higher probability.

How large does the subsample have to be? The smaller the subsample, the less likely it is representative for the population. The more complex a model is, the more parameters have to be estimated, and the more data is necessary to achieve robust estimates. But on the other hand, if things seem to converge after observing several thousand cases, considering more and more data does not add any further value. In some cases, when the parameters to be estimated can be represented as a sum (like the mean), a theoretical bound for the size of the subsample can be provided: In case of the mean value $\bar{x}$ of all x_i, $1 \le i \le n$, and a given (almost certain) variable range $[a, b]$, that is, $x_i \in [a, b]$, the probability that the mean estimate $\bar{x}$ deviates from the true value $E[x]$ only by ε is given by

$$P\left(|\bar{x} - E[x]| \ge \varepsilon\right) \le 2\exp\left(-\frac{2n\varepsilon^2}{(b-a)^2}\right),$$

which is known as the Hoeffding inequality. For some desired probability, the number of required samples can be derived directly. However, this conservative estimate requires quite a large number of samples.

The choice of the modeling technique is always a compromise of interpretability, runtime, stability, etc. (see Sect. 3.3). If runtime becomes a problem, why not reduce the dataset to those cases that are really important for the modeling technique? Why not remove the bulk of highly similar, uninformative records and just keep a few representatives instead? This **instance selection** works surprisingly well for many data mining tasks. If class boundaries exist in a classification task (see Sect. 5.4), the instances that influence the class boundaries most are usually those close to the boundaries—we get the same resulting classifier if some of the *interior* cases are removed. This is exploited by classifiers called support vector machines

(Sect. 9.3). Instance selection for clustering tasks often involves some condensation or compression by choosing one record that represents a number of very similar records in its neighborhood. Often the number of records it represents is included as an additional *weight* attribute, which gets higher the more data it substitutes. Many clustering algorithms (such as k-means clustering, Sect. 7.3) are easily extended to process this additional weight information appropriately. The instance selection can be considered as some kind of preclustering as it identifies representative cases that may stand for a small group of data (a small cluster). The small clusters may then be aggregated further in subsequent steps.

6.2 Clean Data

6.2.1 Improve Data Quality

Data scrubbing involves the correction of simple errors, such as inconsistencies, spelling mistakes, spelling variants, abbreviations, varying formats (time, date, currency, phone). Such distorted values are frequently called **noise**. We have to correct these values in order to recognize differently spelled but semantically identical values as such (which includes text entries and numerical values). Appropriate steps include

- Turn all characters into capital letters to level case sensitivity.
- Remove spaces and nonprinting characters.
- Fix the format of numbers, date, and time (including decimal point, sign, currency signs). Statistical methods have been used in the data understanding phase to identify potential outliers in numerical fields (e.g., limit the deviation from mean value to 2–3 standard deviations), which may indicate a missed decimal point resulting in a much too high value. Such errors are easily corrected, and others may require careful inspection.
- Split fields that carry mixed information into two separate attributes, e.g., *"Chocolate, 100g"* into *"Chocolate"* and *"100.0"*. This is known as **field overloading**.
- Use spell-checker or stemming[1] to normalize spelling in free text entries.
- Replace abbreviations by their long form (with the help of a dictionary).
- Normalize the writing of addresses and names, possibly ignoring the order of title, surname, forename, etc. to ease their reidentification.
- Convert numerical values into standard units, especially if data from different sources (and different countries) are used.
- Use dictionaries containing all possible values of an attribute, if available, to assure that all values comply with the domain knowledge.

---

[1]The reduction of inflected or derived words to their root (or stem) is called stemming. So-called *stemmers* are computer programs that try to automate this step.

While humans are usually very good in spotting and correcting noise, the problem is that it may occur in many variants which makes an automatic recovery very difficult. It is therefore extremely helpful to define rules and patterns the *correctly spelled values* have to match (e.g., regular expressions for text and valid ranges for numerical values), so that the data can be filtered and conversion rules may be defined case-by-case whenever a new rule violation occurs. If additional data will be considered later, these rules prevent us from repeating a manual analysis over and over again.

While the steps mentioned so far were concerned with single records only, other errors may be recognized only if multiple records are investigated. The most prominent example is the discovery of duplicate records, e.g., in the table of customers when creating customer profiles. This problem is known as **record linkage** or **entity resolution**. Methods for detecting duplicates utilize some measure of similarity between records (which will be discussed in Sect. 7.2); two records that are highly similar are than considered as variants of the same entity and are merged to a single record. The greatest difficulty is to avoid that two different entities are merged mistakenly (false positive).

It is always a good idea to document changes, such as the correction of an outlier, because it may turn out later that the entry is not defective at all but points out some rare, special case. An artificial attribute *degree of trustworthiness* may be introduced, which is 1.0 by default but decreased whenever some correction is applied to the record that involves heuristic guesswork, be it outlier correction or missing value treatment, which will be discussed in the next section. Methods which are numerical in nature frequently offer the possibility to take an additional weight into account, such that suspicious, less trustworthy records get less influence on the resulting model (see also robust regression and M-estimation in Sect. 8.3.3.3).

Correcting all possible errors is a very time-consuming task; and if the data turns out to be useless for the task at hand, it is definitely not worth the effort. If potentially useful attributes have been identified early, efforts in data cleaning can be geared towards those attributes that pay back.

6.2.2 Missing Values

Why is it necessary to treat missing data, anyway? If the data is not in the database, this is because we do not know anything about the true value—so there is not much we can do about it—besides *guessing* the true value, but this seems to be even worse than keeping the missing data, because it almost certainly introduces errors.

The reason why we nevertheless may want to change the missing fields is that the implementations of some methods simply cannot deal with empty fields. Then the imputation of estimated values is a simple mean to get the program run and deliver some result, which is often, although affected by estimation errors, better than having no result at all. Another option would be to remove records with missing values completely, so that only complete records remain, and the program does not run into any problems either. In case of nominal attributes there is a third possibility: we may simply make the missing value explicit by adding another possible outcome, say MISSING, to its range. Let us consider all these options individually.

Ignorance/Deletion   In **case deletion** all records that contain at least one missing value are removed completely and are thus not considered in subsequent steps. Although this procedure is easy to carry out, it also removes a lot of information (from the nonempty attributes), especially if the number of attributes is large. It may be safely applied in case of missing completely at random (see Sect. 4.6), as long as the remaining dataset is still sufficiently large. As soon as we deal with missing at random (MAR), we may start to seriously distort the data. To continue the example from Sect. 4.6, where missing temperature values due to empty batteries are less likely to get fixed when it is raining, case deletion will remove more rainy, cold days than sunny, warm days, which may lead to biased estimates elsewhere. A removal of records may therefore threaten the representativeness of the dataset.

Imputation   The missing values may be replaced by some estimate. As a trivial example consider an attribute *"number of pregnancies"* in a database, which is for men either 0 or missing (in the sense of "not applicable"). In such cases *we know* that the true value must be 0 and may impute it safely. For arbitrary attributes, perhaps the most simple (but not advisable) guess is the mean (for numerical data) or the mode (for categorical data). As such a guess clearly does not add any new information to the data, it should not do any harm either: it should leave the characteristics of the dataset unaltered. Indeed, by replacing all missing values of one variable by the mean of the measured values, the mean of the modified set will remain unchanged—however, its variance will decrease. This replacement procedure affects the variance of the attributes to different degrees (depending on the portion of missing values). Of course, it is also possible to impute a value that preserves the variance (but affects the mean). In one way or the other, any imputed value will affect some characteristics of the known dataset. To decide which imputation is better, it is useful to know which statistic will play an important role in the subsequent steps, and this statistic should be least distorted.

If the values are missing at random (but not completely at random), other attributes can provide hints towards the true value. If the horse power is missing in a car database, we would estimate quite different values for a compact car and an off-road vehicle. If an attribute such as fuel consumption or engine displacement are given, their correlation to horsepower can be exploited to achieve better estimates. How to develop models that help us to predict numerical values (regression) or categorical values (classification) will be discussed in Chaps. 8 and 9.

Replacing all missing values with the very same constant number always reduces the variability of the dataset, which often influences derived values such as correlation coefficients or fitting errors. This is less pronounced with substitutions that come from a predictor, but the problem remains. Adding a random value to each imputed value is not a general solution but helps to reduce the effects.

Table 6.4 Treatment of missing values

Treatment of missing values for ...	
Operational reasons	
Case deletion	Removal of all records where missing values occur, only MCAR
Imputation	Estimate the missing value (MAR: with help from other attributes)
Global constant	Turn all missing values into a special, newly introduced global constant
Data quality reasons	
Harmonize	Select a single representation for missing values and change all other representations to it
Separate	Use background knowledge and other attributes to distinguish between *there is a real value, but it is unknown* and *there is no value (yet)* (e.g., grade in exams: unknown grade because the test got lost vs. absence of grade because student was ill)
Lossless preparation	
Construct data	Construct new attributes that indicate imputation, as this may carry important information itself

Explicit Value or Variable A very simple approach is to replace the missing values by some new value, say MISSING. This is only possible for nominal attributes, as any kind of distance to or computation with this value is undefined and meaningless. If the fact that the value is missing carries important information about the value itself (nonignorable missing values), the explicit introduction of a new value may actually be advisable, because it may express an intention that is not recoverable from the other attributes. If we suppose that the absence of the value and the intention correlate, we may luckily capture this situation by a new constant, but the problem is that we cannot assure this from the data itself. If the values are missing completely at random, there is no need to introduce a new constant. The cases that are marked by a constant do not exhibit any deviation from those cases that are not marked. If we introduce it nevertheless, the models will have to deal with an additional, useless value, which makes the models unnecessary complicated and estimates less robust.

A better approach is to introduce a new (binary) variable that simply indicates that the field was missing in the original dataset (and then impute a value). In those cases where neither the measured nor the estimated values really help, but the fact that the data was missing represents the most important bit of information, this attribute preserves all chances of discovering this relationship: the newly introduced attribute will turn out to be informative and useful during modeling. On the other hand, if no such missing value pattern is present, the original variable (with imputed values) can be used without any perturbing MISSING entries.

6.3 Construct Data

By now we have selected and cleaned the data; next we consider the transformation of existing and construction of new attributes that may replace the original attributes. There are at least three reasons for this step: (1) the tool implementations require a transformation because they will not work otherwise, (2) in absence of any background knowledge, we want to make sure that every attribute will play the same role in the model, and no particular attribute is preferred over or has larger influence than another, (3) given some domain knowledge, we want to give helpful hints to the modeling techniques to improve the expected results.

6.3.1 Provide Operability

Scale Conversion   If the modeling algorithm handles only certain types of data, we have to transform the respective attributes such that they fit the needs of the modeling tool. Some techniques assume that all attributes are numerical (e.g., regression in Sect. 8.3, neural networks in Sect. 9.2), so we either ignore the categorical attributes or have to apply some transformation first. Simply assigning a numerical value to each of the possible (categorical) values is not an option because typical operations on numerical values (such as averaging or comparison) are not necessarily meaningful for the assigned numerical values. A typical solution is to convert nominal and ordinal attributes into a set of binary attributes, as it will be discussed in Sect. 7.2.

Other methods prefer categorical attributes (like Bayesian classifiers[2] in Sect. 8.2) or perform superior when a discretization is carried out beforehand [2, 17]. **Discretization** means that the range of possible numerical values is subdivided into several contiguous, nonoverlapping intervals and a label of the respective interval replaces the original value. There are many (more or less intelligent) ways to define the boundaries of the intervals. An **equi-width** discretization selects the boundaries such that all intervals are of the same width (see page 41 for selecting a reasonable number of bins). As we can see from Fig. 6.1(a) for the sepal length attribute of the Iris dataset, it may happen that some intervals contain almost no data. An **equi-frequency** discretization tries to assure that all intervals contain the same number of data objects. Of course an equally distributed frequency is not always achievable. For example, if the dataset with 100 cases contains only 4 distinct numerical values and a partition into 5 intervals is desired, there are not enough split points. Again, such a partition does not necessarily represent the underlying data distribution appropriately: in the partition of the sepal length attribute in Fig. 6.1(b), the compact group of similar data (left) is separated into two intervals; in particular the region of highest density is split into two intervals.

---

[2]Bayesian classifiers can handle numerical data directly by imposing some assumptions on their distribution. If such assumptions cannot be justified, discretization may be a better alternative.

Fig. 6.1 Results of different discretization strategies of the sepal length attribute of the Iris dataset. The class distribution of the three Iris flowers is shown by the *stacked bars*. (**a**) Equi-width partition, (**b**) equi-frequency partition, (**c**) V-optimal partition, (**d**) minimal entropy partition

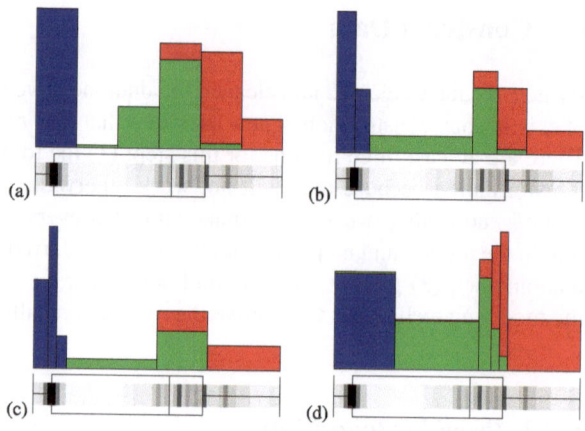

The discretization problem is essentially the same as the histogram problem (mentioned already in Sect. 4.3.1). A so-called **V-optimal** histogram minimizes the quantity $\sum_i n_i V_i$, where n_i is the number of data objects in the ith interval, and V_i is the variance of the data in this interval. Minimization of variances leads to intervals with a roughly uniform data density, as can be seen in Fig. 6.1(c), which makes it quite similar to clustering (Sect. 7.3) where we also search for areas of higher data density to define compact groups of similar data. Clustering algorithms may therefore also be used to derive partitions for discretization.

If additional information is available (such as the labels of some target variable as in classification tasks), it can provide valuable hints for finding a discretization that supports the classification task best. Note that the plots in Fig. 6.1(a)–(c) already show this class label information but that it has not been used by the discretization so far. We will discuss the idea of discretization via entropy minimization in the context of binary splitting in Sect. 8.1, which can be extended to an arbitrary number of splits (or intervals) [3, 4]. The underlying ideas are similar to those of the V-optimal histogram, but now we evaluate a partition by the (im)purity of its intervals: If an interval contains only cases with the same target label, such an interval appears informative wrt. the target variable. On the other hand, if the distribution of the target variable within one interval is identical to the overall distribution, such an interval is useless for the prediction. Just as in feature selection, we may now select the partition that utilizes the classification task most, that is, leads to the purest distribution of target values. This is the same objective we had in feature selection, so that we can reuse the evaluation measures mentioned there (such as information gain, gain ratio, Gini index, etc.). The partition that evaluates best among all possible partitions with k intervals can be found quite efficiently (see, e.g., [3]). An example for the sepal length attribute is shown in Fig. 6.1(d). Having fixed the number of intervals to 6, note that most of the data are already covered by three pure bins, whereas the remaining three bins are located where the class distribution is ambiguous.

Dynamic Domains We have already tackled the problem of dynamic domains with nominal attributes in Chap. 4. In order to apply our models to future data with dynamic domains, we must be prepared to handle values unseen before. One solution is to introduce a new variable, which is an abstraction of the original one. By mapping the original values to more abstract or general values we can hope that newly occurring entries can also be mapped to the generalized terms so that the model remains applicable. In a data warehouse such a granularization scheme may already be installed, otherwise we have to introduce it and install means that newly arriving values will be mapped into the hierarchy first.

Problem Reformulation   Transformations also help to reformulate a problem so that the range of applicable techniques is enlarged. The logit-transformation maps the unit interval to $\mathbb{R}$, enabling standard linear regression techniques to be used for (binary) classification problems (see Sect. 8.3.3.1). Some classification techniques handle binary class variables only, so that multiclass attributes have to be transformed first (e.g., one class against all others).

6.3.2 Assure Impartiality

Another reason for transforming attributes is to ensure that the influence or importance of every attribute is a priori the same. This is not automatically the case, as the next examples will show.

When looking for two natural groups in the two datasets of Fig. 6.2 (which is the task of clustering, see Chap. 7), most people would probably suggest the groups indicated by dotted lines. The rationale for both groupings is that the elements within each group appear to be more similar than elements of other groups. However, any notion of similarity or distance strongly depends on the used *scaling*: Both graphs in Fig. 6.2 actually display the same dataset, only the scaling of the x-axis has been changed from hour to minute. If the distance between the elements is taken as an indication of their similarity, a different scaling may change the whole grouping. In practical applications with considerable background knowledge, one may be in the comfortable position to conclude that a difference of 1.0 in attribute A is as important or relevant as a difference of 2.5 in attribute B. In such a case, all attributes should be rescaled accordingly so that in the rescaled variables a distance of 1.0 is

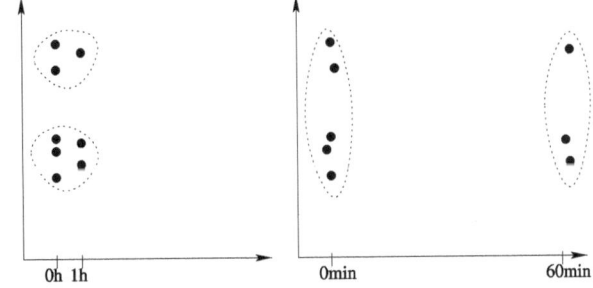

Fig. 6.2 Distance measures are influenced by scaling factors (here: hours vs. minutes)

of equal importance in any attribute. In absence of such knowledge, typical transformations include the min-max-**normalization** and the **z-score standardization**.

- **min-max normalization**: For a numerical attribute $X \in \mathcal{A}$ with $\min_X$ and $\max_X$
  being the minimum and maximum value in the sample, the min-max normaliza-
  tion is defined as

$$n : \text{dom}\, X \rightarrow [0, 1], \qquad x \mapsto \frac{x - \min_X}{\max_X - \min_X}.$$

- **z-score standardization**: For a numerical attribute X with sample mean $\hat{\mu}_X$ and
  empirical standard deviation $\hat{\sigma}_X$, the z-score standardization is defined as

$$s : \text{dom}\, X \rightarrow \mathbb{R}, \qquad x \mapsto \frac{x - \hat{\mu}_X}{\hat{\sigma}_X}.$$

- **robust z-score standardization**: As the sample mean and empirical standard deviation are easily affected by outliers, a more robust alternative is to replace the
  mean by the median $\tilde{x}$ and the standard deviation by the interquartile range IQR_X
  (difference between third and first quartile, see also boxplots in Sect. 4.3.1):

$$s : \text{dom}\, X \rightarrow \mathbb{R}, \qquad x \mapsto \frac{x - \tilde{x}}{IQR_X}.$$

- **decimal scaling**: For a numerical attribute X and the smallest integer value s
  larger than $\log_{10}(\max_X)$, the decimal scaling is defined as

$$d : \text{dom}\, X \rightarrow [0, 1], \qquad x \mapsto \frac{x}{10^s}.$$

While it may be considered as advantageous if the resulting range is the unit
interval $[0, 1]$, the min-max normalization and the decimal scaling are very sensitive
to outliers: even a single outlier can force the majority of the data to concentrate in a
very small interval (and thus being very similar in comparison to other cases). This
effect is less prominent in z-score normalization, and the robust z-score is unaffected
by a moderate number of outliers.

Such as distance measures are used to capture the similarity of data, other
measures are used for different purposes, such as the evaluation of attributes (see
Sect. 6.1.1 on feature selection). Those measures try to identify how much an attribute contributes to the prediction of a given binary target attribute. Suppose that
we have 100 cases in the database and three attributes with 2, 10, and 50 values,
respectively. For the sake of simplicity, let us assume that none of these attributes
can help in predicting the target attribute and that the values are equally likely. It is
probably easy to see that the 2-valued attribute cannot help very much in predicting
the target value, but what about the 50-valued attribute? On average, we have only
two cases in our database where the same value will occur. If (by chance) these two
cases carry the same target value, some algorithm may find it easy to predict the
target value (at least in this case). As we have discussed in Sect. 5.4, as the number of cases gets smaller and smaller, the danger of memorizing (overfitting) rises.
Of course, this effect takes place gradually, and it will already be recognizable with
the 10-valued attribute but to a lesser extent. Some measures (such as information

gain) are sensitive to such effects and thus implicitly prefer attributes with a larger number of values (just because memorizing is then much easier). A model, such as a decision tree, is very likely to evaluate poorly if such a variable is chosen. Rather than removing such attributes completely, their range of values should be reduced to a moderate size. This corresponds to selecting a coarser **granularity** of the data. For instance, the very large number of individual (full) zip-codes should be reduced to some regional or area information. The website access timestamp (which even includes seconds) may be reduced to day-of-week and am/pm. The binning techniques mentioned earlier for discretizing numerical attributes may also be helpful. If there is useful information contained in these attributes, it is very likely that it is retained in the coarsed representation, but the risk of overfitting is greatly reduced.

Another typical assumption is that some variables obey a certain probability distribution, which is not necessarily the case in practice. If the assumption of a Gaussian distribution is not met, we may transform the variable to better suit this requirement. This should be a last resort, because the interpretability of the transformed variable suffers. Typical transformations include the application of the square root, logarithm, or inverse when there is moderate, substantial, or extreme skewness. (If the variable's domain includes the zero, it has to be shifted first by adding an appropriate constant before applying the inverse or logarithm.) Another choice is the **power transform**: Given the mean value $\bar{y}$, the power transform

$$y \mapsto \begin{cases} \frac{y^\lambda - 1}{\lambda \bar{y}^{\lambda-1}} & \text{if } \lambda \neq 0, \\ \bar{y} \log y & \text{if } \lambda = 0, \end{cases}$$

transforms the data monotonically to better approximate a normal distribution. The parameter λ is optimized for the problem at hand, e.g., such that the sum of squares of residuals in regression problems is minimized.

6.3.3 Maximize Efficiency

As already mentioned in Sect. 6.1.1, the performance of many standard learning algorithms, like those we will discuss in Chaps. 7ff, degrades if irrelevant or redundant attributes are supplied. The purpose of feature selection is the identification and removal of such variables to prevent a performance decline. Sometimes the attributes do contain all necessary information, but it is hard for the chosen modeling technique to extract this information from the data. As an example, suppose that the chances for some diagnosis are especially high if at least n out of m symptoms can be observed. In case of 4 available observations a, b, c, and d, this would mean that any combination of two of these symptoms can be taken as an indication of this diagnosis (typical case in the medical domain [11]). Is this relationship difficult to capture for a given modeling technique? Let us consider two possible models, a hyperplane model and a rule-based model. If the four symptoms are modeled by binary 0–1 values (absence–presence), the diagnosis corresponds to the following condition:

$$a + b + c + d \geq 2,$$

Fig. 6.3 Simple example for the usefulness of derived features: for a number of journeys, the travel time and distance are shown; the color indicates whether the driver was ticketed or not. Discriminating both classes with axis-parallel rectangles is laborious, but easy with a new attribute for travel speed

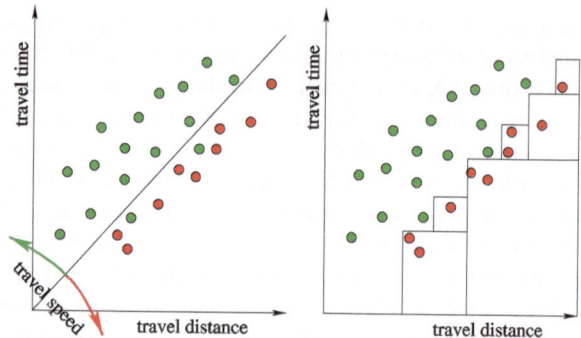

which is already the definition of the respective hyperplane: the hyperplane $a + b + c + d = 2$ divides the space into two half-spaces, on one side the sum is less, and on the other side the sum is larger than 2. If a new case lies on the hyperplane or the latter side, it satisfies the "*at least two symptoms*" condition. However, a rule-based learner delivers rules such as

if (a and b) or (a and c) or (a and d) or ... **then** diagnosis $= \cdots$.

Apparently, the rule-based model requires to include all possible combinations of 2 out of 4 symptoms, which makes the rule quite lengthy and more laborious to discover. Regardless of the chosen model and thus the model representation, it will always be well-suited for *some* real-world problems and perform poorly on others.

Accepting this fact, the idea of feature construction is to overcome the limits of the representation (representational bias) by offering new features that were otherwise difficult to represent by the original model. Such features include:

- **Derived features**. Given some background knowledge, new attributes may be defined from existing attributes, such as *average speed* from *travel time* and *travel distance*. If the true relationship in the data depends on some threshold on the average travel speed but the model requires to condition on the individual attributes, the exact relationship may only be approximated by various pairwise conditions on the individual variables (see Fig. 6.3). Such models will perform worse and are difficult to interpret.
- **Change the baseline**. For an important variable to stand out clearly, selecting the right baseline is often crucial. For instance, to estimate the spending power "*income per household*" (with a thoughtful definition of the size of a household) may outperform "*income per head.*"
- **Grouping**. If it is conjectured that the data form some natural groups or clusters, then several attributes may strongly correlate within the same group, but not necessarily across groups. Knowing to which group a particular case belongs to can be helpful to find a concise model. In such cases, a cluster analysis may be carried out first, and the resulting cluster labels are added as a new attribute. (Make sure that the cluster analysis uses only attributes that will also be available for new data when applying the model.)

- **Hyperplanes**. A linear combination of various attributes may define a new numerical attribute (where a threshold has still to be found by the learner) or, if a threshold is already supplied (cf. the example above), a binary attribute. A useful example is a binary attribute that becomes true if at least (or exactly) n out of m preselected features hold. Such conditions are hard to identify for models that work on the symbolic level (such as rules).

- **Level of granularity**. When looking for interdependencies in the data, it is a priori not clear on which level of granularity they may unveil. Grouping the values of a variable to meaningful aggregates or even a hierarchy or taxonomy would allow us to test for dependencies at various levels of detail.

- **Nonlinearities**. A linear model $y = \alpha x + \beta y + \gamma$ is linear in its parameters α, β, and γ, which is where the name *linear model* comes from, and also linear in x and y. If it is known or conjectured that the true dependencies *are not linear in x or y*, the input data may be transformed beforehand: new attributes x^2 and y^2 may be constructed, and a more complex linear model $y = \alpha x + \beta y + \gamma x^2 + \delta y^2 + \varepsilon$ can be built.

- **Add metadata**. Attributes that indicate where we have altered the data (either because it was missing or because we supposed it was an outlier) can be helpful in situations where our assumptions were wrong (the data is not missing completely at random).

New attributes are often constructed by applying arithmetic or logic operators to existing attributes. Instead of deriving them manually with the help of background knowledge, it is in principle also possible to automatically *search* for helpful attributes by constructing them randomly or heuristically and identifying the best generated features. This idea has been successfully applied in the context of decision tree learning [10].

We have seen in Sect. 5.4.5 that the model error can be decomposed into machine learning bias and variance. The variance is caused by sample errors and causes overfitting. In principle, the sample error can be reduced by acquiring a larger dataset (or by feature selection, as discussed before). The bias is a systematic error that expresses a lack of fit between the model and the real-world relationship. In this sense, feature construction can help to reduce the limitations of the learning bias by giving it some new direction. However, we must be aware that the introduction of newly constructed attributes may also simultaneously increase the sample error again (see [16] for a case study). If attributes are constructed that take a large number of different values, an increase of the sample error is inevitable. But it is also possible to construct features that reduce the sample error: whenever we condense information (from multiple attributes) to get a coarser view, the robustness of our estimates increases. Examples for such transformations are aggregations (one attribute *total number (or price) of equipment* rather than many individual attributes for each type of equipment), introducing hierarchies (one attribute *fruit juice* rather than individual attributes for *apple juice*, *orange juice*, etc.), or grouping. Although we have discussed feature selection first, this does not mean that we have to apply the same principles and requirements to our newly constructed attributes.

6.4 Complex Data Types

In this book we mainly focus on simple data types such as numerical and nominal attributes. However, the majority of data collected has much more structure: there are archives of documents, images, and videos. There are databases containing molecular structure and gene sequences. And there are, of course, the increasing number of network-style information repositories. It would be outside of the scope of this book to discuss the analysis of such complex data types in detail, but in almost all cases one of two approaches is used: either the complex data is somehow transformed into a simpler representation which can be used to apply standard data analysis methods to, or existing methods are adopted to deal with specific types of more complex data types. The following brief overview does not even attempt to be a complete list of complex data types but rather tries to show a few prominent examples of these two approaches.

Text Data Analysis Textual data is somewhat difficult to represent in a data table. Among the possible representations, the vector model is most frequently used. The individual words are reduced to their stem and very frequent and very rare (stemmed) words are removed. The remaining words are then represented by a number of boolean attributes, indicating that the word occurred in the text or not. This approach is also known as bag-of-words representation because it is usually easier to simply enumerate the words contained in the vector instead of storing the entire (usually very sparse) vector itself. Once we have reduced texts/documents to such a vector-based representation, we can apply the methods presented later in the book without change. A good summary can be found in [5]. Under the umbrella of information retrieval, a lot of work has been undertaken as well.

Graph Data Analysis The analysis of large graphs and also the analysis of many collections of graphs poses interesting challenges: from finding common patterns of similar structure in social networks that can help to identify groups of similar interest to the identification of reoccurring elements in molecular databases that could potentially lead to the discovery of a new molecular theme for a particular medication. Network analysis is often performed on abstractions, describing degrees of connectivity, and then proceeds by using standard techniques. However, a number of approaches also exist that operate on the network structure directly. A prominent example is frequent subgraph mining approaches which identify graph patterns that occur sufficiently often or are sufficiently discriminative. These methods follow the frequent itemset mining approaches described later in this book (see Chap. 7) but adopt those methods to directly operate on the underlying graph or graph database. In [1] a good introduction to the analysis of graph data is given. We also discuss one example of a molecular graph algorithm in more detail in Sect. 7.6.3.4 when we introduce item mining algorithms.

Image Data Analysis Analyzing large numbers of images often requires a mix of both methods. In order to find objects or regions of interest, one needs to apply sophisticated image segmentation approaches. Often already here analysis steps are involved to semiautomatically find the best segmentation technique. Once this is done, various types of image feature computations can be applied to describe brightness, shape, texture, or other properties of the extracted regions. On these features classical data analysis routines are then applied. The analysis of these types of data emphasizes an interesting additional problem in data analysis, which is often ignored: each object can be described in many different ways, and it is not always clear from the beginning which of these representations is best suited for the analysis at hand. Finding out automatically which descriptors to use goes further than feature selection discussed earlier in this chapter (Sect. 6.1.1) as we cannot select from various subsets of features accompanied by different semantics. [12] offers a good summary over the different aspects of image mining.

Other Data Types   Of course, plenty other types of data are around: temporal data from stock indices, process control applications, etc. Lots of work has been done in the analysis of movies and other multimedia data such as music and speech. But complex data is also generated by various ways to record data such as weblogs and other types of user interactions recordings. Complex data also arise during observations of various types of populations (e.g., cattle herds). From an analysis point of view, the next big challenge lies in the combined analysis of separate information sources and types.

6.5 Data Integration

Almost all of the discussions in the remainder of this book focus on the modeling or analysis of data that arrives in one nice, well-defined table. In the previous sections of this chapter we have discussed how we can prepare the structure of this table by selecting, cleaning, or constructing data, but what if our data is spread out over numerous repositories? In reality, the data recording has begun over time, and various departments have chosen their own recording mechanism. Worse, a merger in the life time of our corporate database may have thrown two completely independently developed database setups together that contain similar, but not quite the same, data. If we now want to analyze the joint customer base, we may want to merge the two (or more) customer databases into a uniform one first. This type of data integration is called vertical data integration since we are really interested in concatenating two tables holding essentially the same information. But already in this simple case we will run into many annoying problems as we will see in the following section. The other type of integration asks for combining different types of information and is called horizontal data integration. Here we aim to enrich an existing table of, e.g., customers with information from another database such as purchase behavior which was recorded independently. Essentially we are aiming to concatenate the entries from one database with the entries of the second one, somehow smartly matching which entries should be appended to each other.

6.5.1 Vertical Data Integration

An example of vertical data integration could be the combination of two customer databases from one company which recently acquired a second one. There usually are two main obstacles to overcome:

Firstly we need to make sure that the structures of our databases are unified. One database may have stored last and first name in one field, separated by a comma, whereas the second database uses two separate fields for this. So we need to define a joint format which we can then transform the entries of both tables into. This often generates missing values, since very often not all fields will be obtainable from both tables. If one table holds gender and the other not, we can either drop this information altogether or replace it with missing value identifiers for the entries of the second table. This structure unification will also often involve the definition of a derived identifier because very likely the two databases used two different types of identifiers.

Secondly, we need to deal with duplicates. This is a problem which often also occurs within one single database which has grown over the years. But joining two formerly completely independent databases will almost certainly create duplicate entries if we not implement counter measures. The problem is that this is usually not simply a problem of detecting duplicate identifiers. If both of our databases contain exactly the same, correct information duplicate detection is simple. But if our identifier is, for example, generated based on the last and first name, we may sometimes have a middle initial and sometimes not, and then two entries are left in our database. Slight spelling mistakes (Mayer vs. Meyer) or abbreviations (Sesame St. vs. Sesame Street) can make this a formidable task. In fact, for postal addresses, commercial companies exist whose only task is to help clean and merge duplicates in a corporate database. And for those of us who receive lots of advertisements, we can easily see how well such systems work.

Vertical data integration therefore comes back to data cleaning. We need to define a proper structure of our new, combined database, and we need to make sure to unify the representation in such a way that we can eliminate duplicate entries and clean incorrect ones.

6.5.2 Horizontal Data Integration

An example of horizontal data integration was discussed before, using our example customer/shopping data. If we have separate customer and purchase databases but are, for instance, interested in which purchases were actually done by females, we need to pull information from these datasets together. In order to do such a join, we will require some identifier which allows us to identify rows in both tables that "belong together."

Table 6.5 illustrates how this works. Here the identifier used for the join is the customer ID. Note that it is actually named differently in the two tables, so we

Table 6.5 The two datasets on top contain information about customers and product purchases. The joint dataset at the bottom combines these two tables. Note how we loose information about individual customers and how a lot of duplicate information is introduced. In reality this effect is, of course, far more dramatic

id	Last name	First name	Gender		Shopper id	Item id	Price
p2	Mayer	Susan	F		p2	i254	12.50
p5	Smith	Walter	M		p5	i4245	1.99
p7	Brown	Jane	F	+	p5	i32123	1.29
…	…	…	…		p5	i254	12.50
					p5	i21435	5.99
					p7	i254	12.50
					…	…	…

Item id	Price	Last name	First name	gEnder
i254	12.50	Mayer	Susan	F
i4245	1.99	Smith	Walter	M
i32123	1.29	Smith	Walter	M
i254	12.50	Smith	Walter	M
i21435	5.59	Smith	Walter	M
i254	12.50	Brown	Jane	F
…	…	…	…	…

will need to inform the joint operator which two columns to use to join the tables. There are, of course, other ways to join tables which we will discuss in a minute. However, let us first discuss the two most problematic issues related to joins in general: overrepresentation and data explosion.

- overrepresentation of items is (although only mildly) already visible in our example table above. If we were to use the resulting table to determine the gender distribution of our shoppers, we would have a substantially higher degree of male shoppers that is evident from our table of shoppers. So here we would really answer the question "what is the number of items purchased by male (female) shoppers" and not a question related to shoppers.
- data explosion is also visible above: we already see a few duplicate entries which are clearly not necessary to keep all essential information. Finding a more compact database setup, i.e., splitting a big database into several smaller ones, to avoid such redundancies is an important aspect in database normalization where the goal is to ensure dependency of database entries onto the key only.

Joins, as illustrated above, are only a very simple case of what is possible in real systems. Most database systems allow one to use predicates to define which rows are to be joined and more generic data analysis systems often allow one to at least choose one or more attributes as join identifiers and select the four main join types: inner, left, right, and outer joins:

- inner join: creates a row in the output table only if at least one entry in the left and right tables can be found with matching identifiers.
- left join: creates at least one row in the output table for each row of the left input table. If no matching entry in the right table can be found, the output table is filled with missing or null values. So we could make sure that each customer appears at least once, even if the customer has not made any purchase.
- right join: similar to the left join, but at least one row is created for each row in the *right* table.
- outer join: creates at least one row for every row in the left *and* right table. If no matching entries can be found in the other tables, the corresponding entries are filled with null or missing values.

Implementing join operations is not an easy task, and various efficient algorithms have been developed. Especially on large tables when all data does not fit into main memory anymore, smart methods need to be used to allow for such out-of-memory joining. Databases are usually optimized to pick the right algorithm for the tables at hand so the data analyst does not need to worry about this. However being aware of the risk of complex SQL statements, launching hidden joins does not hurt.

Integrating data is, of course, a brute force way to transforming a problem into one that we can then deal with using methods developed for single tables. Mining relational databases, which should actually be called "mining multirelational databases," is a research area which deals with finding information directly in such distributed tables. However, the methods developed so far are rather specialized, and in the following we will continue to assume the existence of one nice, well-formed table holding all the relevant data. This is, of course, a drastic simplification, and quite often, the integration and cleaning of data can easily require much more time and effort than the analysis itself.

6.6 Data Preparation in Practice

Data Preparation is the most ignored aspect in data analysis, and this is also visible in many data analysis software packages. They often assume the existence of a nice file representation or a flat table in database. Visual data-flow-oriented tools such as KNIME offer the ability to prepare the data intuitively, whereas command line or script-based tools are often a lot harder to use for data preparation. One of the important aspects here is not so much to possibility of doing data preparation in the data analysis tool of choice but the repeatability or reproducibility. If a data analysis process is to be used in a production environment, it is often necessary to rerun the analysis periodically or whenever new data becomes available. Having a clear and intuitive way to model and document the data, preparation phase in conjunction with the actual analysis is then a serious benefit.

6.6.1 Data Preparation in KNIME

KNIME offers a vast array of nodes for all kinds of data preparation. Most of them can be found in the category *Data Manipulation* where there are nodes for binning, converting/replacing, filtering, and transforming columns, rows, and entire tables. Data selection tasks can be modeled by filtering out columns using the *Column Filter* node. For row filtering, many more modules are available which filter rows based on certain criteria applied to individual cells (*Row Filter*). These criteria can be regular expressions for string cells or range or other constraints on numerical attributes. In addition, the *Nominal Value Row Filter* allows one to select from the range of nominal values to determine which rows should be filtered out. Those nodes also come in a splitter-variant which does not filter out rows but instead has a second output port containing the table with the remaining elements. Those—and more— functions are all contained within individual nodes, so we skip the corresponding very simple examples.

However, automated versions of column or feature selection such as the ones described in Sect. 6.1.1 are worth being discussed in a bit more detail as they demonstrate the interesting and very powerful loop-feature of KNIME. KNIME brings along the framework to run wrapper approaches for feature selection using arbitrary models. Figure 6.4 shows a workflow performing backward feature elimination using a decision tree as the underlying models. The flow starts with a loop node which creates different subsets of features by filtering out all other columns. The reduced data is then fed into a data partitioner, which splits it into training and validation data. Afterwards we learn a model (here a decision tree) on the training data and apply it to the validation data. The Loop end node (*Backward Feature Elimination End*) calculates the accuracy of the resulting prediction and remembers this for all features that were involved in this run. The loop is now executed several times (the number of runs and number of features per run can be adjusted in the dialog of the loop-start node), and finally a column filter model is produced, which can be used to filter all but the top k features/columns from the second table using the *Backward Feature Elimination Filter* node.

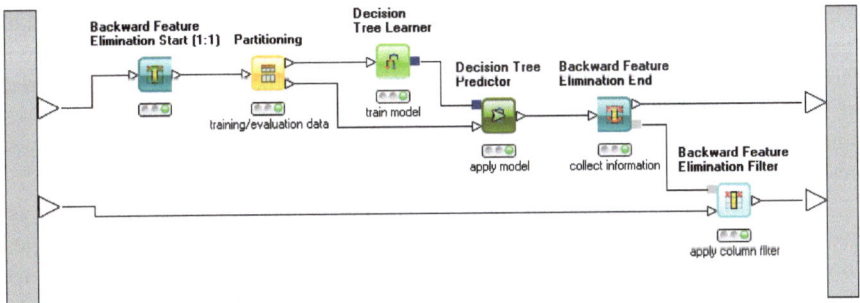

Fig. 6.4 An example for feature selection in KNIME using a decision tree as the underlying model to evaluate feature importance

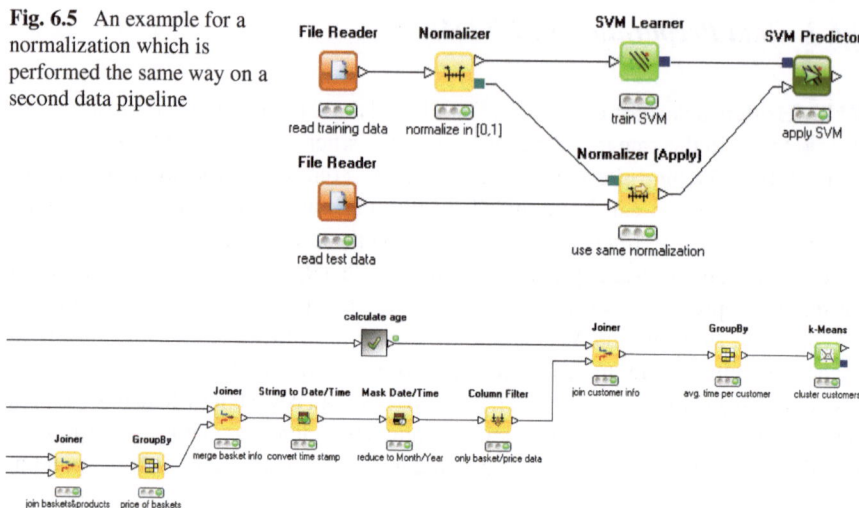

Fig. 6.5 An example for a normalization which is performed the same way on a second data pipeline

Fig. 6.6 An example for data preparation in KNIME. We join a number of tables and compute aggregate information in the *Group By* nodes. The resulting table is finally fed into a clustering node

Nodes for constructing and converting data are available in KNIME as well. Different types of normalizations are available via the *Normalizer* node. Note that this node has a second outport which carries the normalization model. This allows one to easily apply the same normalization to a second data pipeline. An often overlooked problem in data analysis is that users tend to normalize their data using, e.g., min/max normalization and then normalize their test data using this same technique! The resulting errors are hard to find since the min/max ranges of the training and test data can, but are not guaranteed to be exactly the same. Especially, if the ranges are almost equal, the resulting small deviations can cause a lot of confusion. Figure 6.5 shows an example of such a flow. Here a different type of model (support vector machine, see Chap. 9) is trained on normalized training data, and the resulting model is applied to testing data which underwent the same normalization procedure.

KNIME also offers nodes for different types of binning or discretization of data which are available in the "Data Manipulation-Column-Binning" category. Additional nodes allow one to concatenate and join tables, enabling vertical and horizontal data integration. Many other functionalities are available in special nodes. Instead of showing individual examples we conclude this section by showing (Fig. 6.6) a small example of a workflow performing more complex data integration, preprocessing steps to create the data we first discussed in Chap. 2.

The workflow fragment shows how information on products, basket/product associations, additional basket information (date, customer), and customers is merged to produce information on the average basket price of individual customers. Since this information is spread out over several different tables, we first need to join the two tables containing product information (most notably product ID and price) and the map between product and basket ID. Once we have this, we can aggregate the table (second node from the left *Group By*) to obtain the price of each basket. We join this table with the table containing (among others) customer and basket ID and the date of purchase. The latter field is only available as a string, so we need to convert it to a date/time representation before reducing this to only contain year and month. We can then again aggregate the resulting table to obtain the total and average basket price per customer and month. Joining this with the customer data allows one to also add the age of the customer. Note that the age needs to be computed based on the customer birth dates first. We aggregate this again to compute the average prices per month and finally obtain a table containing the features we are interested in (age, average purchase per month, average basket size) which can be fed into the clustering algorithm.

Modeling data integration, transformation, and aggregation in such a graphical way has two advantages. Firstly, this process can be documented and communicated to others. Secondly—and even more importantly—this process can be easily executed again whenever the underlying data repositories change. The documentation aspect is often a problem with script-based platforms. In contrast, reproducibility is an issue with table-based tools, which do allow for an intuitive way to integrate and transform data but do not offer the ability to rerun this process reliably over and over again in the future.

6.6.2 Data Preparation in R

6.6.2.1 Dimensionality Reduction

The use of PCA as technique for dimensionality reduction has already been explained in Sect. 4.3.2.1.

6.6.2.2 Missing Values

The logical constant NA (not available) is used to represent missing values in R. There are various methods in R that can handle missing values directly.

As a very simple example, we consider the mean value. We create a data set with one attribute with four missing values and try to compute the mean:

```
> x <- c(3,2,NA,4,NA,1,NA,NA,5)
> mean(x)
[1] NA
```

The mean value is in this case also a missing value, since R has no information about the missing values and how to handle them. But if we explicitly say that missing values should simply be ignored for the computation of the mean value (na.rm=T), then R returns the mean value of all nonmissing values:

```
> mean(x,na.rm=T)
[1] 3
```

Note that this computation of the mean value implicitly assumes that the values are missing completely at random (MCAR).

6.6.2.3 Normalization and Scaling

Normalization and standardization of numeric attributes can be achieved in the following way. The function is.factor returns true if the corresponding attribute is categorical (or ordinal), so that we can ensure with this function that normalization is only applied to all numerical attributes, but not to the categorical ones. With the following R-code, z-score standardization is applied to all numerical attributes:

```
> iris.norm <- iris

> for (i in c(1:length(iris.norm))){
    if (!is.factor(iris.norm[,i])){
      attr.mean <- mean(iris.norm[,i])
      attr.sd <- sd(iris.norm[,i])
      iris.norm[,i] <- (iris.norm[,i]-attr.mean)/attr.sd
    }
  }
```

Other normalization and standardization techniques can carried out in a similar manner. Of course, instead of the functions mean (for the mean value) and sd (for the standard deviation), other functions like min (for the minimum), max (for the maximum), median (for the median), or IQR (for the interquartile range) have to be used.

References

1. Cook, D.J., Holder, L.B.: Mining Graph Data. Wiley, Chichester (2006)
2. Dougherty, J., Kohavi, R., Sahami, M.: Supervised and unsupervised discretization of continuous features. In: Proc. 12th Int. Conf. on Machine Learning (ICML 95, Lake Tahoe, CA), pp. 115–123. Morgan Kaufmann, San Mateo (1995)
3. Elomaa, T., Rousu, J.: Efficient multisplitting revisited: optima-preserving elimination of partition candidates. Data Min. Knowl. Discov. 8(2), 97–126 (2004)
4. Fayyad, U., Irani, K.: Multi-interval discretization of continuous-valued attributes for classification learning. In: Proc. 10th Int. Conf. on Artificial Intelligence (ICML'93, Amherst, MA), pp. 1022–1027. Morgan Kaufmann, San Mateo (1993)
5. Feldman, R., Sanger, J.: The Text Mining Handbook: Advanced Approaches in Analyzing Unstructured Data. Cambridge University Press, Cambridge (2007)
6. Guyon, I., Elisseeff, A.: An Introduction to Variable and Feature Selection. J. Mach. Learn. Res. 3, 1157–1182 (2003)

7. Hall, M.A., Smith, L.A.: Feature subset selection: a correlation based filter approach. In: Proc. Int. Conf. on Neural Information Processing and Intelligent Information Systems, pp. 855–858. Springer, Berlin (1997)

8. Kolodyazhniy, V., Klawonn, F., Tschumitschew, K.: A neuro-fuzzy model for dimensionality reduction and its application. Int. J. Uncertain. Fuzziness Knowl. Based Syst. **15**, 571–593 (2007)

9. Lowe, D., Tipping, M.E.: Feed-forward neural networks topographic mapping for exploratory data analysis. Neural Comput. Appl. **4**, 83–95 (1996)

10. Markovitch, S., Rosenstein, S.: Feature generation using general constructor functions. Mach. Learn. **49**(1), 59–98 (2002)

11. Murphy, P., Pazani, M.: ID2-of-3: constructive induction of m-of-n concepts for discriminators in decision trees. In: Proc. 8th Int. Conf. on Machine Learning (ICML'91, Chicago, IL), pp. 183–188. Morgan Kaufmann, San Mateo (1991)

12. Petrushin, V.A., Khan, L. (eds.): Multimedia Data Mining and Knowledge Discovery. Springer, New York (2006)

13. Rehm, F., Klawonn, F., Kruse, R.: POLARMAP—efficient visualisation of high dimensional data. In: Information Visualization, pp. 731–740. IEEE Press, Piscataway (2006)

14. Rehm, F., Klawonn, F.: Improving angle based mappings. In: Advanced Data Mining and Applications, pp. 3–14. Springer, Berlin (2008)

15. Robnik-Sikonja, M., Kononenko, I.: Theoretical and empirical analysis of ReliefF and RReliefF. Mach. Learn. **53**(1–2), 23–69 (2004)

16. van der Putten, P., van Someren, M.: A bias-variance analysis of a real world learning problem: the COIL challenge 2000. Mach. Learn. **57**, 177–195 (2004)

17. Yang, Y., Webb, G.I.: Discretization for naive Bayes learning: managing discretization bias and variance. Mach. Learn. **74**(1), 39–74 (2009)

References

Chapter 7
Finding Patterns

In this chapter we are concerned with summarizing, describing, or exploring the dataset as a whole. We do not (yet) concentrate on a particular target attribute, which will be the focus of Chaps. 8 and 9. Compact and informative representations of (possibly only parts of) the dataset stimulate the data understanding phase and are extremely helpful when exploring the data. While a table with mean and standard deviation for all fields also summarizes the data somehow, such a representation would miss any interaction between the variables. Investigating two fields in a scatter plot and examining their correlation coefficients could reveal linear dependencies among two variables, but what if more variables are involved, and the dependency is restricted to a certain part of the data only? We will review several techniques that try to group or organize the data intelligently, so that the individual parts are meaningful or interesting by themselves. For the purpose of getting a quick impression of the methods we will address in this section, consider a dataset of car offers from an internet platform. A number of interactions and dependencies, well known to everyone who has actively searched for a car, can be found in such a dataset.

Clustering For instance, we may want to *group* all cars according to their similarities. Rather than investigating the offers car by car, we could then look at the groups as a whole and take a closer look only at the group of cars we are particularly interested in. This aggregation into similar groups is called clustering. (From our background knowledge we expect to recover the well-known classes, such as luxury class, upper and lower middle-sized class, compact cars, etc.) **Cluster analysis** (or clustering) looks for groups of similar data objects that can naturally be separated from other, dissimilar data objects. As a formal definition of clusters turns out to be quite challenging, depending on the application at hand or the algorithmic approach used to find the structure (e.g., search, mathematical optimization, etc., see Chap. 5), algorithms with quite different notions of a cluster have been proposed. Their strengths (and weaknesses) lie in different aspects of the analysis: some focus on the data constellation (such as compactness and separation of clusters), others concentrate on the detection of changes in data density, and yet others aim at an easily understandable and interpretable summary.

M.R. Berthold et al., *Guide to Intelligent Data Analysis,*
Texts in Computer Science 42,
DOI 10.1007/978-1-84882-260-3_7, © Springer-Verlag London Limited 2010

We have selected the following clustering methods for this chapter:

Hierarchical Clustering As the name suggests, hierarchical clustering constructs a full-fledged hierarchy for all data that can be inspected visually. As every datum is represented by its own leaf of the tree, this technique is only feasible for small data sets but provides a good overview (Sect. 7.1).

Prototype-Based Clustering Alternatively, the data may be condensed to a small, fixed number of *prototypical records*, each of them standing for a full subgroup of the dataset (Sect. 7.3). This technique is efficient, and the results are interpretable even for much larger datasets. Such techniques assume, however, that the groups roughly follow some regular shape.

Density-Based Clustering Clusters may also be defined implicitly by the regions of higher data density, separated from each other by regions of lower density (Sect. 7.4). Most techniques from this category of density-based clustering allow for arbitrary shaped groups—but the price for this flexibility is a difficult interpretation of the obtained clusters.

Self-organizing Maps We could even arrange all cars into a two-dimensional *map* where similar car offers are placed close together so that we explore cars similar to some selected car by examining its neighborhood in the map. Such an overview is generated by so-called self-organizing maps (Sect. 7.5). By means of color coding it becomes immediately clear where many similar records reside and where we have a sparsely populated area. Self-organizing maps were not intended as clustering techniques but can deliver similar insights.

While clustering methods seek for similarities among cases or records in our database, we may also be interested in relationships between different variables or in subgroups that behave differently with respect to some target variable.

Association Rules Rather than grouping or organizing the cars, we may be interested in interdependencies among the individual variables. Several automobile manufacturers, for instance, offer certain packages that contain a number of additional features for a special price. If the car equipment is listed completely, it is possible to recover those features that frequently occur together. The existence of some features increases the probability of others, either because both features are offered in a package or simply because people frequently select a certain set of features in combination. One technique to find associations of this kind are **association rules** (Sect. 7.6): For every feature that can be predicted confidently from the occurrence of some other features, we obtain a rule that describes this relationship.

Deviation Analysis The last method we are going to investigate in this chapter is deviation analysis. Usually expensive cars offer a luxury equipment and consume significantly more petrol than standard cars. However, new cars with new fuel-saving engines, hybrid technology, etc. may represent exceptions from this general rule. Typically, there is a negative correlation with price and mileage, however, the group of oldtimer cars represents an exception. Typically a domain expert is already familiar with the more global dependencies and is more interested in deviations from the standard case or exceptional rules. The discovery of deviating subgroups of the population is discussed in Sect. 7.7.

7.1 Hierarchical Clustering

We start the discussion of clustering algorithms with one of the oldest and most frequently used clustering techniques: **agglomerative hierarchical clustering**. The term "agglomerative" indicates that this type of clustering operates bottom-up. Consider a dataset as shown in Fig. 7.1. Given a database consisting of many records, we are interested in a descriptive summary of the database, organizing all cases according to their properties and mutual similarity. Before comparing and arguing about *groups of records*, we must first solve a much simpler problem, namely the comparison of single cases. Given the first two rows in Fig. 7.1, how can their similarity be expressed by a single number of proximity or (dis)similarity? We do not just want to know whether two entries are identical, but we want to express *how different they are* in terms of a single number. Without loss of generality, let us concentrate on dissimilarity or distance (the smaller the value, the more similar the data) rather than similarity or proximity (the higher the value, the more similar the data).

In the case of a two-dimensional dataset (only two relevant attributes X and Y are given), we can visualize the dataset in a scatterplot and immediately see that, for instance, a and c are more similar than a and b—simply because a and c are *closer* together. Thus, the length of a straight line segment between two records appears to be a meaningful measure of dissimilarity (or distance). For the remainder of this section, we will use this notion of Euclidean distance but will discuss other options in detail in Sect. 7.2.

Given n records in our dataset, we need all pairwise distances, that is, a distance matrix $[d_{i,j}] \in \mathbb{R}_{\geq 0}^{n \times n}$, where $d_{i,j}$ denotes the dissimilarity between record #i and #j (or #j and #i for reasons of symmetry, that is, $d_{i,j} = d_{j,i}$), as shown on the right in Fig. 7.1. For very small datasets, such a matrix could be acquired manually from experts (e.g., wine tasting), but for larger datasets, it is more practical to obtain the values directly from a distance metric $d(\mathbf{x}, \mathbf{y})$ that can be applied to arbitrary records $\mathbf{x}, \mathbf{y} \in \mathcal{D}$ and defines the distance matrix by $d_{i,j} := d(\mathbf{x}_i, \mathbf{x}_j)$.

Hierarchical clustering requires either the provision of a distance metric $d(\mathbf{x}, \mathbf{y})$ or a distance matrix $[d_{i,j}]$ and builds a series of partitions to organize the data. Formally, a **partition** $\mathcal{P}$ is defined as a set of clusters $\{\mathcal{C}_1, \ldots, \mathcal{C}_c\}$, where each cluster $\mathcal{C}_i \subseteq \mathcal{D}$ contains a subset of the data. Any partition $\mathcal{P}$ must contain all the

Fig. 7.1 A small two-dimensional data table, its graphical representation, and distance matrix	<table><tr><td>id</td><td>X</td><td>Y</td></tr><tr><td>a</td><td>1</td><td>2</td></tr><tr><td>b</td><td>3</td><td>3</td></tr><tr><td>c</td><td>2</td><td>1</td></tr></table> 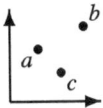	<table><tr><td>$[d_{i,j}]$</td><td>a</td><td>b</td><td>c</td></tr><tr><td>a</td><td>0.00</td><td>2.23</td><td>1.41</td></tr><tr><td>b</td><td>2.23</td><td>0.00</td><td>2.23</td></tr><tr><td>c</td><td>1.41</td><td>2.23</td><td>0.00</td></tr></table>

data from $\mathcal{D}$ (i.e., $\bigcup_{\mathcal{C} \in \mathcal{P}} \mathcal{C} = \mathcal{D}$) and assign it exclusively to a single cluster (i.e., $\forall \mathcal{C}, \mathcal{C}' \in \mathcal{P} : \mathcal{C} \cap \mathcal{C}' = \emptyset$). For practical reasons, it is usually required that none of the clusters is empty (i.e., $\forall \mathcal{C} \in \mathcal{P} : \mathcal{C} \neq \emptyset$).

7.1.1 Overview

Suppose that we have $\mathbf{x}, \mathbf{y} \in \mathcal{D}$ with $\delta = d(\mathbf{x}, \mathbf{y})$ and that from our background knowledge we conclude that both records should belong to the same cluster. As distance is our only measure to decide upon belongingness to a cluster, to be consistent, any other $\mathbf{z} \in \mathcal{D}$ with $d(\mathbf{x}, \mathbf{z}) \leq \delta$ should *also* belong to the same cluster, as $\mathbf{z}$ is even more similar to $\mathbf{x}$ than $\mathbf{y}$ is. Given some distance threshold δ, we may define clusters *implicitly* by requiring that any data $\mathbf{x}, \mathbf{y}$ must belong to the same cluster $\mathcal{C}$ if they are closer together than δ:

$$\forall \mathbf{x}, \mathbf{y} \in \mathcal{D} : \qquad (d(\mathbf{x}, \mathbf{y}) \leq \delta \quad \Rightarrow \quad \exists \mathcal{C} \in \mathcal{P} : \mathbf{x}, \mathbf{y} \in \mathcal{C}). \qquad (7.1)$$

Apparently, for different values of δ, we will obtain different partitions. Deferring how to find the partition for the moment, one difficult question is then how to come up with the *right* distance threshold δ. The choice must not be arbitrary, because the resulting clusters should be *stable* and *robust* in the sense that a slightly changed threshold should not lead to completely different clusters (which would render our clusters arbitrary and useless). The idea is to overcome this problem by exploring the full range of thresholds and deciding on the best threshold on the basis of the obtained results.

Exploring the full range of thresholds sounds like a costly task delivering many different partitions. Comparing all these partitions against each other appears to be even more expensive. The key observation is, however, that the solutions form a hierarchy, where clusters obtained from a threshold δ_1 are always completely contained in clusters obtained from δ_2 if $\delta_1 < \delta_2$. This follows directly from (7.1): Suppose $\mathbf{x}, \mathbf{y} \in \mathcal{C}$, where cluster $\mathcal{C}$ has been obtained using δ_1. Then, $d(\mathbf{x}, \mathbf{y}) \leq \delta_1$, and thus $d(\mathbf{x} - \mathbf{y}) \leq \delta_2$ (as $\delta_1 \leq \delta_2$), so $\mathbf{x}$ must be contained in some cluster $\mathcal{C}'$ obtained from δ_2 that contains $\mathbf{y}$.

The evolution of the partitions (as δ increases) can therefore be summarized in a single hierarchy, called **dendrogram**. On the horizontal axis, all data objects are listed in a suitable order, and the vertical axis shows δ. Consider the example in Fig. 7.2: For the smallest possible value of $\delta = 0$, we obtain 7 clusters (each of size 1, represented by leaves in the hierarchy). Records d and g are closest, so at the level $\delta = d(\mathsf{d}, \mathsf{g})$, these two points unite to a cluster, while all others still remain singletons. For $\delta_1 = d(\mathsf{a}, \mathsf{c})$ (green color), all four points on the right side can be reached within this distance from d, so that all points belong to the same cluster already. The three points on the left are still separated into two groups ($\{\mathsf{a}, \mathsf{c}\}$ and $\{\mathsf{b}\}$), because the distance $d(\mathsf{a}, \mathsf{b}) = \delta_2$ (blue color) is larger than δ_1. If we cut the dendrogram just above δ_1, the set of leaves of the remaining subtrees represent the currently obtained clusters: $\{\mathsf{a}, \mathsf{c}\}$, $\{\mathsf{b}\}$, and $\{\mathsf{d}, \mathsf{e}, \mathsf{f}, \mathsf{g}\}$. At δ_2 we obtain two clusters

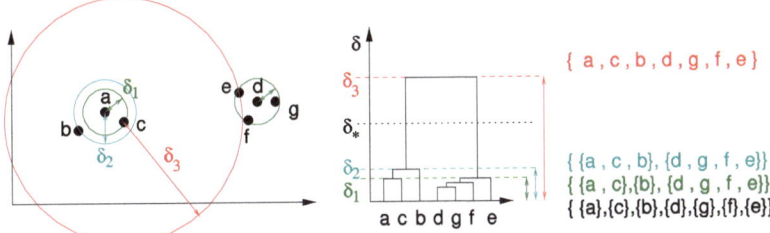

Fig. 7.2 Hierarchical clustering (single linkage) applied to a small example dataset

$\{a, c, b\}$ and $\{d, e, f, g\}$, and at a very large distance $\delta_3 = d(c, e)$, the two clusters become connected, and all data belong to a single cluster.

Coming back to the initially stated problem of finding the best distance threshold δ, the dendrogram provides useful information: As already mentioned, for robust clusters, a slightly changed δ should not lead to completely different results. If we select δ_1, it makes a difference whether we cut the tree slightly above or slightly below δ_1, because we obtain different clusters in both cases. However, for δ_*, we obtain clusters that are stable in the sense that we get the same clusters if we choose δ somewhat smaller or larger. By using $\delta_* = \frac{1}{2}(\delta_2 + \delta_3)$ this *stability* is maximized, and we end up with two clusters. Looking for the largest difference between δ_i and δ_{i+1}, which corresponds to the largest gap between in the dendrogram, helps us in finding the most stable partition.

Figure 7.3 shows two examples of performance (so-called single linkage cluster- ing, see next section) to illustrate the interpretation of a dendrogram. Example (a) consists of three well-separated clusters, which are easily discovered by the algo- rithm. The data belonging to each of the clusters unites at relatively low values of δ already, whereas quite a large value of δ is required to merge two of the three clus- ters into one. Therefore, the vertical gap is large, and the existence of three clusters is easily recognized by visual inspection. From hierarchy (b) we immediately see that we have two clusters, one being a bit larger (left cluster in tree occupies more horizontal space and thus has more leaves) and less dense than the other (higher δ-values compared to the right cluster). The fact that the shape of the clusters is very different (cf. scatterplot) does not make a difference for the algorithm (and is likewise not reflected in the dendrogram).

Although example (c) is somewhat similar to (a), just some random noise has been added, the dendrogram is very different. To belong to a cluster, it is sufficient to have a high similarity to just one of the cluster members, noise points get eas- ily *chained*, thereby building bridges between the clusters that cause cluster merg- ing at a relatively low δ-value. These *chaining* capabilities were advantageous for case (b) but prevent us from observing a clear cluster structure in case (c). Hierar- chical (single-linkage) clustering is extremely sensitive to noise—adding just a few noise points to dataset (a) quickly destroys the clear structure of the dendrogram. We will discuss solutions to this problem in Sect. 7.1.3.

In addition to the dendrogram, a **heatmap** can also help to illustrate the cluster- ing result. For heatmap, hierarchical clustering is carried out for the data records,

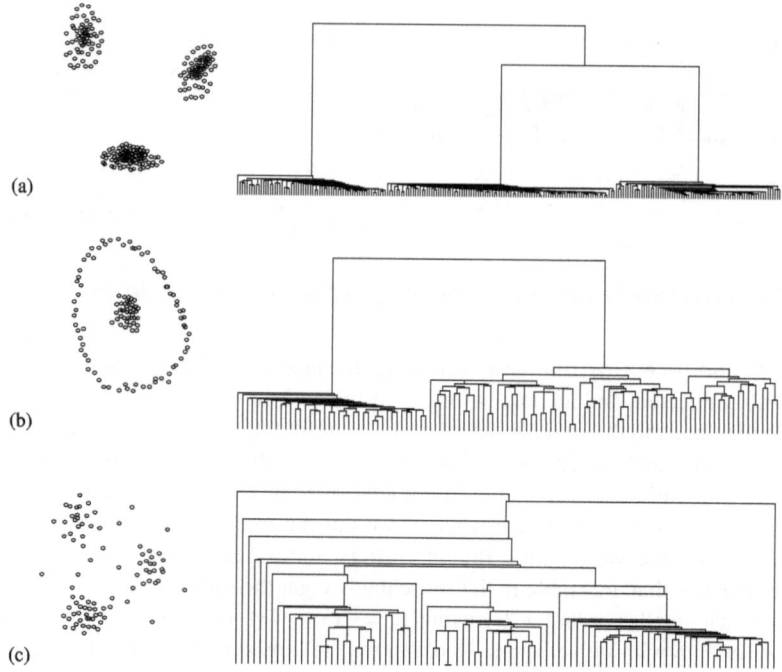

(a)

(b)

(c)

Fig. 7.3 Hierarchical clustering (single linkage) for three example data sets

i.e., the rows of a data table, and also for the attributes by transposing the data table, so that clustering is applied to the columns of the data table. Then the records and the attributes are reordered according to the clustering result. Instead of showing the numerical entries in the data table, a color scale is used to represent the numerical values.

Figure 7.4 shows a heatmap for clustering the numerical attributes of the Iris data set (after z-score standardization). The dendrogram for the record clustering is shown on the left, and the dendrogram for the attribute clustering on top. One can easily see that in the lower part of the diagram, records are clustered based on lower values for sepal width, the petal length, and the petal width (red color) and on higher values for the sepal length (yellow or orange color).

7.1.2 Construction

Table 7.1 shows an algorithm to perform hierarchical clustering. Starting from $\delta_0 = 0$, we have $n = |\mathcal{D}|$ clusters of size 1. In line 4 we find the two closest clusters in the current partition $\mathcal{P}_t$. These two clusters are removed from the current partition, and their union is reinserted as a new cluster (line 6). In this way, the number of clusters reduces by 1 for each iteration, and after n iterations the algorithm terminates.

Note that the distance measure d' used in Table 7.1 measures the distance between clusters $\mathcal{C}$ and $\mathcal{C}'$ and not between data objects. While the Euclidean distance

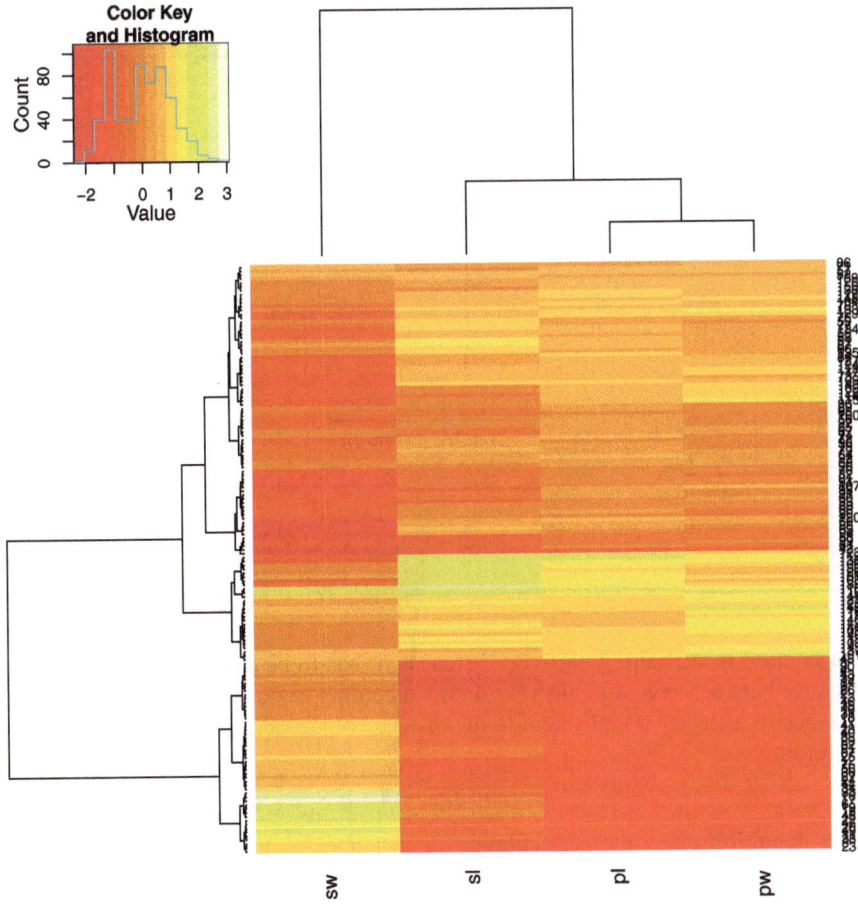

Fig. 7.4 A heatmap for the Iris data set

Table 7.1 Hierarchical clustering

Algorithm $\text{HC}(\mathcal{D}) : (\mathcal{P}_t)_{t=0..n-1}, (\delta_t)_{t=0..n-1}$

input: data set $\mathcal{D}, |\mathcal{D}| = n$
output: series of hierarchically nested partitions $(\mathcal{P}_t)_{t=0..n-1}$
        series of hierarchy levels $(\delta_t)_{t=0..n-1}$

1	$\mathcal{P}_0 = \{\{\mathbf{x}\} \mid \mathbf{x} \in \mathcal{D}\}$
2	$t = 0, \delta_t = 0$
3	**while** current partition $\mathcal{P}_t$ has more than one cluster
4	find pair of clusters $(\mathcal{C}_1, \mathcal{C}_2)$ with minimal distance $d'(\mathcal{C}_1, \mathcal{C}_2)$
5	$\delta_{t+1} = d'(\mathcal{C}_1, \mathcal{C}_2)$
6	construct $\mathcal{P}_{t+1}$ from $\mathcal{P}_t$ by removing $\mathcal{C}_1$ and $\mathcal{C}_2$ and inserting $\mathcal{C}_1 \cup \mathcal{C}_2$
7	$t = t+1$
8	**end while**

Table 7.2 First five iterations of hierarchical clustering on records a–g with given distance matrix. The cluster pairs $(\mathcal{C}_1, \mathcal{C}_2)$ are selected by minimal distances (bold font)

	a	b	c	d	e	f	g
a	0	50	63	17	72	81	12
b		0	49	41	42	54	37
c			0	52	13	16	61
d				0	56	66	15
e					0	**11**	64
f						0	74
g							0

	a	b	c	d	ef	g
a	0	50	63	17	72	**12**
b		0	49	41	42	37
c			0	52	13	61
d				0	56	15
ef					0	74
g						0

	ag	b	c	d	ef
ag	0	37	61	15	72
b		0	49	41	42
c			0	52	**13**
d				0	56
ef					0

	ag	b	d	efc
ag	0	37	**15**	61
b		0	41	42
d			0	52
efc				0

	agd	b	efc
agd	0	**37**	52
b		0	54
efc			0

can be used to measure point-wise distances, it has to be extended to measure the distance between sets of points (clusters). According to our initial idea in (7.1), the distance to a cluster is equivalent to the distance of the nearest point of the cluster; therefore,

$$d'(\mathcal{C}, \mathcal{C}') := \min\{d(x, y) \mid x \in \mathcal{C}, y \in \mathcal{C}'\}. \tag{7.2}$$

This definition eases the recalculation of distances of merged clusters, because $d'(\{\mathcal{C} \cup \mathcal{C}'\}, \mathcal{C}'')$ can be obtained directly from the previously used distances: $d'(\{\mathcal{C} \cup \mathcal{C}'\}, \mathcal{C}'') = \min\{d'(\mathcal{C}, \mathcal{C}''), d'(\mathcal{C}', \mathcal{C}'')\}$. This recursive formulation is helpful if the distances are given by a distance matrix, because then the distances to the merged cluster are obtained by simply looking up two distances from the old distance matrix and taking their minimum.

As an example, consider the data set from Table 7.2, which shows the distance matrix for seven objects a–g. In each iteration of the algorithm, the smallest distance in the matrix is identified, and the corresponding objects (clusters, resp.) are merged. Afterwards, a new distance matrix is constructed, where the new cluster replaces the two merged clusters, and the entries in the matrix are determined according to (7.2). The process eventually stops at a 1×1 matrix.

7.1.3 Variations and Issues

As long as the data set size is moderate, the dendrogram provides useful insights, but a tree with several thousand leaves is quite difficult to manage. Furthermore, at least in the single linkage clustering, one cannot tell anything about how the data distributes in the data space, since the cluster may be shaped arbitrarily. A brief characterization of each cluster is thus difficult to provide. To get an impression of the cluster, one has to scan through all of its members. On the other hand, hierarchical clustering may be carried out even if only a distance matrix is given, whereas the approaches in the next two sections require an explicitly given distance function (often even a specific function, such as the Euclidean distance).

The runtime and space complexity of the algorithm is quite demanding: In a naïve implementation we need to store the full distance matrix and have a cubic runtime complexity (n times find the minimal element in an $n \times n$ matrix). At least the runtime complexity may be reduced by employing data structures that support querying for neighboring data. Further possibilities to reduce time and space requirements are subsampling (using only part of the dataset), omission of entries in the distance matrix (transformation into a graph, application of, e.g., minimum spanning trees), and data compression techniques (use representatives for a bunch of similar data, e.g., [60]).

7.1.3.1 Cluster-to-Cluster Distance

The so-called **single-linkage** distance (7.2) is responsible for the high sensitivity to noise we have observed already in Fig. 7.3(c). Several alternative distance aggregation heuristics have been proposed in the literature (cf. Table 7.3). The **complete-linkage** uses the maximum rather than the minimum and thus measures the diameter of the united cluster rather than the smallest distance to reach one of the cluster points. It introduces a bias towards compact cloud-like clusters, which enables hierarchical clustering to better recover the three clusters in Fig. 7.2(c) but fails in case of Fig. 7.2(b) because any cluster consisting of the ring points would also include all data from the second (interior) cluster.

The single-linkage and complete-linkage approaches represent the extremes in measuring distances between clusters. Calculating the exact average distance be-

Table 7.3 Alternative cluster distance aggregation functions

$$d'(\{\mathcal{C} \cup \mathcal{C}'\}, \mathcal{C}'') = \cdots$$

$$\text{single linkage} = \min\{d'(\mathcal{C}, \mathcal{C}''), d'(\mathcal{C}', \mathcal{C}'')\} \tag{7.3}$$

$$\text{complete linkage} = \max\{d'(\mathcal{C}, \mathcal{C}''), d'(\mathcal{C}', \mathcal{C}'')\} \tag{7.4}$$

$$\text{average linkage} = \frac{|\mathcal{C}|d'(\mathcal{C}, \mathcal{C}'') + |\mathcal{C}'|d'(\mathcal{C}', \mathcal{C}'')}{|\mathcal{C}| + |\mathcal{C}'|} \tag{7.5}$$

$$\text{Ward} = \frac{(|\mathcal{C}| + |\mathcal{C}''|)d'(\mathcal{C}, \mathcal{C}'') + (|\mathcal{C}'| + |\mathcal{C}''|)d'(\mathcal{C}', \mathcal{C}'') - |\mathcal{C}''|d'(\mathcal{C}, \mathcal{C}')}{|\mathcal{C}| + |\mathcal{C}'| + |\mathcal{C}''|} \tag{7.6}$$

$$\text{centroid (metric)} = \frac{1}{|\mathcal{C} \cup \mathcal{C}'||\mathcal{C}''|} \sum_{x \in \mathcal{C} \cup \mathcal{C}'} \sum_{y \in \mathcal{C}''} d(\mathbf{x}, \mathbf{y}) \tag{7.7}$$

tween data from different clusters via (7.7) is computationally more expensive than using the heuristics (7.5) or (7.6). If the data stems from a vector space, the cluster means may be utilized for distance computation. The cluster means can be stored together with the clusters and easily updated at the merging step, making approach (7.7) computationally attractive. For hyperspherical clusters, all these measures perform comparable but otherwise may lead to quite different dendrograms.

7.1.3.2 Divisive Clustering

Rather than agglomerating small clusters to large ones, we may start with one single cluster (full data set) and subsequently subdivide clusters to find the best result. Such methods are called **divisive hierarchical clustering**. In agglomerative clustering we have merged two clusters, whereas in divisive clustering typically a single cluster is subdivided into two subclusters (**bisecting strategy**), which leads us again to a binary tree of clusters (a hierarchy). Two critical questions have to be answered in every iteration: (1) which cluster has to be subdivided, and (2) how to divide a cluster into subclusters. For the first problem, validity measures try to evaluate the quality of a given cluster, and the one with the poorest quality may be the best candidate for further subdivision. Secondly, if the number of subclusters is known or assumed to be fixed (e.g., 2 in a bisecting strategy), clustering algorithms from Sect. 7.3 may be used for this step, as these algorithms subdivide the data into a partition with a fixed number of clusters. In the bottom-up approach of agglomerative clustering, the decisions are based on local information (neighborhood distances), which give poor advice in case of diffuse cluster boundaries and noisy data. In such situations, top-down divisive clustering can provide better results, as the global data distribution is considered from the beginning. The computational effort is higher for divisive clustering, though, which can be alleviated by stopping the construction of the hierarchy after a few steps, as the desired number of clusters is typically rather small.

7.1.3.3 Categorical Data

Providing distances for categorical data leads to a coarse view on the data, for instance, the number of distinct distance values is limited (as we will see in the next section). For the moment, just consider six binary features and let us compare the cases: Visual inspection of the cases a–h in Fig. 7.5 reveals two clusters, one group of cars (a–d) equipped with features u and v (plus one optional feature) and the second group (e–h) equipped with y and z (plus one optional features).

Applying the Jaccard measure (a well-suited distance measure for binary data, which will be discussed in the next section) to this dataset leads to the distance matrix in Fig. 7.5. A distance of 0.5 occurs both within groups (e.g., $d(a, b)$, $d(e, g)$) and across groups ($d(c, g)$). The Jaccard distance is therefore not helpful to discriminate between these two groups.

full data set

car	u	v	w	x	y	z
a	1	1	1	0	0	0
b	1	1	0	1	0	0
c	1	1	0	0	1	0
d	1	1	0	0	0	0
e	0	0	0	1	1	1
f	0	0	1	0	1	1
g	0	1	0	0	1	1
h	0	0	0	0	1	1

Jaccard distance

	a	b	c	d	e	f	g	h
a	0	.50	.50	.33	1	.80	.80	1
b	.50	0	.50	.33	.80	1	.80	1
c	.50	.50	0	.33	.80	.80	.50	.80
d	.33	.33	.33	0	1	1	.75	1
e	1	.80	.80	1	0	.50	.50	.33
f	.80	1	.80	1	.50	0	.50	.33
g	.80	.80	.50	.75	.50	.50	0	.33
h	1	1	.80	1	.33	.33	.33	0

common neighbours

	a	b	c	d	e	f	g	h
a		2	2	2	0	0	1	0
b	2		2	2	0	0	1	0
c	2	2		2	0	1	0	1
d	2	2	2		0	0	1	0
e	0	0	0	0		2	2	2
f	0	0	1	0	2		2	2
g	1	1	0	1	2	2		2
h	0	0	1	0	2	2	2	

Fig. 7.5 Grouping of categorical data: A Jaccard distance of 0.5 is achieved within the groups (e.g., $d(a, b)$, $d(e, g)$) and also across groups ($d(c, g)$)

Distance measures usually compare only two records from the dataset, but they do not take any further information into account. We have seen that single-linkage clustering reacts sensitive to single outliers, which may drastically reduce the distance between clusters. Knowing the outliers can thus help to improve the clustering results we obtain from distances alone. One approach in this direction is to use link information between data objects: The number of links that exist between data x_1 and x_2 is the number of records $x \in \mathcal{D}$ that are close to both, x_1 and x_2. Based on some threshold ϑ, we define the neighborhood $N_\vartheta(x)$ of x as $\{y \in \mathcal{D} \mid d(x, y) \leq \vartheta\}$. In our example, $N_{0.5}(c) = \{a, b, c, d, g\}$. The number of links or common neighbors of x and y is then $(N_\vartheta(x) \cap N_\vartheta(y)) \backslash \{x, y\}$. In the example we obtain a single common neighbor for a and g: $(\{a, b, c, d\} \cap \{c, e, f, g, h\}) \backslash \{a, g\} = \{c\}$. The matrix on the right of Fig. 7.5 shows the number of links for all pairs of data objects. Using the number of links as a measure of similarity rather than the distance now allows us to discriminate both clusters, and we have no longer the same degree of similarity within and across groups. This approach is used in the **ROCK** algorithm (robust clustering using links) [26].

7.2 Notion of (Dis-)Similarity

In the previous section we have simply taken the length of the line segment between two data points as the distance or dissimilarity between records. This section is solely dedicated to distance measures, as they are crucial for clustering applications. We have already seen in Sect. 6.3.2 that by a different scaling of individual attributes, the clustering result may be completely different. If we have clearly pronounced clusters in two dimensions (as in Fig. 7.3(a)), but the distance measure takes also a third variable into account, which is just random noise, this may mess up the clear structure completely, and in the end, all groups contain data from the original two-dimensional clusters. We have discussed the negative influence of irrelevant and redundant variables already in Sect. 6.1.1 but assumed a predictive task such that we had some reference variable to measure a variable's informativeness—we have no such target variable in explorative clustering tasks. Instead, we can only observe

Table 7.4 Small excerpt from a database of cars offered on an internet platform

id	Manufacturer	Name	Type	yofr	cap	fu	fxu	used	air	abs	esp	hs	ls
1	Volkswagen	Golf Variant	Estate Car	2004	1.9	6.6	4.3	Y	Y	Y	Y	Y	N
2	Dacia	Logan	Limousine	2006	1.4	9.4	5.5	Y	N	Y	N	N	N
3	Ford	Galaxy	Van	2004	1.9	12.2	8.3	Y	Y	Y	N	N	N

Legend: yofr = year of first registration, cap = engine capacity, fu = fuel consumption (urban), fxu = fuel consumption (extra urban), used = used car, air = air conditioned, abs = anti blocking system, esp = electronic stability control, hs = heated seats, ls = leather seats

how good the data gets clustered in two and three dimensions. A clear grouping with respect to variables A and B means that both attributes correlate somehow (certain combinations of attribute values occur more frequently than others). There is no reason why a similar co-occurrence pattern should be observable if we add irrelevant or random variables. Thus, whether we will observe clusters or not strongly depends on the choice of the distance measure—and that includes the choice of the variable subset.

For this section, we return to the example of car offers as those shown in Table 7.4. This time we have a mixed dataset with binary, categorical, and numerical attributes. How can we define the similarity between two cars?

Numerical Attributes At the first glance, turning attributes such as manufacturer, car type, or optional equipment into a numerical value seems a bit more challenging; therefore we start with numerical attributes, such as horse power, fuel consumption, or year of first registration. Given that the differences of the numerical values are meaningful (that is, the data is at least on an interval scale), we can simply take the absolute difference to measure the dissimilarity in terms of, say, fuel consumption. Let x_1 be the first, and x_2 the second row in Table 7.4; then based on fuel consumption (fu) alone, we have $d(x_1, x_2) = |x_{1,fu} - x_{2,fu}| = |6.6 - 9.4| = 2.8$. If we want to include the extra-urban fuel consumption into our comparison, we actually compare tuple $(x_{1,fu}, x_{1,fxu})$ against $(x_{2,fu}, x_{2,fxu})$ rather than just numbers and have the additional freedom of how to combine the dissimilarity of the individual attributes (see Table 7.5): we could use the sum of the individual distances (Manhattan distance)

$$d(x_1, x_2) = |x_{1,fu} - x_{2,fu}| + |x_{1,fxu} - x_{2,fxu}| = |6.6 - 9.4| + |4.3 - 5.5| = 4.0,$$

the length of the shortest line between x_1 and x_2 (Euclidean distance)

$$d(x_1, x_2) = \sqrt{(x_{1,fu} - x_{2,fu})^2 + (x_{1,fxu} - x_{2,fxu})^2} = \sqrt{7.84 + 1.44} \approx 3.04,$$

or even the maximum of all distances

$$d(x_1, x_2) = \max\{|x_{1,fu} - x_{2,fu}|, |x_{1,fxu} - x_{2,fxu}|\} = \max\{2.8, 1.2\} = 2.8.$$

Under the maximum norm, the Logan (car #2) is equally dissimilar to the Golf (car #1) and the Galaxy (car #3), whereas for the other measures, the dissimilarity of the Galaxy is 1.4 times higher for the Manhattan distance and 1.3 times

Table 7.5 Dissimilarity of numeric vectors. The graph on the right shows (for selected measures) all two-dimensional points that have a distance of 1.0 from the small bullet in the centre

Minkowksi L_p	$d_p(x,y) = \frac{1}{p}\sqrt{\sum_{i=1}^{n}|x_i - y_i|^p}$
Euclidean L_2	$d_E(x,y) = \sqrt{(x_1 - y_1)^2 + \cdots + (x_n - y_n)^2}$
Manhattan L_1	$d_M(x,y) = |x_1 - y_1| + \cdots + |x_n - y_n|$
Chebyshev L_∞	$d_\infty(x,y) = \max\{|x_1 - y_1|, \ldots, |x_n - y_n|\}$
Cosine	$d_C(x,y) = 1 - \frac{x^\top y}{\|x\|\|y\|}$
Tanimoto	$d_T(x,y) = \frac{x^\top y}{\|x\|^2+\|y\|^2-x^\top y}$
Pearson	Euclidean of z-score transformed $\mathbf{x}$, $\mathbf{y}$

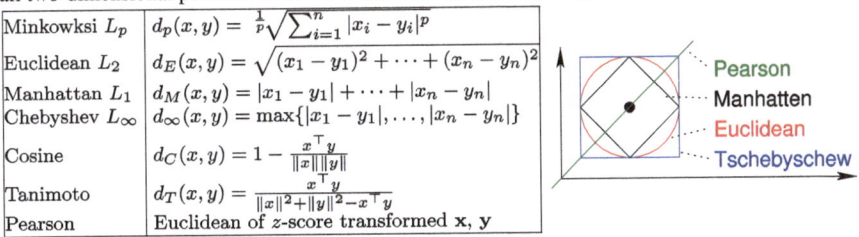

higher for the Euclidean distance. We have already seen that these distance values directly influence the order in which the clusters get merged, so we should carefully check that our choice corresponds to our intended notion of dissimilarity.

Nonisotropic Distances The term **isotropic** means that the distance grows in all directions equally fast, which is obviously true for, e.g., the Euclidean distance. All points with distance 1.0 from some reference point $\mathbf{x}$ form a perfect, symmetric surface of a sphere. We use the distance to capture the dissimilarity between records, and a large distance may even be used as an indication for an outlier. (In particular, if we have z-score transformed data, which can be interpreted as "*how many standard deviations is this value away from the mean?*", see Sect. 6.3.2). If we consider a pair of correlated variables, say urban and extra-urban fuel consumption, we may perceive isotropic distances and counterintuitive: The differences of the fuel consumptions $\mathbf{x}_1 = (8.1, 5.3)$ and $\mathbf{x}_2 = (5.1, 5.3)$ from the fuel consumption $\mathbf{x} = (6.6, 4.3)$ is $\Delta_1 = (1.5, 1.0)$ and $\Delta_2 = (-1.5, 1.0)$. The length of the difference vectors are identical, so are the distances when using an isotropic distance function. However, we would be very surprised if the difference in the urban fuel consumption is negative while at the same time the difference in the extra-urban fuel consumption is positive—this deviates vastly from our expectations.

Nonisotropic distance functions help us to respect such cases. We have already seen that the joint distribution of multiple variables is captured by the covariance matrix C (see also Sect. 4.3.2.1). The nonisotropic **Mahalanobis distance** is defined as

$$d(\mathbf{x}, \mathbf{y}) = \sqrt{(\mathbf{x} - \mathbf{y})^\top C^{-1}(\mathbf{x} - \mathbf{y})}$$

and considers the variance across all variables. Similar to the z-score transformed data, the distance may be interpreted as the number of standard deviations a vector deviates from the mean, but this time the deviation is not measured independently for every variable but jointly for all variables at the same time. Now, all points with distance 1.0 from some reference points $\mathbf{x}$ do no longer form the surface of a symmetric sphere but the surface of an arbitrary hyperellipsoid.

Text Data and Time Series Sometimes not the absolute difference is important but the fact that the data *varies in the same way*. This is often the case with embedded time series ($\mathbf{x}$ contains the values of a variable at consecutive time points) or a bag-of-words text representation (the attributes indicate the number of word occurrences, see Sect. 6.4). When comparing two texts in a bag-of-words representation, they are usually considered similar if the same words occur in comparable frequency. However, if one of the texts is, say, three times longer than the other, the absolute frequencies will be three times higher, so both texts appear dissimilar. By normalizing both word vectors (to unit length), the absolute frequencies become relative frequencies, and both vectors become similar. The cosine measure yields the cosine of the angle between the normalized vectors $\mathbf{x}$ and $\mathbf{y}$ and is well suited for bag-of-words comparison due to the built-in normalization. If we apply a z-score transformation to the vectors $\mathbf{x}$ and $\mathbf{y}$, the Euclidean metric becomes the Pearson metric (up to a constant factor) and is also invariant to shifting (adding a constant). The latter is useful when comparing embedded time series that have different offsets (and scaling).

Binary Attributes   Next, we consider binary attributes, e.g., whether the car was offered by a dealer or not (private offer). If we encode the Boolean values (yes/no) into numbers (1/0), we may apply the same differencing we used for numeric values before. In this case, we get a zero distance if both cars are used cars or both cars are new cars (and a distance of one otherwise).

However, this approach may be inappropriate for other binary attributes. Suppose that we have a feature "offer by local dealer," where "local" refers to the vicinity of your own home. Then, the presence of this feature is much more informative than its absence: The cars may be offered by private persons in the vicinity and by private persons or dealers far away. The presence of this feature tells us much more than the absence; the probability of absence is typically much higher, and therefore the absence should not have the same effect in the dissimilarity measure as the presence. A similar argument applies to the optional car equipment, such as ABS, EPS, heated seats, leather seats, etc. We could easily invent hundred such attributes, probably only a handful of cars having more than ten of these properties. But if we consider the absence of a feature as a *sharing a common property*, most cars will be very similar, simply because they share that 90% of the equipment is not present. In such cases, the measure by Russel and Rao could be used that accounts only for shared *present* features (see Table 7.6). Another possibility is the Jaccard measure, which takes its motivation from set theory, where the dissimilarity of two sets $A, B \subseteq \Omega$ is well reflected by $J - 1 - \frac{|A \cap B|}{|A \cup B|}$. Apparently, this definition is independent of the size of the domain $|\Omega|$. For the cars example restricted to the last six binary attributes, we obtain, for the Jaccard measure, $d_J(\mathbf{x}_1, \mathbf{x}_2) = 1 - \frac{2}{5} = \frac{3}{5}$ and $d_J(\mathbf{x}_2, \mathbf{x}_3) = 1 - \frac{2}{3} = \frac{1}{3}$ versus $d_S(\mathbf{x}_1, \mathbf{x}_2) = 1 - \frac{3}{6} = \frac{1}{2}$ and $d_S(\mathbf{x}_2, \mathbf{x}_3) = 1 - \frac{5}{6} = \frac{1}{6}$ for the simple match.

Nominal Attributes   Nominal attributes may be transformed into a set of binary attributes, each of them indicating one particular feature of the attribute (see example in Table 7.7). Only one of the introduced binary attributes may be active at a time. The measures from Table 7.6 could then be reused for comparison. Clearly, as soon as $n > 2$, we *always* have co-occurring zero fields, and therefore the Jaccard or Dice measure should be used. Equivalently, a new attribute might be introduced that

Table 7.6 Measuring the dissimilarity in case of a vector of predicates (binary variables) $\mathbf{x}$ and $\mathbf{y}$, which may also be interpreted as a set of properties $X, Y \subseteq \Omega$. The Tanimoto measure from Table 7.5 becomes identical to the Jaccard measure when applied to binary variables

	binary attributes	sets of properties
simple match	$d_S = 1 - \frac{b+n}{b+n+x}$	
Russel & Rao	$d_R = 1 - \frac{b}{b+n+x}$	$1 - \frac{|X \cap Y|}{|\Omega|}$
Jaccard	$d_J = 1 - \frac{b}{b+x}$	$1 - \frac{|X \cap Y|}{|X \cup Y|}$
Dice	$d_D = 1 - \frac{2b}{2b+x}$	$1 - \frac{2|X \cap Y|}{|X|+|Y|}$

no. of predicates that...

$b = \ldots$ hold in both records

$n = \ldots$ do not hold in both records

$x = \ldots$ hold in only one of both records

Example:

$\mathbf{x}$	$\mathbf{y}$	set X	set Y	b n x	d_M	d_R	d_J	d_D
101000	111000	$\{A, C\}$	$\{A, B, C\}$	2 3 1	$0.1\bar{6}$	$0.6\bar{6}$	$0.3\bar{3}$	0.20

Table 7.7 Converting a nominal attribute

Manufacturer	...	Binary vector
Volkswagen	...	00001
Dacia	... $\rightarrow$	01000
Ford	...	00100

Table 7.8 Converting an ordinal attribute

Number of previous owners ...	Binary vector		Rank
1 previous owner ...	100		1
2 previous owner ... $\rightarrow$	110	/	2
more than 2 previous owners ...	111		3

becomes true if and only if both records share the same value. Finally, a third alternative is the provision of a look-up table for individual dissimilarity values, where Porsche and BMW are more similar than, say, Mercedes and Dacia.

Ordinal Attributes Ordinal attributes can also be transformed into binary attributes in the same fashion as nominal attributes, but then any two different values appear equally (dis)similar, which does not reflect the additional information contained in an ordinal attribute. A better transformation is shown in Table 7.8, where the Jaccard dissimilarity is proportional to the number of ranks lying between the compared values. This is equivalent to introducing an integer-valued rank attribute and apply, say, the Manhattan distance to measure the dissimilarity of two values. The provision of a dissimilarity matrix remains as a third option.

In a mixed scale situation, such as the comparison of car offerings in Table 7.4, there are two possibilities. First, even interval- and ratio-scaled attributes could be discretized and then transformed into binary attributes, so that only the Jaccard measure needs to be applied on the final set of (only) binary attributes. A second option is the combination of various distance measures (sum, mean, etc.), where each of the measures considers a different subset of attributes.

Attribute Weighting Although we have mentioned this point earlier, we want to emphasize once more that it is important to take care about the relative weight an individual attribute gains. This holds for numeric and nominal-scaled attributes. Suppose that we have compared two cars based on fuel consumption (urban and extra-urban)—now we want to take mileage into consideration. What will be the effect? The huge differences in the additional attribute will completely dominate the small differences obtained earlier. The results would be almost identical to those obtained from mileage only. The variances of the attributes differ far too much. As a second example, consider the attributes capturing the car equipment: If two cars share an additional property, they become more similar. Should they become more similar to the same degree, regardless of the property they share? Compare a drink-holder to an automatic air conditioning (a feature that much more contributes to the value of the car). For the first example (mileage), the standard solution is to standardize the data so that all numeric attributes have unit variance. The second problem is only solved by using additional weights for the aggregation of dissimilarity values when calculating the overall car-to-car distance.

The Curse of Dimensionality While we are used to work with distances in two- or three-dimensional spaces, our intuition does not help us in high-dimensional spaces. By considering more and more attributes in a dataset, we mathematically construct a high-dimensional space. While the generalization of, e.g., the Euclidean distance to an arbitrary number of dimensions is straightforward, the interpretation of the distances may be somewhat counterintuitive.

Suppose that all variables have been normalized to the unit interval $[-1, 1]$. If we select two attributes, the space in which our data resides corresponds to the box of edge length 2 around the origin $\mathbf{0}$ of the coordinate system. Most of its area ($\approx 78\%$) is contained in the circle of radius 1 around the origin; only the four corners are omitted. If we add another attribute, the data space corresponds to a three-dimensional cube. Now, we have already eight corners that are not covered by the a sphere with radius 1 around the origin, and the sphere contains less than 55% of the cube's volume. Figure 7.6 shows the percentage of the data space's hypervolume that is covered by a hypersphere of radius 1: The percentage quickly approximates zero.

If one data object is located at the origin of a two-dimensional space and we take a second data object from $[-1, 1]^2$, there is a chance of only $\approx 22\%$ that its distance to the first object is greater than 1 (assuming the data space is uniformly populated). In case of an eight-dimensional data space, the chances of being further away than a distance of 1 rises to more than 95%. The distances among the data objects increase, and most of the data is in the corners of the hypercube. If we want to identify data similar to some reference object (here: the origin), the higher the dimensionality, the less data we will find. This phenomenon is known as the **curse of dimensionality**, a term coined by Bellmann [9].

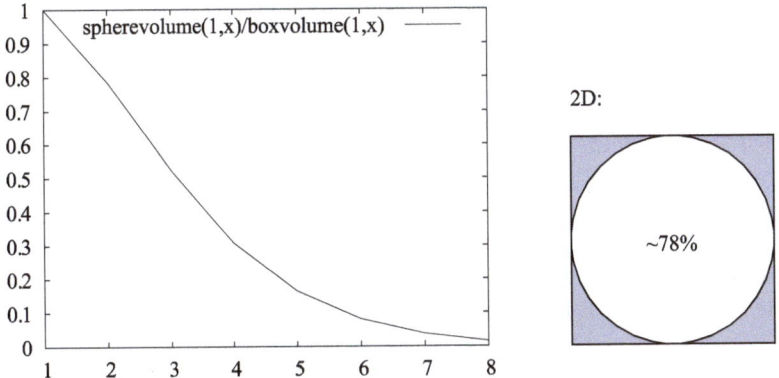

Fig. 7.6 Curse of dimensionality: The higher the dimensionality, the smaller the fraction of the hyperbox that is covered by the embedded hypersphere

Fig. 7.7 The distribution of the length of random vectors in the unit hypercube (Euclidean norm) in 1, 2, 5, 10, and 20 dimensions (from bottom to top). While in 1D any vector length in [0, 1] is equally likely, there were no vectors of length ≤ 1 in the 20-dimensional sample of size 20,000

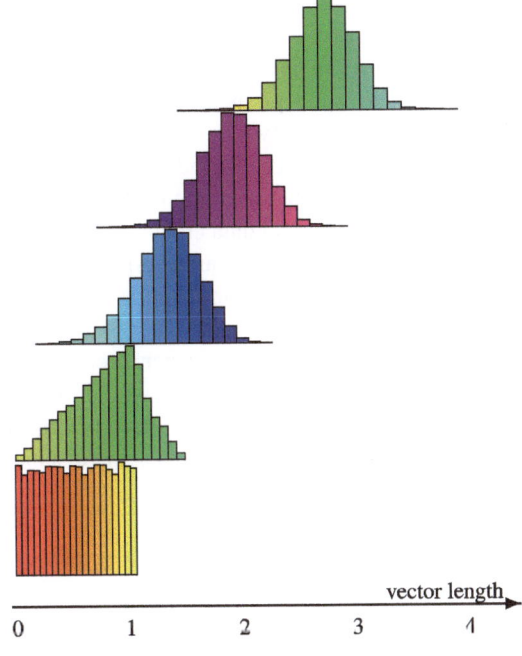

This effect is also shown in Fig. 7.7, which shows the distribution of the vector lengths of 20,000 random vectors in the m-dimensional unit hypercube $[0, 1]^m$. As expected, in the one-dimensional case (bottom row) the unit interval [0,1] is almost uniformly covered. To obtain an Euclidean distance of zero in an m-dimensional space, all m components must be zero at the same time, so the chances of low overall distances (sum of all components) decreases rapidly with the dimensionality. In the 10-dimensional sample there are almost no vectors of length ≤ 1, that is, within a range of 1.0 from the origin.

7.3 Prototype- and Model-Based Clustering

Clusters are represented by subtrees in a dendrogram, and all cluster members are represented individually by leaves. For large datasets, such dendrograms are too detailed, and the clusters themselves somewhat difficult to grasp, because the only way to inspect the cluster is by looking at the members one by one. A condensed representation by, say, the mean and variance of the members may be misleading, because the cluster's shape is arbitrary. The alternative perspective on clustering in this section is that we explicitly search for clusters that can be well represented by simple statistics such as mean and variance.

7.3.1 Overview

In this section, we consider approaches that make (directly or indirectly) assumptions on the shape of the clusters, either because there is actually some evidence about their shape or simply because clusters of a simple shape are easy to grasp. From the summarization point of view, a roughly hyperspherical cluster is well characterized by its mean (where we would expect the bulk of the data), standard deviation (compact or widely spread data), and its size. A table of these values per cluster eases the interpretation of the results (given that the clusters roughly follow this model of hyperspherical data clouds). The question is how to determine the location (or more generally, the parameters) of each cluster. Several approaches in this direction exist—although their motivation is different, the resulting algorithms are quite similar.

As the section title indicates, a cluster is represented by a **model**, which may be as simple as a prototypical data object (in which case it is usually called a **prototype**). For the moment, we assume that the space of models/prototypes $\mathcal{M}$ is identical to the dataspace $\mathcal{X} = \mathbb{R}^m$. We require that some distance measure

$$d : \mathcal{M} \times \mathcal{X} \to \mathbb{R}_{\geq 0}$$

allows us to compare clusters (represented by prototypes) and data directly. As long as $\mathcal{M} = \mathcal{X}$, we may use the Euclidean distance here.

From our assumptions on the cluster shape we can derive the correct cluster representative (e.g., by calculating the mean of the associated data) if we just know which data belongs to the cluster. Here we use a different encoding of cluster membership than in Sect. 7.1: A **membership matrix** $[p_{i|j}] \in [0, 1]^{c \times n}$ contains the degree of belongingness of data $\mathbf{x}_j \in \mathcal{D}$ to cluster $\mathcal{C}_i$ by $p_{i|j} \in [0, 1]$. A one-to-one mapping of a **partition** onto a binary membership matrix $p_{i|j} \in \{0, 1\}$ is obtained from $p_{i|j} = 1 \Leftrightarrow \mathbf{x}_j \in \mathcal{C}_i$. Again, none of the clusters shall be empty ($\forall i : \sum_j p_{i|j} > 0$), and all data must participate at equal shares in the clustering ($\sum_i p_{i|j} = 1$), so that the membership degrees of the full matrix add up to the size of the dataset ($\sum_i \sum_j p_{i|j} = n$). The exclusiveness of the parti-

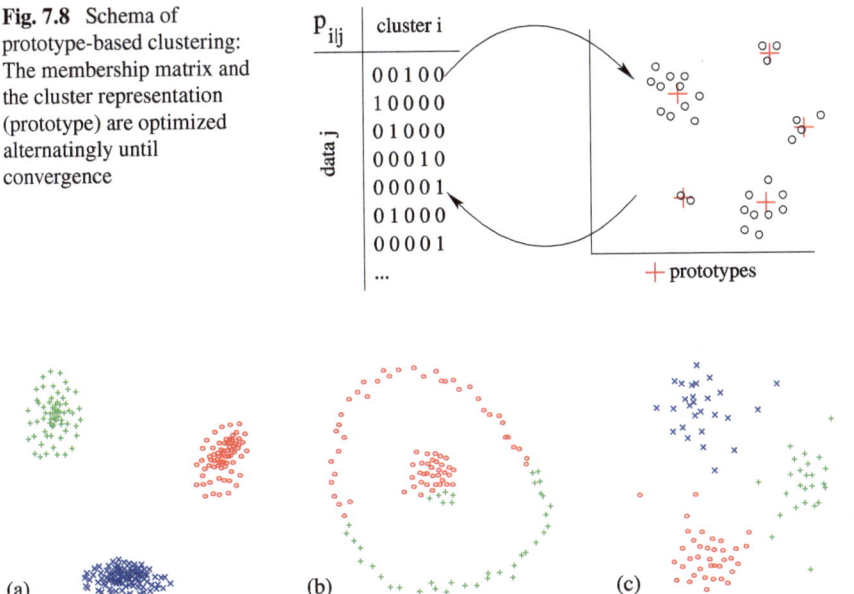

Fig. 7.8 Schema of prototype-based clustering: The membership matrix and the cluster representation (prototype) are optimized alternatingly until convergence

Fig. 7.9 Fuzzy c-Means clustering for three example data sets (cf. Fig. 7.3)

tion ($p_{i|j} \in \{0, 1\}$) is required only in some variants of prototype-based clustering.

As shown in Fig. 7.8, prototype-based clustering starts with some initial guesses of the prototypes and usually alternates between the adjustment of the membership matrix (using the currently given prototypes) and the adjustment of the prototypes (using the currently given membership degrees). It is assumed that the number of clusters c is known beforehand. Actually, all prototype-based clustering procedures optimize an objective-function during this iterative refinement (which could not be formulated without knowing the number of clusters in advance). Although an inspection of the membership matrix might be useful in some cases, the prototypes are considered as the primary output, since they concisely characterize the data associated with them. New data may be associated with the established clusters easily by assigning it to the nearest prototype (according to the distance $d(\mathbf{p}_i, \mathbf{x})$).

If the cluster is represented by a point prototype (assuming hyperspherical clusters), results for the datasets already presented in Fig. 7.3 are shown in Fig. 7.9. Where the model assumption was correct (cases (a) and (c)), the clusters are quickly and robustly recovered. In case (b), however, the ring-shaped outer cluster cannot be discriminated from the inner cluster: a typical prototype for the outer ring would have roughly the same distance to all its members, and thus would have to be placed in the center of the inner cluster. But then, both prototypes compete for the central data, and eventually both clusters are split among the prototypes.

Table 7.9 Algorithm for prototype-based clustering: k-Means (kM), Fuzzy c-Means (FCM), and Gaussian mixture decomposition (GMD)

Algorithm CentroidBasedClustering($\mathcal{D}, \mathbf{p}_1, \ldots, \mathbf{p}_c$) $\rightarrow \mathbf{p}_1, \ldots, \mathbf{p}_c$			
input: data set $\mathcal{D}$, $|\mathcal{D}| = n$,			
initial models/prototypes $\mathbf{p}_1, \ldots, \mathbf{p}_c$,			
output: optimized models/prototypes $\mathbf{p}_1, \ldots, \mathbf{p}_c$			

		kM	FCM	GMD
1	**repeat**			
2	**for all** $1 \leq j \leq n$:			
3	update $p_{i|j}$ according to...	(7.10)	(7.11)	(7.14)
4	**for all** $1 \leq i \leq c$:			
5	update prototype/model $\mathbf{p}_i$ according to...	(7.9)	(7.13)	(7.15)–(7.17)
6	**until** maximum number of iterations reached			
7	or $\mathbf{p}_1, \ldots, \mathbf{p}_c$ converged			

7.3.2 Construction

The general sketch of the algorithm has already been discussed and is depicted in Table 7.9. However, there are still placeholders for the two main steps, the membership update and the prototype update. We will briefly review three common algorithms for hyperspherical clusters, namely the k-Means, Fuzzy c-Means, and Gaussian mixture decomposition.

7.3.2.1 Minimization of Cluster Variance (k-Means Model)

The k-Means model corresponds to our initially stated situation where the prototypes stem from the same space as the data and the membership degrees are binary ($p_{i|j} \in \{0, 1\}$). The objective function of k-Means is given by

$$J_{kM} = \sum_{i=1}^{c} \sum_{j=1}^{n} p_{i|j} \|\mathbf{x}_j - \mathbf{p}_i\|^2 = \sum_{i=1}^{c} \underbrace{\sum_{\mathbf{x} \in C_i} \|\mathbf{x}_j - \mathbf{p}_i\|^2}_{\text{variance within cluster } C_i} \tag{7.8}$$

subject to $\sum_{i=1}^{c} p_{i|j} = 1$ for $1 \leq i \leq c$. If we think of a prototype $\mathbf{p}_i$ as the center of a cluster, the term $\sum_{j=1}^{n} p_{i|j} \|\mathbf{x}_j - \mathbf{p}_i\|^2$ calculates (up to a constant) the variance within the cluster (only data associated with the cluster is considered because $p_{i|j}$ is zero otherwise). By minimizing J_{kM} the prototypes will be chosen such that the variance of all clusters is minimized, thereby seeking for compact clusters. From the necessary condition for a minimum of J_{kM} ($\nabla_{\mathbf{p}_i} J_{kM} = 0$) the optimal prototype location is derived:

$$\mathbf{p}_i = \frac{\sum_{j=1}^{n} p_{i|j} \mathbf{x}_j}{\sum_{j=1}^{n} p_{i|j}}, \tag{7.9}$$

which corresponds to the mean of the data associated with the prototype. All prototypes are updated using (7.9) in the prototype-update step of Algorithm 7.9. As to
the membership update step, optimization of J_{kM} is simple: for data $\mathbf{x}_j$, we have to
select a single cluster i with $p_{i|j} = 1$ (and $p_{k|j} = 0$ for all $k \neq i$). Since we seek to
minimize the objective function, we select the prototype where $\|\mathbf{x}_j - \mathbf{p}_i\|^2$ becomes
minimal (which constitutes the membership update step):

$$p_{i|j} = \begin{cases} 1 & \text{if } \|\mathbf{p}_i - \mathbf{x}_j\| \text{ becomes minimal for } i; \text{ ties are broken arbitrarily;} \\ 0 & \text{otherwise.} \end{cases} \quad (7.10)$$

7.3.2.2 Minimization of Cluster Variance (Fuzzy c-Means Model)

The k-Means model can be extended to a membership matrix that contains continuous membership degrees within $[0, 1]$ rather than only binary memberships from
$\{0, 1\}$ (the cluster representation remains unchanged). A gradual membership allows one to distinguish data close to the prototype (very typical for the cluster) from
incidentally associated data (far away from prototype). To guarantee a meaningful
comparison of membership values, they should be normalized ($\sum_{i=1}^{c} p_{i|j} = 1$) and
proportional to the distance to the prototype (if $\mathbf{x}_1$ has twice the distance to the prototype than $\mathbf{x}_2$, $\mathbf{x}_1$ should receive only one half of the membership that $\mathbf{x}_2$ receives).
This determines the membership degrees up to a constant [32]:

$$p_{i|j} = \frac{1}{\sum_{k=1}^{c} \frac{\|\mathbf{x}_j - \mathbf{p}_i\|^2}{\|\mathbf{x}_j - \mathbf{p}_k\|^2}}. \quad (7.11)$$

These membership degrees can also be obtained by minimizing the objective function of Fuzzy c-Means

$$J_{FcM} = \sum_{i=1}^{c} \sum_{j=1}^{n} p_{i|j}^2 \|\mathbf{x}_j - \mathbf{p}_i\|^2, \quad (7.12)$$

which includes an exponent of 2 and is usually called **fuzzifier**. Without the introduction of a fuzzifier, the minimization of J_{FcM} would still lead to binary membership degrees. Other choices than 2 are possible but are less frequently used in
practice. The cluster centroid update is very similar to k-Means; the only difference
is due to the introduction of the fuzzifier:

$$\mathbf{p}_i = \frac{\sum_{j=1}^{n} p_{i|j}^2 \mathbf{x}_j}{\sum_{j=1}^{n} p_{i|j}^2}. \quad (7.13)$$

7.3.2.3 Gaussian Mixture Decomposition (GMD)

In Gaussian mixture decomposition a probabilistic model is estimated that may
explain the observed data. It is assumed that a single cluster roughly follows an

isotropic[1] Gaussian distribution, which is parameterized by its mean μ and variance σ:

$$g(\mathbf{x}; \mu, \sigma) = \frac{1}{\sqrt{2\pi}\sigma} e^{-\frac{1}{2}\frac{\|\mathbf{x}-\mu\|^2}{\sigma}}.$$

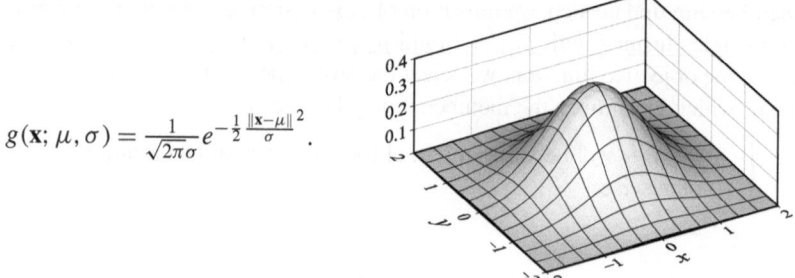

Since such a distribution is **unimodal** (i.e., there is only a single maximum) but each cluster represents its own peak in the density itself, we assume the overall density being a **mixture of Gaussians**. Given that the number of clusters c in the data is a priori known, the **generative model** of a dataset of size n is repeating the following two steps n times:

- First, draw a random integer between 1 and c with probability p_i of drawing i. The data generated next will then belong to cluster #i.
- Second, draw a random sample from the Gaussian distribution $g(\mathbf{x}; \mu_i, \sigma_i)$ with the parameters μ_i, σ_i taken from model #i.

The overall density function is thus given by

$$f(\mathbf{x}; \theta) = \sum_{i=1}^{c} p_i g(\mathbf{x}; \mu_i, \sigma_i),$$

where $\theta = (\theta_1, \ldots, \theta_c)$ and $\theta_i = (p_i, \mu_i, \sigma_i)$. For some given data $\mathbf{x}_j$, we interpret the probability of being generated by model #i as the cluster membership

$$p_{i|j} = p_i g(\mathbf{x}_j; \mu_i, \sigma_i). \tag{7.14}$$

Given a mixture θ, the likelihood of the full dataset being generated by this model is

$$L = \prod_{j=1}^{n} f(\mathbf{x}_j; \theta).$$

To find the most likely cluster configuration θ, the following **maximum likelihood estimate** can be derived:

$$\mu_i = \frac{\sum_{j=1}^{n} p_{i|j} \mathbf{x}_j}{\sum_{j=1}^{n} p_{i|j}}, \tag{7.15}$$

---

[1]Isotropic means that the density is symmetric around the mean, that is, the cluster shapes are roughly hyperspheres. Using the Mahalanobis distance instead would allow for ellipsoidal shape, but we restrict ourselves to the isotropic case here.

$$\sigma_i = \sqrt{\frac{1}{m} \frac{\sum_{j=1}^{n} p_{i|j} \|x_j - \mu_i\|^2}{\sum_{j=1}^{n} p_{i|j}}}, \tag{7.16}$$

$$p_i = \frac{1}{n} \sum_{i=1}^{n} p_{i|j}. \tag{7.17}$$

(Note the similarity of (7.15) with (7.9) and (7.13).) The so-called EM-algorithm performs a gradient descent search for the maximum likelihood estimate and as such is prone to local minima. If the model is poorly initialized and the clusters interact, convergence may be very slow.

7.3.3 Variations and Issues

If a good initialization is provided, the model- or prototype-based clustering algorithms are faster than hierarchical clustering, because only a few iterations are necessary. They are therefore better suited for larger datasets, also because the resulting models are easier to grasp than a complex dendrogram.

7.3.3.1 How many clusters?

One drawback of all prototype-based methods is that the number of clusters needs to be known (or suspected) in advance. If nothing is known whatsoever about the correct number of clusters, additional effort is necessary to apply the methods from this section. One may think of at least three different approaches: (1) top-down, divisive clustering: start with a relatively small number of clusters and split the cluster in case it does not fit the associated data well (e.g., bisecting k-means; see, e.g., [61]); (2) bottom-up, agglomerative clustering: overestimate the number of clusters and merge similar clusters; and (3) run the algorithm for a full range of possible numbers of cluster and evaluate each partition w.r.t. the overall goodness of fit. From the plot of obtained results the *optimal number of clusters* is determined by finding an extremum.

To find the *optimal* result, all these methods make use of so-called **validity measures** (see, e.g., [19, 27]). Approaches (1) and (2) employ a local validity measure that evaluate a single cluster only, such as the data density within the cluster or the distribution of membership degrees (unambiguous memberships are preferable as they indicate a successful separation of clusters). Another example is the **silhouette coefficient** [46] of a cluster $\mathcal{C}$, which is the average of silhouette coefficients $s(\mathbf{x})$ of its members $\mathbf{x} \in \mathcal{C}$:

$$s(\mathbf{x}) = \frac{b(\mathbf{x}) - a(\mathbf{x})}{\max\{a(\mathbf{x}), b(\mathbf{x})\}} \in [-1, 1],$$

where $a(\mathbf{x}) = d(\mathbf{x}, \mathcal{C})$ is the average distance of $\mathbf{x}$ to members of the same cluster $\mathcal{C}$, and $b(\mathbf{x})$ is the average distance to the members of the nearest cluster $\mathcal{C}'$ other than $\mathcal{C}$ ($b(\mathbf{x}) = \min_{\mathcal{C}' \neq \mathcal{C}} d(\mathbf{x}, \mathcal{C}')$). Well-clustered data $\mathbf{x}$ is close to members of its own cluster (small $a(\mathbf{x})$) but far away from members of other clusters (large $b(\mathbf{x})$) and thus receives a high silhouette coefficient near 1. A good cluster receives a high average silhouette coefficient from its members.

In contrast, approach (3) requires a global validity measure that evaluates all clusters at the same time. Regarding the global validity measures, we could employ the objective functions themselves to determine the optimal number of clusters. However, all the objective functions tend to become better and better the larger the number of cluster is (and at the end, there would be n singleton clusters). Validity measures such as the Aikake information criterion compensate this effect by reducing the obtained likelihood by the number of necessary parameters. Other typical measures, such as the **separation index** [20], identify *compact and well-separated clusters*:

$$D = \min_{i=1..c} \min_{j=i+1..c} \frac{\min_{x \in \mathcal{C}_i, y \in \mathcal{C}_j} d(\mathbf{x}_i, \mathbf{x}_j)}{\max_{k=1..c} \text{diam}_k},$$

where the numerator represents the distances between clusters $\mathcal{C}_i$ and $\mathcal{C}_j$ (which should be large), and the denominator $\text{diam}_k = \max_{x,y \in \mathcal{C}_k} \|x - y\|$ expresses the extension of cluster $\mathcal{C}_k$ (which should be small for compact clusters). Since the separation index considers neither fuzzy membership nor probabilities, it is well applicable to k-Means. There are also fuzzified versions of the separation index [58].

Admittedly these heuristic approaches fail to find the number of clusters if the model assumptions do not hold (e.g., cluster shape is not hyperspherical, cluster sizes are very different, cluster density is very different) or there is too much noise in the data.

Although finding the correct number of clusters using model-based clustering may be difficult, sometimes it is very handy to specify the number of groups in advance. Clustering algorithms may also be used as a tool for other tasks, such as discretization or allocation. If customers have to be assigned to k account managers, k-Means clustering could help to assign a homogeneous group of customers to each manager. In such settings, the number of groups is usually known beforehand.

7.3.3.2 Cluster Shape

We have only considered hyperspherical, **isotropic** clusters, that is, clusters that extend uniformly from the centre into the dataspace. With real data this assumption is seldom satisfied. The Euclidean distance used in GMD and FcM can be generalized to the Mahalanobis distance, which adapts to the correlations between dimensions and accounts for ellipsoidal clusters. There are also many variants of Fuzzy c-Means that consider somewhat exotic cluster shapes such as circles, ellipses, rectangles, or quadrics [33] (shell clustering).

Integrating flexible cluster shapes into the used distance measure greatly increases the number of parameters; typically the resulting algorithms react very sensitively to poor initializations. Another approach is to subdivide the data into a pretty large number of small clusters (using, e.g., k-Means) and then agglomeratively join these prototypes to clusters. In a sense, k-Means clustering is then used like a preprocessing phase to compress the data volume before then agglomerative hierarchical clustering is applied. (On the other hand, such a condensed representation can help to speed up k-Means and its variants significantly [31, 60].)

7.3.3.3 Noise Clustering

A very simple extension of fuzzy c-means clustering to cope with outliers is noise clustering [18]. We have already discussed that, to consider a different cluster shape, we only have to define an appropriate distance function. The idea of a noise cluster is that it has a fixed (usually large) distance d_{noise} to any data point. As soon as the distance of some data $\mathbf{x} \in X$ to the nearest prototype $\mathbf{p}_i$ comes close to d_{noise}, the noise cluster gains a considerable fraction of the total membership degree, thereby reducing the influence of $\mathbf{x}$ with respect to $\mathbf{p}_i$. Noise clustering simply requires to exchange the membership update (7.11) by

$$p_{i|j} = \frac{1}{\left(\frac{\|\mathbf{x}_j - \mathbf{p}_i\|^2}{d_{noise}^2}\right) + \sum_{k=1}^{c} \left(\frac{\|\mathbf{x}_j - \mathbf{p}_i\|^2}{\|\mathbf{x}_j - \mathbf{p}_k\|^2}\right)} \tag{7.18}$$

and represents and effective mean to reduce the influence of noise and extract cluster prototypes more clearly.

The noise cluster allows for a better analysis of the clustering results. What typically happens when no noise cluster is used is that the memberships of an outlier tend to $\approx \frac{1}{k}$ for the k closest clusters. This situation may also occur for data that is well within the usual data range (and thus not an outlier) but half way between these k clusters. By inspection of membership degrees these two cases are undistinguishable. When a noise cluster is present, the former case is easily detected, as it has gained most of the membership degree (and only a small remainder is equally distributed among the closest prototypes).

7.4 Density-Based Clustering

We have seen in Sect. 7.1 that single-linkage clustering is flexible in terms of recognizable cluster shapes, but sensitive to outliers, which may easily be misused as *bridges* between clusters. These bridges are particularly counterintuitive if the core of the clusters consists of a considerable amount of data, but the bridge is established by a few data objects only. To establish a substantial connection between the clusters, the data density of the bridge should be comparable to that of the cluster itself. Density-based clustering goes into this direction by requiring a minimum data density within the whole cluster.

7.4.1 Overview

The underlying idea of density-based clustering is to measure the data density at a certain spot or region of the data space and then declare those regions as being clusters, whose density exceeds a certain density threshold. The final clusters are then obtained by *connecting* neighboring dense spots or regions. This is easy to illustrate for so-called grid-clustering algorithms, where an artificial (multidimensional) grid is laid over the data space and the number of data objects falling into a particular cell is counted. Figure 7.10(a) shows an illustrative example for the two-dimensional case. While the two larger groups are recognized as clusters, there is no bridge between both clusters as the data density is too low.

However, with this grid two other small data agglomerations are not recognized as a cluster (blue regions) but would have been if the grid was slightly shifted. Another problem is that in a naïve implementation the total number of cells increases exponentially with the number of attributes (but not the number of cells with high data density). A simple solution to both problems is to center the cells directly on the data rather than defining the grid independently from the data. Figure 7.10(b) shows a spherical area centered on four data points where each of these *neighborhoods* contains at least 3 data objects, similarly to the approach of the **DBScan** algorithm [21]. If the threshold for establishing a cluster (that is, a dense sphere) is four, we have thereby found the first four (distinct) clusters in Fig. 7.10(b). All data within the neighborhoods is considered as belonging to the same cluster. Using this data as starting points, the area occupied by the cluster is expanded as long as the minimal density threshold is satisfied. At the end, we arrive at the clusters shown in Fig. 7.10(c). Note that at the cluster border some data may belong to a cluster although their density does not satisfy the threshold.

The result of density-based clustering algorithms is a set of clusters, together with all data belonging to each cluster (a partition). The data within the cluster may be

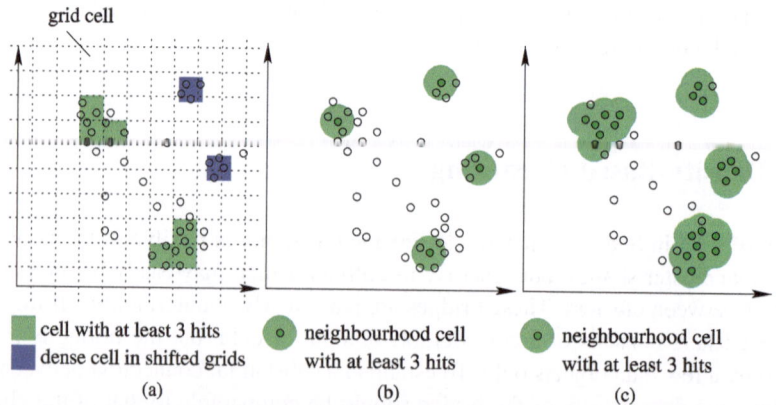

Fig. 7.10 Density-based clustering using (**a**) grid cells or (**b**)/(**c**) data neighborhoods to estimate the density

summarized by statistical measures, but—just as with hierarchical clustering—has to be interpreted with care as the shape of the clusters may be arbitrary.

7.4.2 Construction

We consider the DBScan algorithm given in Table 7.10. The density threshold is encoded by two parameters, the radius of the (hyperspherical) neighborhood and the number *MinPts* of data objects that is needed in the neighborhood to consider it as *dense*. The actual density at some location $\mathbf{x}$ is measured within the ε-**neighborhood** $N_\varepsilon(\mathbf{x}) = \{\mathbf{y} \in \mathcal{D} \mid \|\mathbf{x} - \mathbf{y}\| \leq \varepsilon\}$ of $\mathbf{x}$. If this neighborhood contains at least *MinPts* elements, $\mathbf{x}$ is located *within* a cluster and is called a **core object**. All data in the ε-neighborhood of a core object also belong to the same cluster as the core object. Other core objects within $N_\varepsilon(\mathbf{x})$ may likewise be core objects (they are **density-reachable**) such that those neighboring elements are also included in the same clus-

Table 7.10 The DBScan algorithm

Algorithm dbscan($\mathcal{D}, \varepsilon, MinPts$)

input: data set $\mathcal{D}$, neighbourhood radius ε, density threshold *MinPts*
output: labels the data with cluster id (or NOISE)

1.	label all data $\mathbf{x} \in \mathcal{D}$ as UNCLASSIFIED
2.	initialize cluster-counter $cid = 0$
3.	**for all** $\mathbf{x} \in \mathcal{D}$
4.	**if** $\mathbf{x}$ is labelled as UNCLASSIFIED
5.	**if** expand($\mathcal{D}, \mathbf{x}, \mathbf{cid}, \varepsilon, \mathbf{MinPts}$)
6.	increment cluster counter $cid = cid + 1$

Algorithm expand($\mathcal{D}, \mathbf{x}, \mathbf{cid}, \varepsilon, \mathbf{MinPts}$) : bool

input: data set $\mathcal{D}$, $x \in D$, currently unused cluster-id *cid*, neighbourhood radius ε,
        density threshold *MinPts*
output: returns true iff a new cluster has been found

1.	let $S = \{\mathbf{y} \in \mathcal{D} \mid \|\mathbf{x} - \mathbf{y}\| \leq \varepsilon\}$ (range query)
2.	**if** not enough data in neighborhood of $\mathbf{x}$ ($|S| < MinPts$)
3.	re-label $\mathbf{x}$ as NOISE and **return** false
4.	**forall** $\mathbf{x}' \in S$: re-label $\mathbf{x}'$ with current cluster-ID cid
5.	remove $\mathbf{x}$ from S
6.	**forall** $\mathbf{x}' \in S$
7.	$T = \{\mathbf{y} \in \mathcal{D} \mid \|\mathbf{x}' - \mathbf{y}\| \leq \varepsilon\}$ (range query)
8.	**if** enough data in neighborhood of $\mathbf{x}'$ ($|T| \geq MinPts$)
9.	**forall** $\mathbf{y} \in T$
10.	**if** $\mathbf{y}$ does not belong to a cluster (labeled as NOISE or UNCLASSIFIED)
11.	**if** $\mathbf{y}$ is labelled UNCLASSIFIED: insert $\mathbf{y}$ into S
12.	re-label $\mathbf{y}$ with cluster-counter cid
13.	remove $\mathbf{x}'$ from S
14.	**return** true

ter. This expansion strategy is repeated until no further expansion is possible. Then, eventually, the full cluster has been determined.

The main routine *dbscan* passes once through the database and evokes the sub-routine *expand* for every data x unlabeled so far (line 5). Whenever this subroutine returns true, a new cluster has successfully been extracted from the seed point **x**.

The *expand* subroutine firstly identifies the data in the neighborhood of **x** (set S in line 1). In case the desired density is not reached ($|S| < MinPts$), **x** is relabeled as NOISE, and no cluster has been found at **x**. Otherwise, a (potentially new) cluster has been found, and all data from the neighborhood are assigned to it (relabeled with cluster-ID *cid*, line 4). At this point, the processing of **x** has been completed, and the function *expand* tries to extend the cluster further. For all the data in the neighborhood, it is consecutively checked if they also satisfy the density threshold (line 8). If that is the case, they also belong to the *core* of the cluster (and are called core-points). All data in the neighborhood of a core-point are then also added to the cluster, that is, they are relabeled and inserted into the set S of data that has to be tested for further cluster expansion. If the data are organized in a data structure that supports neighborhood queries (as called in line 1), the runtime complexity is low ($O(n \log n)$). However, the performance of such data structures (e.g., R^*-trees) degrades as the data dimensionality increases.

To obtain optimal results, two parameters need to be adjusted, *MinPts* and ε. An alternative set of parameters is the desired *density threshold* and the *resolution* of the analysis: Any choice of *MinPts* and ε corresponds to the selection of a lower bound on the data density. But for some given density threshold ϱ, we have the freedom to choose *MinPts* twice as large and double at the same time the volume of N_ε without affecting the density threshold. The larger the volume, the more robust the density estimation, but at the same time we loose the capability of recognizing clusters smaller than this volume. Besides that, a single best choice may not exist for a given dataset if clusters of different densities are present. As a rule of thumb, *MinPts* may be set to $2 * dimension - 1$, and then ε is determined from visually inspecting a sorted *MinPts*-distance plot: For all data x_i, the radius r_i is determined for which $N_{r_i}(x) = MinPts$. These radii r_i are then plotted in descending order (cf. Fig. 7.11). Under the assumption that the clusters have roughly comparable densities, the r_i-values of data belonging to clusters should roughly be the same (in the ordered plot this corresponds to a marginal negative slope). In contrast, outside the clusters

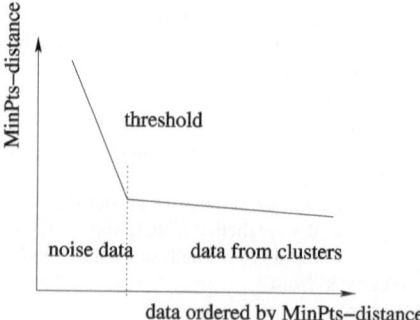

Fig. 7.11 Determination of the ε-parameter by visual inspection (identification of knee point)

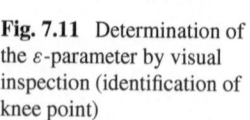

the distance to the *MinPts*th-nearest neighbor strongly depends on the surrounding noise points, and thus the *MinPts*-distances will vary much stronger (leading to a deeper descend in the plot). A reasonable choice for ε is then given by the knee point in the *MinPts*-distance plot.

7.4.3 Variations and Issues

7.4.3.1 Relaxing the density threshold

The **OPTICS** algorithm [5] is an extension of DBScan that solves the problem of determining *MinPts* and ε differently. It can also be considered as a hierarchical clustering algorithm, as it provides the resulting partition for a full range of possible values ε. For each data object, OPTICS determines the so-called reachability- and core-distance. The core-distance is basically the smallest value of ε under which the data object becomes a core point. The reachability distance is the smallest distance under which the data objects becomes a cluster member, that is, belongs to the neighborhood of the closest core object. The reachability plot of all data objects in a specific order determined during clustering can be interpreted as a cluster hierarchy, as shown in Fig. 7.12. Valleys in this plot indicate clusters in the dataset: the broader the valley, the larger the cluster, and the deeper the valley, the higher the density.

Another possibility is to define the data density over the full data space $\mathcal{X}$ by the superposition of influence functions centered at every data object:

$$\varrho(\mathbf{x}) = \sum_{\mathbf{y} \in \mathcal{D}} f_{\mathbf{y}}(\mathbf{x}),$$

where $f_{\mathbf{y}}(\cdot)$ tells us how well $\mathbf{x}$ is represented (or influenced) by $\mathbf{y}$. In the case of DB-Scan, we may define $f_{\mathbf{y}}(\mathbf{x}) = 1 \Leftrightarrow \|\mathbf{x} - \mathbf{y}\| \le \varepsilon$ and 0 otherwise. For a core object of

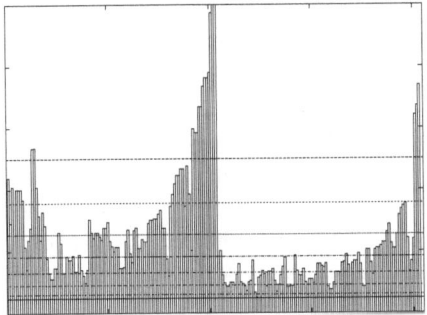
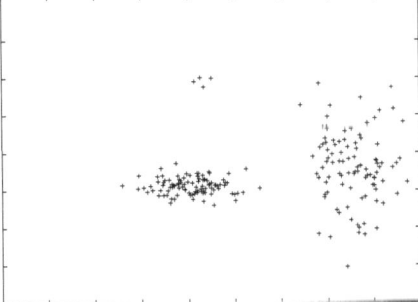

Fig. 7.12 The reachability plot of the dataset on the right. The data objects have been ordered, and for each data object, its reachability value is shown. Contiguous areas on the *horizontal axis*, where the reachability drops below a *horizontal line*, represent a cluster. The *horizontal lines* represent certain data density levels; the lower the line, the higher the density

a DBScan-cluster, we have $\varrho(\mathbf{x}) \geq MinPts$, because at least *MinPts* data objects exist around $\mathbf{x}$ within the ε-neighborhood. The more data exist around $\mathbf{x}$, the higher the density function $\varrho(\mathbf{x})$. The **DENCLUE** algorithm [29] uses a Gaussian distribution for $f_{\mathbf{y}}(\mathbf{x})$ instead and concentrates on local maxima of ϱ (called density attractors) via a hill-climbing technique. Such local maxima can be considered as point-like clusters as in prototype-based clustering. Alternatively, several points with densities above some threshold can be considered as representing jointly a cluster of arbitrary shape. The DENCLUE algorithm preprocesses the data and organizes it into grid cells. The summation of ϱ is restricted to some neighboring cells, thereby keeping the overall runtime requirements low. Since the neighborhood relationship between grid cells is known, no expensive neighborhood queries are necessary as with DB-Scan or OPTICS. The regular grid of DENCLUE is adapted to the data at hand in the OptiGrid approach [30].

7.4.3.2 Subspace Clustering

We have emphasized in Sect. 7.2 that the choice of the (dis)similarity measure has major impact on the kind and number of clusters the respective algorithm will find. A certain distance function may ignore some attributes; therefore we may consider the feature selection problem being part of the distance function design. All the clustering algorithms we have discussed so far assume that the user is capable of providing the right distance function—and thus, to select the right features or, in other words, the right subspace that contains the clusters.

The term **subspace clustering** refers to those methods that try to solve the problem of subspace selection simultaneously while clustering. (This short section on subspace clustering has been located in the section on density-based clustering, because most subspace clustering techniques base their measurements on a density-based clustering notion.) As in feature selection, the problem with a naïve approach, where we simply enumerate all possible subspaces (i.e., subsets of the attribute set) and run the clustering algorithm of our choice, is the exponential complexity: the number of potential subspaces increases exponentially with the dimensionality of the dataspace. There are at least three ways to attack this problem.

In a bottom-up approach, we first seek for clusters in the individual one-dimensional spaces (1D). From those we can construct two-dimensional spaces via the cross-product of dense one-dimensional clusters (blue shaded regions in Fig. 7.13, right). We thereby safe ourselves from investigating those 2D-areas whose 1D-projections are not dense (white areas in Fig. 7.13, right), as they have no chance of becoming dense in higher dimensions anyway. The algorithm **CLIQUE** [4] follows this approach and takes the basic idea of dimension-wise cluster construction from frequent pattern mining, which will be discussed in Sect. 7.6.

In a top-down solution we seek for a few data points with high data density in the full dataset (anchor points) and then identify those variables that lead to the highest data density. This can be done cluster-wise, such that every cluster may live in a different subspace. A member of this family of algorithms is **PROCLUS** (projected

Fig. 7.13 The stylized histograms along the axes indicate density estimates. The *red line* corresponds to a threshold for dense intervals. By analyzing the density along the individual attributes, areas in the 2D-space can be excluded from further cluster search (areas *left white*)

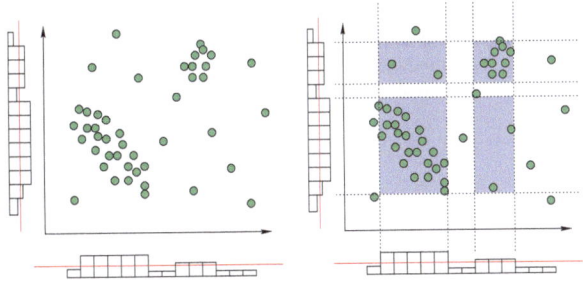

clustering) [1]. Each dimension gets a weight that indicates whether the dimension belongs to the cluster's subspace, and this weight is iteratively adjusted to increase the cluster validity measure.

The third option is to rank subspaces according to their suspected suitability for clustering and use the top-ranked subspace(s) for a *conventional* clustering algorithm (similar to feature selection). The basic idea of the **SURFING** (subspaces relevant for clustering) approach [7] is to look at the distribution of distances to the kth nearest neighbor, as it has already been used to determine ε in the DBScan algorithm, cf. Fig. 7.11. If the data is distributed on a regular grid, this rather uninteresting dataset wrt. clustering is recognized by an almost straight line in this graph. If clusters are present, we suspect varying distances to the kth neighbor. The more the distances deviate from a constant line, the more promising is the subspace. Similar to CLIQUE, the potential subspaces are then explored levelwise to identify those subspaces that appear most promising.

7.5 Self-organizing Maps

Maps of all kinds are used to distill a large amount of information into a handy, small, and comprehensible visualization. People are used to maps; being able to transform the (potentially high-dimensional) dataset into a two-dimensional representation may therefore help to get an intuitive understanding of the data.

7.5.1 Overview

A map consists of many tiny pictograms, letters, and numbers, each of them meaningful to the reader, but the most important property when using a map is the neighborhood relationship: If two point are close to each other on the map, so they are in the real world. The **self-organizing maps** or **Kohonen maps** [38] start from a mesh of map nodes[2] for which the neighborhood relationship is fixed. One may think

---

[2]The node are usually called neurons, since self-organizing maps are a special form of neural network.

Fig. 7.14 The self-organizing map (*middle*) covers the occupied data space (*left*) over time

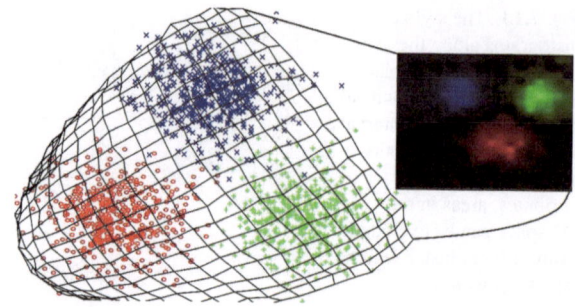

of a fishing net, where the knots represent the nodes, and all neighbors are connected by strings. The structure of the fishing net is two-dimensional, but in a three-dimensional world it may lay flat on the ground, forming a regular grid of mesh nodes, may be packed tightly in a wooden box, or may even float freely in the ocean. While the node positions in the higher-dimensional space varies, the neighborhood relationship remains fixed by the a priori defined connections of the mesh (knots). The idea of **self-organizing maps** is to define the mesh first (e.g., two-dimensional mesh)—which serves as the basic layout for the pictorial representation—and then let it float in the dataspace such that the data gets well covered. Every node $\mathbf{p}_i$ in the pictorial representation is associated with a corresponding reference node $\mathbf{r}_i$ in the high-dimensional space. At the end, the properties of the map at node $\mathbf{p}_i$, such as color or saturation, are taken directly from the properties of the data space at $\mathbf{r}_i$. The elastic network represents a nonlinear mapping of reference vectors (high-dimensional positions in the data space) to the mesh nodes (two-dimensional positions in the mesh).

Figure 7.14 shows an example for a two-dimensional data set. In such a low-dimensional case the self-organizing map provides no advantages over a scatter plot, it is used for illustrating purposes only. The occupied dataspace is covered by the reference vectors of the self-organizing map (overlaid on the scatter plot). The color and data density information is taken from the respective locations and gets mapped to the two-dimensional display coordinates (right subfigure).

7.5.2 Construction

Learning a self organizing map is done iteratively (see Table 7.11): for any data vector $\mathbf{x}$ from the dataset, the closest reference vector $\mathbf{r}_w$ (in the high-dimensional dataspace) is identified. The node w is usually called the **winner neuron**. From the low-dimensional map topology the neighboring nodes of node w are identified. Suppose that node i is one of these neighbors, then the associated reference vector $\mathbf{r}_i$ is moved slightly towards $\mathbf{x}$ so that it better represents this data vector. How much it is shifted is controlled by a heuristic learning rate η, which slowly approaches zero to ensure the convergence of the reference vectors. Thus, any new data vector leads to a local modification of the map around the winner neuron. After a fixed number

Table 7.11 Algorithm for mapping a dataset onto a two-dimensional map

Algorithm SOM($\mathcal{D}, \mathbf{p}_1, \ldots, \mathbf{p}_m, h$) $\rightarrow \mathbf{r}_1, \ldots, \mathbf{r}_m$

input: data set $\mathcal{D}$, $|\mathcal{D}| = n$,
        two-dimensional mesh nodes $\mathbf{p}_1, \ldots, \mathbf{p}_m \in \mathbb{R}^{2 \times m}$,
        neighborhood function $h : \mathbb{R}^2 \times \mathbb{R}^2 \rightarrow [0, 1]$
output: reference vectors $\mathbf{r}_1, \ldots, \mathbf{r}_m$

1	initialize reference vectors $\mathbf{r}_1, \ldots, \mathbf{r}_m$
2	**repeat**
3	**forall** $\mathbf{x} \in \mathcal{X}$
4	find reference vector $\mathbf{r}_w$ closest to $\mathbf{x}$ wrt distance function $d(\mathbf{x}, \mathbf{r}_i)$
5	update all reference vectors $\mathbf{r}_i$:
6	$\mathbf{r}_i = \mathbf{r}_i + \eta \cdot h(\mathbf{p}_w, \mathbf{p}_i) \cdot (\mathbf{x} - \mathbf{r}_i)$
7	slightly decrease η and δ
8	**until** maximum number of iterations reached or $\mathbf{r}_1, \ldots, \mathbf{r}_m$ converged

of iterations or when the changes drop below some threshold, the map adjustment is finished.

To determine the winner node any distance function might be used (see the measures discussed in Sect. 7.2), but the Euclidean and the cosine distances are most frequently employed. As not only the winning reference vector $\mathbf{r}_w$ is shifted towards the new sample, but also neighboring vectors, we need a way to specify which nodes will be affected. This is done by a neighborhood function $h : \mathbb{R}^2 \times \mathbb{R}^2 \rightarrow [0, 1]$: For node i, this function provides a factor $h(\mathbf{p}_w, \mathbf{p}_i)$ that directly influences the reference vector modification. Note that the neighborhood information is taken from the two-dimensional node locations $\mathbf{p}_i$, while the update concerns the high-dimensional reference vectors $\mathbf{r}_i$. The smaller the factor, the smaller the effect of the modification; a factor of zero means that the reference vector is left unaltered. The following, typically used functions share a common parameter, the influence range δ, which controls the size of the neighborhood:

cylinder: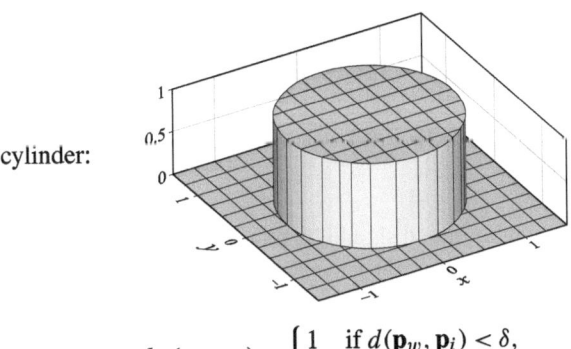

$$h_1(\mathbf{p}_w, \mathbf{p}_i) = \begin{cases} 1 & \text{if } d(\mathbf{p}_w, \mathbf{p}_i) < \delta, \\ 0 & \text{otherwise.} \end{cases}$$

cone:

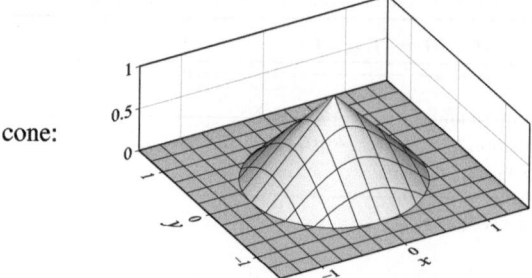

$$h_2(\mathbf{p}_w, \mathbf{p}_i) = \max\{0, 1 - d(\mathbf{p}_w, \mathbf{p}_i)/\delta\}.$$

cosine:

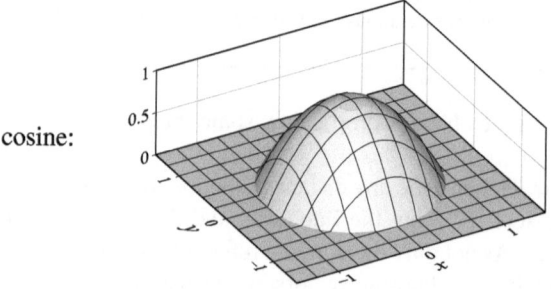

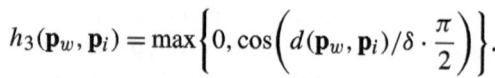

$$h_3(\mathbf{p}_w, \mathbf{p}_i) = \max\left\{0, \cos\left(d(\mathbf{p}_w, \mathbf{p}_i)/\delta \cdot \frac{\pi}{2}\right)\right\}.$$

Gauss:

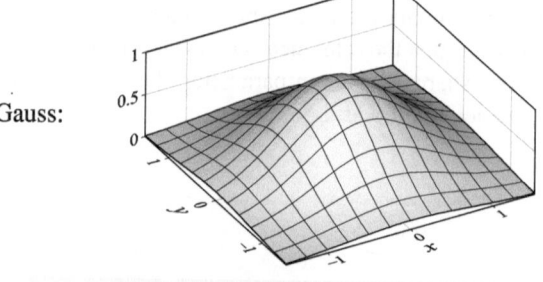

$$h_4(\mathbf{p}_w, \mathbf{p}_i) = \exp\left(-d(\mathbf{p}_w, \mathbf{p}_i)^2/\delta^2\right).$$

Initially, the neighborhood should be relatively large as the map needs to be first unfolded. The random initialization leads to an unorganized initial state, and neighboring nodes i and j in the low-dimensional representation are not associated with neighboring reference vectors. To avoid distorted or twisted maps, the so-called stiffness parameter δ must be relatively large at the beginning and must not decrease too quickly. Figure 7.15 shows some intermediate steps in the evolution of a self-organizing map.

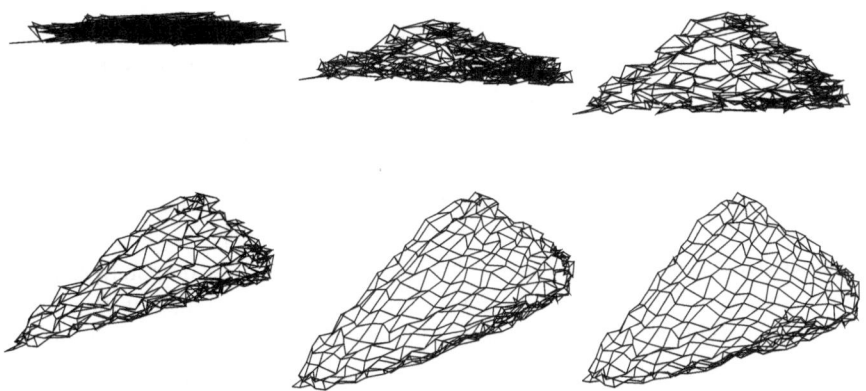

Fig. 7.15 Evolution of a self organizing map

7.6 Frequent Pattern Mining and Association Rules

Frequent pattern mining tackles the problem of finding common properties (patterns) that are shared by all cases in certain sufficiently large subgroups of a given data set. The general approach to find such patterns is to (efficiently) search a space of potential patterns. Based on the type of data to be mined, frequent pattern mining is divided into (1) frequent item set mining and association rule induction, (2) frequent sequence mining, and (3) frequent (sub)graph mining with the special subarea of frequent (sub)tree mining.

7.6.1 Overview

Frequent pattern mining was originally developed—in the special form of **frequent item set mining**—for market basket analysis, which aims at finding regularities in the shopping behavior of the customers of supermarkets, mail-order companies, and online shops. In particular, it tries to identify sets of products that are frequently bought together. Once identified, such sets of associated products may be exploited to optimize the organization of the offered products on the shelves of a supermarket or on the pages of a mail-order catalog or web shop, may be used to suggest other products a customer could be interested in (so-called cross-selling), or may provide hints which products may conveniently be bundled. In order to find sets of associated products, one analyzes recordings of actual purchases and selections made by customers in the past, that is, basically lists of items that were in their shopping carts. Due to the widespread use of scanner cashiers (with bar code readers), such recordings are nowadays readily available in basically every supermarket.

Generally, the challenge of frequent pattern mining consists in the fact that the number of potential patterns is usually huge. For example, a typical supermarket has thousands of products on offer, giving rise to astronomical numbers of sets of potentially associated products, even if the size of these sets is limited to few items: there are about 8 quadrillion $(8 \cdot 10^{12})$ different sets of 5 items that can be selected from a set of 1000 items. As a consequence, a brute force approach, which simply enumerates and checks all possible patterns, quickly turns out to be infeasible.

More sophisticated approaches exploit the structure of the pattern space to guide the search and the simple fact that a pattern cannot occur more frequently than any of its subpatterns to avoid useless search, which cannot yield any output. In addition, clever representations of the data, which allow for efficient counting of the number of cases satisfying a pattern, are employed. In the area of frequent item set mining, research efforts in this direction led to well-known algorithms like Apriori [2, 3], Eclat [59], and FP-growth [28], but there are also several variants and extensions [8, 25].

It is often convenient to express found patterns as rules, for example:
If a customer buys bread and wine, she will probably also buy cheese.

Such rules are customarily called **association rules**, because they describe an association of the item(s) in the consequent (then-part) with the item(s) in the antecedent (if-part). Such rules are particularly useful for cross-selling purposes, because they indicate what other products may be suggested to a customer. They are generated from the found frequent item sets by relating item sets one of which is a subset of the other, using the smaller item set as the antecedent and the additional items in the other as the consequent.

With the algorithms mentioned above, the basic task of frequent item set mining can be considered satisfactorily solved, as all of them are fast enough for most practical purposes. Nevertheless, there is still room for improvement. Recent advances include filtering the found frequent item sets and association rules (see, e.g., [53, 54]), identifying temporal changes in discovered patterns (see, e.g., [10, 11]), and mining fault-tolerant or approximate frequent item sets (see, e.g., [16, 44, 52]).

Furthermore, **sequences**, **trees**, and generally **graphs** have been considered as patterns, thus vastly expanding the possible applications areas and also introducing new problems. In particular, for patterns other than item sets—and especially for general graphs—avoiding redundant search is much more difficult. In addition, the fact that with these types of data a pattern may occur more than once in a single sample case (for example, a specific subsequence may occur at several locations in a longer sequence) allows for different definitions of what counts as frequent and thus makes it possible to find frequent patterns for single instances (for example, mining frequent subgraphs of a single large graph).

The application areas of frequent pattern mining in its different specific forms include market basket analysis, quality control and improvement, customer management, fraud detection, click stream analysis, web link analysis, genome analysis, drug design, and many more.

Fig. 7.16 A simple example database with 10 transactions (market baskets, shopping carts) over the item base $B = \{a, b, c, d, e\}$ (*left*) and the frequent item sets that can be discovered in this database if the minimum support is chosen to be $s_{min} = 3$ (*right*; the numbers state the support of these item sets)

transaction database
1: $\{a, d, e\}$
2: $\{b, c, d\}$
3: $\{a, c, e\}$
4: $\{a, c, d, e\}$
5: $\{a, e\}$
6: $\{a, c, d\}$
7: $\{b, c\}$
8: $\{a, c, d, e\}$
9: $\{b, c, e\}$
10: $\{a, d, e\}$

Frequent item sets (with support)
(minimum support $s_{min} = 3$)

0 items	1 item	2 items	3 items
$\emptyset$: 10	$\{a\}$: 7	$\{a, c\}$: 4	$\{a, c, d\}$: 3
	$\{b\}$: 3	$\{a, d\}$: 5	$\{a, c, e\}$: 3
	$\{c\}$: 7	$\{a, e\}$: 6	$\{a, d, e\}$: 4
	$\{d\}$: 6	$\{b, c\}$: 3	
	$\{e\}$: 7	$\{c, d\}$: 4	
		$\{c, e\}$: 4	
		$\{d, e\}$: 4	

7.6.2 Construction

Formally, the task of **frequent item set mining** can be described as follows: we are given a set B of **items**, called the **item base**, and a database T of **transactions**. Each item may represent a product, a special equipment item, a service option, etc., and the item base represents the set of all products etc. that are offered. The term **item set** refers to any subset of the item base B. Each transaction is an item set and represents a set of products that has been bought by a customer. Since two or even more customers may have bought the exact same set of products, the total of all transactions must be represented as a vector, a bag, or a multiset, since in a simple set each transaction could occur at most once. (Alternatively, each transaction may be enhanced by a unique *transaction identifier*, and these enhanced transactions may then be combined in a simple set.) Note that the item base B is usually not given explicitly but only implicitly as the union of all transactions. An example transaction database over the item base $B = \{a, b, c, d, e\}$ is shown in Fig. 7.16.

The **support** $s_T(I)$ of an item set $I \subseteq B$ is the number of transactions in the database T it is contained in. Given a user-specified **minimum support** $s_{min} \in \mathbb{N}$, an item set I is called **frequent** in T iff $s_T(I) \geq s_{min}$. The goal of frequent item set mining is to identify all item sets $I \subseteq B$ that are frequent in a given transaction database T. In the transaction database shown in Fig. 7.16, 16 frequent item sets with 0 to 3 items can be discovered if $s_{min} = 3$ is chosen. Note that more frequent item sets are found than there are transactions, which is a typical situation.

In order to design an algorithm to find frequent item sets, it is beneficial to first consider the properties of the support of an item set. Obviously we have

$$\forall I : \forall J \supseteq I : \quad s_T(J) \leq s_T(I).$$

That is: *If an item set is extended, its support cannot increase.* This is immediately clear from the fact that each added item is like an additional constraint a transaction has to satisfy. One also says that support is **antimonotone** or **downward closed**. From this property it immediately follows what is known as the **a priori property**:

$$\forall s_{min} : \forall I : \forall J \supseteq I : \quad s_T(I) < s_{min} \rightarrow s_T(J) < s_{min}.$$

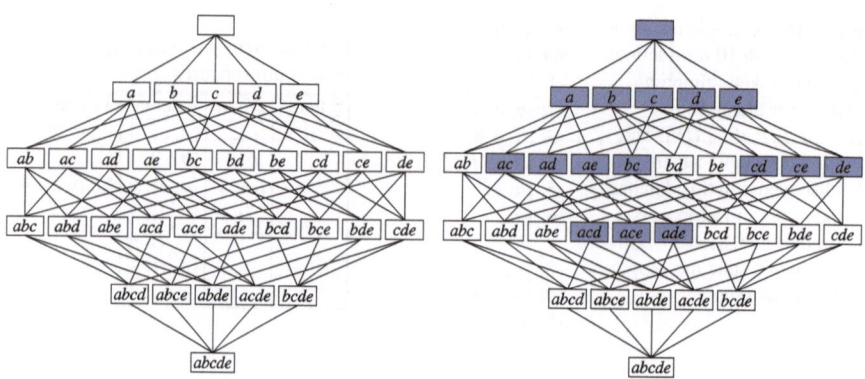

Fig. 7.17 A subset lattice for five items (*left*) and the frequent item sets (*right*, frequent item sets in *blue*) for the transaction database shown in Fig. 7.16

That is: *No superset of an infrequent item set can be frequent.* As a consequence, we need not explore any supersets of infrequent item sets—an insight which can nicely be exploited to prune the search for frequent item sets.

In order to structure the actual search, it is advantageous to consider the **subset lattice** of the item base B. For the simple item base $B = \{a, b, c, d, e\}$, which underlies the transaction database shown in Fig. 7.16, a Hasse diagram of this subset lattice is shown in Fig. 7.17 on the left: two item sets are connected if they differ in size by only one item and if one item set is a subset of the other.

Because of the a priori property, the frequent item sets must be located at the top of this lattice, thus suggesting that we should search the subset lattice top-down. As an illustration, in the right part of Fig. 7.17 the frequent item sets of Fig. 7.16 are highlighted. In this diagram the a priori property is exhibited by the fact that there cannot be a link between a white node (infrequent item set) in a higher level to a blue node (frequent item set) in a lower level. Links leading from a blue node to a white node mark the boundary between frequent and infrequent item sets.

A core problem of searching the subset lattice is that all item sets with at least two items can be explored on multiple paths. For example, the item set $\{a, c, e\}$ may be reached through $\{a\}$ and $\{a, c\}$, or through $\{c\}$ and $\{c, e\}$, or through $\{e\}$ and $\{a, e\}$, etc. As a consequence, a lot of redundant search would ensue if we always extended any frequent item set by all possible other items until we reach the boundary between the frequent and infrequent item sets. In order to avoid this redundancy, a **unique parent** is assigned to each item set, thus turning the subset lattice into a **subset tree**. This is illustrated in Fig. 7.18, where the parents are assigned by the following simple scheme: we choose an order of the items (which is arbitrary but fixed; here we simply use the alphabetically order) and then apply the rule

$$\forall I \neq \emptyset: \quad \mathrm{parent}(I) = I - \left\{\max_{i \in I} i\right\}.$$

This scheme has the important advantage that the permissible extensions of an item set, which only generate proper children, are easy to find: an item set may be extended only by items succeeding all already present items in the chosen item order.

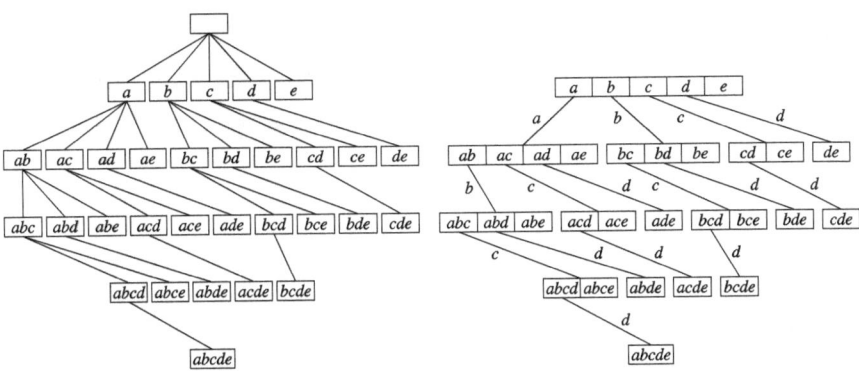

Fig. 7.18 A subset tree that results from assigning a unique parent to each item set (*left*) and a corresponding prefix tree in which sibling nodes with the same prefix are merged (*right*)

Since with this scheme of assigning unique parents, all sibling item sets share the same **prefix** (w.r.t. the chosen item order), namely the item set that is their parent, it is convenient to structure the search as a prefix tree, as shown in Fig. 7.18. In this prefix tree the concatenation of the edge labels on the path from the root to a node is the common prefix of all item sets explored in this node.

A standard approach to find all frequent item sets w.r.t. a given database T and given support threshold s_{min}, which is adopted by basically all frequent item set mining algorithms (except those of the Apriori family), is a **depth-first search** in the subset tree of the item base B. Viewed properly, this approach can be interpreted as a simple **divide-and-conquer** scheme. For the first item i in the chosen item order, the problem to find all frequent item sets is split into two subproblems: (1) find all frequent item sets containing the item i, and (2) find all frequent item sets *not* containing the item i. Each subproblem is then further divided based on the next item j in the chosen item order: find all frequent item sets containing (1.1) both items i and j, (1.2) item i but not j, (2.1) item j but not i, (2.2) neither item i nor j, and so on, always splitting with the next item (see Fig. 7.19 for an illustration).

All subproblems that occur in this divide-and-conquer recursion can be defined by a **conditional transaction database** and a **prefix**. The prefix is a set of items that has to be added to all frequent item sets that are discovered in the conditional database. Formally, all subproblems are tuples $S = (C, P)$, where C is a conditional database, and $P \subseteq B$ is a prefix. The initial problem, with which the recursion is started, is $S = (T, \emptyset)$, where T is the given transaction database, and the prefix is empty. The problem S is, of course, to find all item sets that are frequent in T.

A subproblem $S_0 = (T_0, P_0)$ is processed as follows: Choose an item $i \in B_0$, where B_0 is the set of items occurring in T_0. Note that, in principle, this choice is arbitrary but usually respects a predefined order of the items, which then also defines the structure of the subset tree as discussed above.

If $s_{T_0}(i) \geq s_{min}$ (that is, if the item i is frequent in T_0), report the item set $P_0 \cup \{i\}$ as frequent with the support $s_{T_0}(i)$ and form the subproblem $S_1 = (T_1, P_1)$ with $P_1 = P_0 \cup \{i\}$. The conditional database T_1 comprises all transactions in T_0 that

Fig. 7.19 Illustration of the divide-and-conquer approach to find frequent item sets. *Top*: first split; *bottom*: second split; *blue*: split item/prefix; *green*: first subproblem; *red*: second subproblem

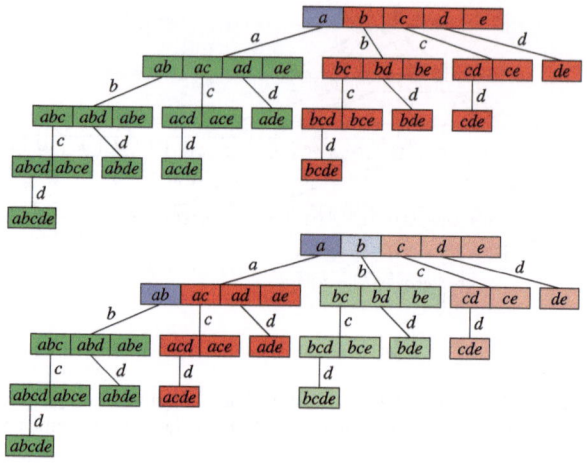

contain the item i, but with the item i removed. This also implies that transactions that contain no other item than i are entirely removed: no empty transactions are ever kept in the search. If T_1 is not empty, S_1 is processed recursively.

In any case (that is, regardless of whether $s_{T_0}(i) \geq s_{min}$ or not), form the subproblem $S_2 = (T_2, P_2)$, where $P_2 = P_0$. The conditional database T_2 comprises all transactions in T_0 (including those that do not contain the item i), but again with the item i removed (and, as before, transactions containing no other item than i discarded). If T_2 is not empty, S_2 is processed recursively.

Eclat, FP-growth and several other frequent item set mining algorithms all rely on the described basic recursive processing scheme. They differ mainly in how they represent the conditional transaction databases. There are two basic approaches: in a **horizontal representation**, the database is stored as a list (or array) of transactions, each of which is a list (or array) of the items contained in it. In a **vertical representation**, on the other hand, a database is represented by first referring with a list (or array) to the different items. For each item, a list (or array) of transaction identifiers is stored, which indicate the transactions that contain the item.

However, this distinction is not pure, since there are many algorithms that use a combination of the two forms of representing a database. For example, while Eclat uses a purely vertical representation and the SaM algorithm presented in the next section uses a purely horizontal representation, FP-growth combines in its FP-tree structure (basically a prefix tree of the transaction database, with links between the branches that connect equal items) a vertical representation (links between branches) and a (compressed) horizontal representation (prefix tree of transactions). Apriori also uses a purely horizontal representation but relies on a different processing scheme, because it traverses the subset tree levelwise rather than depth-first.

In order to give a more concrete idea of the search process, we discuss the particularly simple SaM (Split and Merge) algorithm [14] for frequent item set mining. In analogy to basically all other frequent item set mining algorithms, the SaM algorithm first preprocesses the transaction database with the aim to find a good item

①		②		③		④		⑤

① a d
  a c d e
  b d
  b c d g
  b c f
  a b d
  b d e
  b c d e
  b c
  a b d f

② g: 1
  f: 2
  e: 3
  a: 4
  c: 5
  b: 8
  d: 8

③ a d
  e a c d
  b d
  c b d
  c b
  a b d
  e b d
  e c b d
  c b
  a b d

④ e a c d
  e c b d
  e b d
  a b d
  a b d
  a d
  c b d
  c b
  c b
  b d

⑤
1 → e a c d
1 → e c b d
1 → e b d
2 → a b d
1 → a d
1 → c b d
2 → c b
1 → b d

Fig. 7.20 The example database: original form (1), item frequencies (2), transactions with sorted items (3), lexicographically sorted transactions (4), and the used data structure (5)

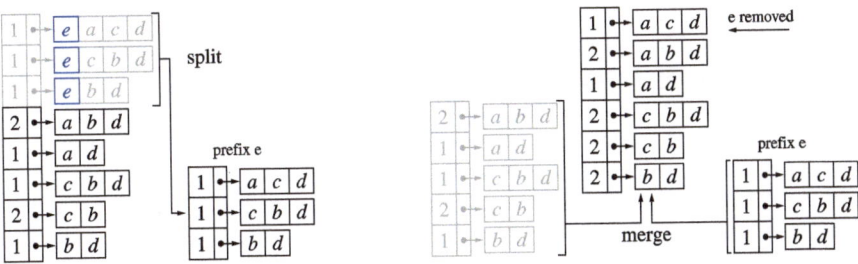

Fig. 7.21 The basic operations of the SaM algorithm: split (*left*) and merge (*right*)

order and to set up the representation of the initial transaction database. The steps are illustrated in Fig. 7.20 for a simple example transaction database. Step 1 shows a transaction database in its original form. In step 2 the frequencies of individual items are determined from this input in order to be able to discard infrequent items immediately. If we assume a minimum support of three transactions for this example, there are two infrequent items, namely f and g, which are discarded. In step 3 the (frequent) items in each transaction are sorted according to their frequency in the transaction database, since experience (also with other algorithms) has shown that processing the items in the order of increasing frequency usually leads to the shortest execution times. In step 4 the transactions are sorted lexicographically into descending order, with item comparisons again being decided by the item frequencies, although here the item with the higher frequency precedes the item with the lower frequency. In step 5 the data structure on which SaM operates is built by combining equal transactions and setting up an array, in which each element consists of two fields: an occurrence counter and a pointer to the sorted transaction. This data structure is then processed recursively, according to the divide-and-conquer scheme discussed in the preceding section, to find the frequent item sets.

The basic operations used in the recursive processing are illustrated in Fig. 7.21. In the **split step** (see the left part of Fig. 7.21) the given array is split w.r.t. the leading item of the first transaction (the *split item*; item e in our example): all array elements referring to transactions starting with this item are transferred to a new array. In this process the pointer (in)to the transaction is advanced by one item, so that the common leading item is removed from all transactions, and empty transactions are discarded. Obviously, the new array represents the conditional database of the first subproblem (see page 183), which is then processed recursively to find all frequent item sets containing the split item (provided that this item is frequent).

The conditional database for frequent item sets *not* containing this item (needed for the second subproblem, see page 184) is obtained with a simple **merge step** (see the right part of Fig. 7.21). The created new array and the rest of the original array (which refers to all transactions starting with a different item) are combined with a procedure that is almost identical to one phase of the well-known *mergesort* algorithm. Since both arrays are obviously lexicographically sorted, one merging traversal suffices to create a lexicographically sorted merged array. The only difference to a *mergesort* phase is that equal transactions (or transaction suffixes) are combined. That is, there is always just one instance of each transaction (suffix), while its number of occurrences is kept in the occurrence counter. In our example this results in the merged array having two elements less than the input arrays together: the transaction (suffixes) *cbd* and *bd*, which occur in both arrays, are combined, and, consequently, their occurrence counters are increased to 2.

As association rules, we formally consider rules $A \rightarrow C$, with $A \subseteq B$, $C \subseteq B$, and $A \cap C = \emptyset$. Here B is the underlying item base, and the symbols A and C have been chosen to represent the antecedent and the consequent of the rule. For simplicity, we confine ourselves in the following to rules with $|C| = 1$, that is, to rules with only a single item in the consequent. Even though rules with several items in the consequent have also been studied, they are usually a lot less useful and only increase the size of the created rule set unnecessarily.

While (frequent) item sets are associated with only one measure, namely their support, association rules are evaluated with two measures:

- **support of an association rule**
  The support of an association rule $A \rightarrow C$ can be defined in two ways:
  - *support of all items appearing in the rule*: $r_T(A \rightarrow C) = \frac{s_T(A \cup C)}{s_T(\emptyset)}$.
    This is the more common definition:
    the support of a rule is the fraction of cases in which it is correct.
  - *support of the antecent of the rule*: $r_T(A \rightarrow C) = \frac{s_T(A)}{s_T(\emptyset)}$.
    This definition is actually more plausible:
    the support of a rule is the fraction of cases in which it is applicable.
  (Note that $s_T(\emptyset)$ is simply the number of transactions in the database T to mine, because the empty item set is contained in all transactions.)
- **confidence of an association rule**
  The confidence of an association rule is the number of cases in which it is correct relative to the number of cases in which it is applicable: $c_T(A \rightarrow C) = \frac{s_T(A \cup C)}{s_T(A)}$. The confidence can be seen as an estimate of the conditional probability $P(C \mid A)$.

Given a transaction database T over some item base B, a **minimum support** $r_{min} \in [0, 1]$, and a **minimum confidence** $c_{min} \in [0, 1]$ (both to be specified by a user), the task of **association rule induction** is to find all association rules $A \rightarrow C$ with $A, C \subset B$, $r_T(A \rightarrow C) \geq r_{min}$, and $c_T(A \rightarrow C) \geq c_{min}$.

Fig. 7.22 The association rules that can be discovered with a minimum support $s_{min} = 3$ and a minimum confidence $c_{min} = 80\%$ in the transaction database shown in Fig. 7.16. Note that in this example it is irrelevant which rule support definition is used

association rule	support of all items	support of antecedent	confidence
$b \to c$	$3 \triangleq 30\%$	$3 \triangleq 30\%$	100.0%
$d \to a$	$5 \triangleq 50\%$	$6 \triangleq 60\%$	83.3%
$e \to a$	$6 \triangleq 60\%$	$7 \triangleq 70\%$	85.7%
$a \to e$	$6 \triangleq 60\%$	$7 \triangleq 70\%$	85.7%
$d, e \to a$	$4 \triangleq 40\%$	$4 \triangleq 40\%$	100.0%
$a, d \to e$	$4 \triangleq 40\%$	$5 \triangleq 50\%$	80.0%

The usual approach to this task consists in finding, in a first step, the frequent item sets of T (as described above). For this step, it is important which rule support definition is used. If the support of an association rule is the support of all items appearing in the rule, then the minimum support to be used for the frequent item set mining step is simply $s_{min} = \lceil s_T(\emptyset) r_{min} \rceil$ (which explains why this is the more common definition). If, however, the support of an association rule is only the support of its antecendent, then $s_{min} = \lceil s_T(\emptyset) r_{min} c_{min} \rceil$ has to be used. In a second step the found frequent item sets are traversed, and all possible rules (with one item in the consequent, see above) are generated from them and then filtered w.r.t. r_{min} and c_{min}.

As an example, consider again the transaction database shown in Fig. 7.16 and the frequent item sets that can be discovered in it (for a minimum support, $r_{min} = 30\%$, that is, $s_{min} = 3$). The association rules that can be constructed from them with a minimum confidence $c_{min} = 0.8 = 80\%$ are shown in Fig. 7.22. Due to the fairly high minimum confidence requirement, the number of found association rules is relatively small compared to the number of transactions.

7.6.3 Variations and Issues

A general scheme to enhance the efficiency of frequent pattern is perfect extension pruning (Sect. 7.6.3.1). Since usually many frequent patterns can be found in a data set, reducing the output is an important task (Sect. 7.6.3.2). The same holds for association rules (Sect. 7.6.3.3), which are usually filtered based on evaluation measures. Data types other than item sets need special methods to eliminate redundant search (Sect. 7.6.3.4). With this last section we go, in one special instance, beyond the usual scope of this book, because we consider data that has a more complex structure than a record in a data set. The idea is to convey at least some idea of how data analysis of such data can look like, while we cannot provide a comprehensive treatment of all more complex data types.

7.6.3.1 Perfect Extension Pruning

The basic recursive processing scheme for finding frequent item sets described in Sect. 7.6.2 can easily be improved with so-called **perfect extension pruning** (also called **parent equivalence pruning**), which relies on the following simple idea: given an item set I, an item $i \notin I$ is called a **perfect extension** of I iff I and $I \cup \{i\}$ have the same support, that is, if i is contained in all transactions that contain I. Perfect extensions have the following properties: (1) if the item i is a perfect extension of an item set I, then it is also a perfect extension of any item set $J \supseteq I$ as long as $i \notin J$, and (2) if I is a frequent item set and K is the set of all perfect extensions of I, then all sets $I \cup J$ with $J \in 2^K$ (where 2^K denotes the power set of K) are also frequent and have the same support as I. These properties can be exploited by collecting in the recursion not only prefix items but also, in a third element of a subproblem description, perfect extension items. Once identified, perfect extension items are no longer processed in the recursion but are only used to generate all supersets of the prefix that have the same support. Depending on the data set, this can lead to a considerable acceleration of the search.

7.6.3.2 Reducing the Output

Often three types of frequent item sets are distinguished by additional constraints:

- **frequent item set**
  A frequent item set is merely frequent:
  I is a frequent item set $\Leftrightarrow s_T(I) \geq s_{\min}$.
- **closed item set**
  A frequent item set is called *closed* if no proper superset has the same support:
  I is a closed item set $\Leftrightarrow s_T(I) \geq s_{\min} \wedge \forall J \supset I : s_T(J) < s_T(I)$.
- **maximal item set**
  A frequent item set is called *maximal* if no proper superset is frequent:
  I is a maximal item set $\Leftrightarrow s_T(I) \geq s_{\min} \wedge \forall J \supset I : s_T(J) < s_{\min}$.

Obviously, all maximal item sets are closed, and all closed item sets are frequent. The reason for distinguishing these item set types is that with them one can achieve a compressed representation of all frequent item sets: the set of all frequent item sets can be recovered from the maximal (or the closed) item sets by simply forming all subsets of maximal (or closed) item sets, because each frequent item set has a maximal superset. With closed item sets, one even preserves the knowledge of their support, because any frequent item set has a closed superset *with the same support*. Hence the support of a nonclosed frequent item set I can be reconstructed with

$$\forall s_{\min} : \forall I : \quad s_T(I) \geq s_{\min} \quad \Rightarrow \quad s_T(I) = \max_{J \in C_T(s_{\min}), J \supseteq I} s_T(J),$$

where $C_T(s_{\min})$ is the set of all closed item sets that are discovered in a given transaction database T if the minimum support $s_{\min}$ is chosen.

Note that closed item sets are directly related to perfect extension pruning, since a closed item set can also be defined as an item set that does not possess a perfect extension. As a consequence, identifying perfect extensions is particularly useful and effective if the output is restricted to closed item sets.

Fig. 7.23 The decimal
logarithms of the numbers of
frequent, closed, and maximal
item sets (*vertical axis*) for
the BMS-Webview-1
database and different
minimum support values
(*horizontal axis*)

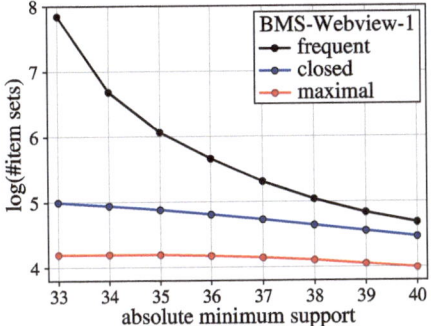

For the simple example database shown in Fig. 7.16, all frequent item sets are closed, with the exception of $\{b\}$ (since $s_T(\{b\}) = s_T(\{b, c\}) = 3$) and $\{d, e\}$ (since $s_T(\{d, e\}) = s_T(\{a, d, e\}) = 4$). The maximal item sets are $\{b, c\}$, $\{a, c, d\}$, $\{a, c, e\}$, and $\{a, d, e\}$. Note that indeed any frequent item set is a subset of at least one of these four maximal item sets, demonstrating that they suffice to represent all frequent item sets. In order to give an impression of the relative number of frequent, closed, and maximal item sets in practice, Fig. 7.23 displays the decimal logarithms of the numbers of these item sets that are found for different minimum support values for a common benchmark data set, namely the BMS-Webview-1 database. Clearly, restricting the output to closed or even to maximal item sets can reduce the size of the output by orders of magnitude in this case.

Naturally, the notions of a closed and a maximal item set can easily be transfered to other pattern types, leading to closed and maximal sequences, trees, and (sub)graphs (with analogous definitions). Their purpose is the same in these cases: to reduce the output to a more manageable size. Nevertheless, however, the size of the output remains a serious problem in frequent pattern mining, despite several research efforts, for example, to define relevance or interestingness measures by which the best patterns can be singled out for user inspection.

7.6.3.3 Assessing Association Rules

In practice the number of found association rules is usually huge, often exceeding (by far) the number of transactions. Therefore considerable efforts have been devoted to filter them in order to single out the interesting ones, or at least to rank them. Here we study only a few simple methods, though often effective. More sophisticated approaches are discussed, for instance, in [53, 54].

The basic idea underlying all approaches described in the following is that a rule is interesting only if the presence of its antecedent has a sufficient effect on the presence of its consequent. In order to assess this, the confidence of a rule is compared to its expected confidence under the assumption that antecedent and

Fig. 7.24 A contingency table for the antecedent A and the consequent C of an association rule $A \to C$ w.r.t. the transactions t of a given transaction database T. Obviously, $n_{11} = s_T(A \cup C)$, $n_{1*} = s_T(C)$, $n_{*1} = s_T(A)$, and $n_{**} = s_T(\emptyset)$

	$A \nsubseteq t$	$A \subseteq t$	
$C \nsubseteq t$	n_{00}	n_{01}	n_{0*}
$C \subseteq t$	n_{10}	n_{11}	n_{1*}
	n_{*0}	n_{*1}	n_{**}

consequent are independent. If antecedent and consequent are independent, then $c_T(A \to C) = \hat{P}(C \mid A)$ and $c_T(\emptyset \to C) = \hat{P}(C)$ should not differ much (the true probabilities should actually be the same, but their estimates may differ slightly due to noise in the data), and thus we can assess association rules by comparing them.

The most straightforward measure based on this idea is the absolute difference of the two confidences, that is, $d_T(A \to C) = |c_T(A \to C) - c_T(\emptyset \to C)|$. However, it is more common to form their ratio, that is,

$$l_T(A \to C) = \frac{c_T(A \to C)}{c_T(\emptyset \to C)}.$$

This measure is known as the **lift value**, because it describes by what factor our expectation to observe C is lifted if we learn that A is present. Clearly, the higher the lift value, the more influence A has on C, and thus the more interesting the rule.

More sophisticated approaches evaluate the dependence of A and C based on the contingency table shown in Fig. 7.24, which lists the numbers of transactions which contain or do not contain A and C, alone and in conjunction. (Note that a star as a first index means that the value is a sum over a column, and a star in the second index means that the value is a sum over a row. Two stars mean a sum over both rows and columns.) Measures based on such a table are the well-known statistical χ^2 measure

$$\chi^2(A \to C) = \sum_{i=0}^{1} \sum_{j=0}^{1} \frac{(p_{i*}p_{*j} - p_{ij})^2}{p_{i*}p_{*j}} = \frac{n_{**}(n_{1*}n_{*1} - n_{**}n_{11})^2}{n_{1*}(n_{**} - n_{1*})n_{*1}(n_{**} - n_{*1})}$$

and **mutual information** or **information gain**

$$I_{\mathrm{mut}}(A \to C) = \sum_{i=0}^{1} \sum_{j=0}^{1} p_{ij} \log_2 \frac{p_{ij}}{p_{i*}p_{*j}} = \sum_{i=0}^{1} \sum_{j=0}^{1} \frac{n_{ij}}{n_{**}} \log_2 \frac{n_{ij}n_{**}}{n_{i*}n_{*j}},$$

where $p_{kl} = n_{kl}/n_{**}$ for $k, l \in \{0, 1, *\}$. For both of these measures, a higher value means a stronger dependence of C on A and thus a higher interestingness of the rule.

It should be noted that such measures are only a simple aid, as they do not detect and properly assess situations like the following: suppose that $c_T(\emptyset \to C) = 20\%$ but $c_T(A \to C) = 80\%$. According to all of the above measures, the rule $A \to C$ is highly interesting. However, if we know that $c_T(A - \{i\} \to C) = 76\%$ for some item $i \in A$, the rule $A \to C$ does not look so interesting anymore: the change in confidence that is brought about by the item i is fairly small, and thus we should rather consider a rule with a smaller antecedent.

While this problem can at least be attacked by comparing a rule not only to its counterpart with an empty antecedent but to all rules with a simpler antecedent (even though this is more difficult than it may appear at first sight), semantic aspects of interestingness are completely neglected. For example, the rule "pregnant → female" has a confidence of 100%, while "∅ → female" has a confidence of only 50%. Nevertheless the rule "pregnant → female" is obviously completely uninteresting, because it is part of our background knowledge (or can at least be derived from it).

7.6.3.4 Frequent Subgraph Mining

In analogy to frequent item set mining, with which item sets are found that are contained in a sufficiently large number of transactions of a given database (as specified by the minimum support), **frequent subgraph mining** tries to find (sub)graphs that are contained in a sufficiently large number of (attributed or labeled) graphs of a given graph database. Since the advent of this research area around the turn of the century, several clever algorithms for frequent subgraph mining have been developed. Some of them rely on principles from inductive logic programming and describe the graph structure by logical expressions [23]. However, the vast majority transfers techniques developed originally for frequent item set mining (see Sect. 7.6.2). Examples include MolFea [39], FSG [40], MoSS/MoFa [12], gSpan [56], CloseGraph [57], FFSM [34], and Gaston [42]. A related, but slightly different approach, is used in Subdue [17], which is geared toward graph compression with common subgraphs rather than frequent subgraph mining.

Here we confine ourselves to a brief outline of the core principles of frequent subgraph mining, because frequent sequence and tree mining can obviously be seen as special cases of frequent subgraph mining. Nevertheless, however, it should be noted that specific optimizations are possible due to the restricted structure of trees and those graphs to which different types of sequences can be mapped. Note also that this is the only place in this book where we consider the analysis of data that does not have a tabular structure. While we cannot provide a comprehensive coverage of all complex data types, we strive in this section to convey at least some idea of the special requirements and problem encountered when analyzing such data.

Formally, frequent subgraph mining works on a database of labeled graphs (also called *attributed graphs*). A **labeled graph** is a triple $G = (V, E, l)$, where V is the set of vertices, $E \subseteq V \times V - \{(v, v) \mid v \in V\}$ is the set of edges, and $l : V \cup E \to L$ assigns labels from some label set L to vertices and edges. The graphs we consider are **undirected** and **simple** (that is, there is at most one edge between two given vertices) and contain **no loops** (that is, there are no edges connecting a vertex to itself). However, graphs without these restrictions (that is, directed graphs, graphs with loops and/or multiple edges) can be handled as well with properly adapted methods. Note also that several vertices and edges may carry the same label.

Fig. 7.25 Examples of subgraph isomorphisms with two molecular fragments and a molecule containing them in different ways

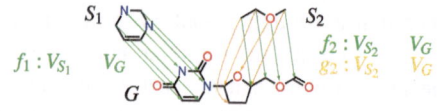

The **support** $s_{\mathcal{G}}(S)$ of a (sub)graph S wrt. a given graph database $\mathcal{G}$ is the number of graphs $G \in \mathcal{G}$ it is contained in. What is meant by a graph being contained in another is made formally precise by the notion of a *subgraph isomorphism*. Given two labeled graphs $G = (V_G, E_G, l_G)$ and $S = (V_S, E_S, l_S)$, a **subgraph isomorphism** of S to G is an injective function $f : V_S \rightarrow V_G$ satisfying (1) $\forall v \in V_S : l_S(v) = l_G(f(v))$ and (2) $\forall (u, v) \in E_S : (f(u), f(v)) \in E_G \wedge l_S((u, v)) = l_G((f(u), f(v)))$. That is, the mapping f preserves the connections and the labels of both vertices and edges.

Note that there may be several ways to map a labeled graph S to a labeled graph G so that the connection structure and the labels are preserved. For example, the graph G may possess several subgraphs that are isomorphic to S. It may even be that the graph S can be mapped in several different ways to the same subgraph of G. This is the case if there exists a subgraph isomorphism of S to itself (a so-called *graph automorphism*) that is not the identity. Examples of subgraph isomorphism wrt. a molecule G and two molecular fragments S_1 and S_2, all three of which are modeled as labeled graphs, are shown in Fig. 7.25. The fragment S_2 is contained in several different ways in the molecule, two of which are indicated in the figure.

Given a database $\mathcal{G}$ of labeled graphs and a user-specified **minimum support** $s_{\min} \in \mathbb{N}$, a (sub)graph S is called **frequent** in $\mathcal{G}$ if $s_{\mathcal{G}}(S) \geq s_{\min}$. The task of frequent subgraph mining is to identify all subgraphs that are frequent in the given graph database $\mathcal{G}$. However, usually the output is restricted to connected subgraphs for two reasons: in the first place, connected subgraphs suffice for most applications. Secondly, restricting the result to connected subgraphs considerably reduces the search space, thus rendering the search actually feasible.

Also in analogy to frequent item set mining, we consider the **partial order** of subgraphs of the graphs of a given database that is induced by the subgraph relationship. A Hasse diagram of an example partial order, for a simple database of three molecule-like graphs (no chemical meaning attached, to be seen at the bottom) is shown in Fig. 7.26 on the top: two subgraphs are connected if one possesses an additional edge (and an additional vertex, since there are no cycles) compared to the other. The frequent subgraphs are located at the top of this Hasse diagram, thus suggesting a top-down search, just as for frequent item set mining.

Like for frequent item set mining (see Sect. 7.6.2), a core problem is that the same graph can be reached on different paths, as is easily visible from the Hasse diagram of subgraphs shown in Fig. 7.26. Hence we also face the task to prevent redundant search. The solution principle is the same as for frequent item sets: to each subgraph, a **unique parent** is assigned, which turns the subgraph partial order into a **subgraph tree**. This is illustrated in Fig. 7.26 on the bottom (the principle underlying this assignment of unique parents is discussed below).

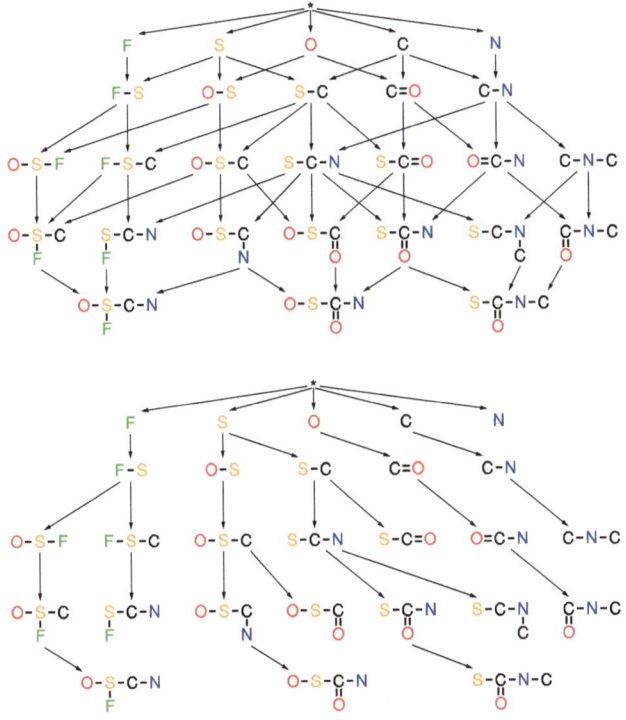

Fig. 7.26 A Hasse diagram of subgraphs of three molecule-like graphs, to be seen at the bottom (*top*) and an assignment of unique parents, which turns the Hasse diagram into a tree (*bottom*)

With a unique parent for each subgraph, we can carry out the search for frequent subgraphs according to the following simple recursive scheme, which parallels the divide-and-conquer scheme described in Sect. 7.6.2: in a base loop, all possible vertex labels are traversed (their unique parent is the empty graph). All vertex labels (and thus all single vertex subgraphs) that are frequent are processed recursively. A given frequent subgraph S is processed recursively by forming all possible extensions R of S by a single edge and a vertex for which S is the chosen unique parent. All such extensions R that are frequent (that is, for which $s_G(R) \geq s_{min}$) are processed recursively, while infrequent extensions are discarded.

Whereas for frequent item sets, it was trivial to assign unique parents, namely by simply defining an (arbitrary but fixed) order of the items, ordering the labels in the set L, though also necessary, is not enough. The reason is mainly that several vertices (and several edges) may carry the same label. Hence the labels do not uniquely identify a vertex, thus rendering it impossible to describe the graph structure solely with these labels. We rather have to endow each vertex with a unique identifier (usually a number), so that we can unambiguously specify the edges of the graph.

Given an assignment of unique identifiers, we can describe a graph with a **code word** which specifies the vertex and edge labels and the connection structure, and from which the graph can be reconstructed. Of course, the form of this code word depends on the chosen numbering of the vertices: each numbering leads to a different code word. In order to single out one of these code words, the **canonical code word**, we simply select the lexicographically smallest code word.

With canonical code words, we can easily define unique parents: the canonical code word of a (sub)graph S is obtained with a specific numbering of its vertices and thus also fixes (maybe with some additional stipulation) an order of its edges. By removing the last edge according to this order, which is not a bridge or is incident to at least one vertex with degree 1 (which is then also removed), we obtain a graph that is exactly one level up in the partial order of subgraphs and thus may be chosen as the unique parent of S. Details about code words, together with a discussion of the important **prefix property** of code words, can be found, for example, in [13].

7.7 Deviation Analysis

The search for new and interesting association rules can be seen as a discovery of subgroups of the dataset that share common properties, e.g., buy a certain product much more often than the average customer. Deviation analysis is another form of subgroup discovery. In contrast to association analysis, there is usually some target property given. The goal is to find subgroups of the populations that are statistically most interesting, that is, they are as large as possible and deviate from the whole population with respect to the property of interest as much as possible [37, 55]. For example, in a dataset consisting of pricy car offers, most of them will have powerful engines with respectable cylinder capacity, so their fuel consumption will be quite high. If we look for a subgroup that deviates from the high-average consumption as much as possible, we may find a (relatively small) subgroup of pricy cars with significantly less fuel consumption having hybrid engine, aluminium body, etc.

7.7.1 Overview

In deviation analysis a pattern describes a divergence of some target measure in a subgroup of the full dataset. The target measure is usually specified by the analyst, and subgroups with the largest deviations are sought by the respective method. Cluster analysis may be considered as deviation analysis in the sense that the cluster represents a subgroup for which the data density deviates from the data density of the whole dataset. Other possibilities are deviations in the distribution of a nominal target variable, the mean of a target variable, or more complex targets such as regression functions. While the methods described in Chaps. 8 and 9 also require a target attribute, their primary goal is the prediction of the target attribute for new data, which requires to build a model that covers the whole dataspace (rather than just subgroups). Furthermore, the successful discovery of a deviation in the distribution of some target attribute does not necessarily imply that we can easily predict the target value now (cf. a change in the distribution of 10:50:40 to 40:45:15).

To gain insights from the discovery, the respective subgroup should be easy to grasp by the analyst and is therefore best characterized by a conjunction of features (conditions on attributes like *fuel = liquid petrol gas, offeredby ≠ dealer, mileage > 100000*, etc.). All possible combinations of constraints define jointly the set of possible subgroups. A search method is then applied to systematically explore this subgroup space. The target measure of each candidate subgroup is compared against the value obtained for, say, the whole dataset, and a statistical test decides about the significance of a discovered deviation (under the null hypothesis that the subgroup does not differ from the whole dataset). Only significant deviations are considered as interesting and reported to the analyst. The found subgroups are ranked to allow the search method to focus on the most interesting patterns and to point the user to the most promising findings. This ranking involves a quality function, which is usually a trade off between size of the subgroup and its unusualness.

One of the major practical benefits of deviation analysis is that we do not even try to find a model for the full dataset, because in practice such a model is either coarse and well known to the expert or very large and thus highly complex. Concentrating on the (potentially most interesting) highest deviation from the expected average and deriving short, understandable rules delivers invaluable insights.

7.7.2 Construction

The ingredients of deviation analysis are (1) the target measure and a verification test serving as a filter for irrelevant or incidental patterns, (2) a quality measure to rank subgroups, and (3) a search method that enumerates candidates subgroups systematically.

Any attribute of the dataset may serve as the target measure, and any combination or derived variable, and even models themselves, may be considered (e.g., regression models [41]). Depending on the scale of the target attribute, an appropriate test has to be selected (see Fig. 7.27). For instance, for some binary target flag, we have a probability of $p_0 = 0.7$ in the full population of observing the flag F. Now, in a subgroup of size $n = 100$ we observe $m = 80$ cases where the flag is set, so $p = P(F) = 0.8$. Before we claim that we have found a deviation (and thus an interesting subgroup), we want to get confident that 80 cases or more are unlikely to occur by chance (say, only in 1%). We state the null hypothesis that $p = p_0$ and apply a one-sample z-test[3] with a significance level of 0.01. We have to compute the z-score

$$z = \frac{m - \mu}{\sigma} = \frac{80 - 70}{0.0158} \approx 2.18,$$

---

[3] As n is relatively large, we approximate the binomial distribution by a normal distribution. We use the z-test for reasons of simplicity—typically p_0 is not known but has to be estimated from the sample, and then Student's t-test is used.

data type/scale	target measure	possible test
binary	single share, e.g., percentage of females	binomial test, χ^2 test
nominal	vector of percentages, e.g. $(0.3, 0.2, 0.5)$	χ^2 goodness-of-fit
ordinal	median or probability of better value in subgroup	ridit analysis
ratio	mean, median, variance	median test, t-test

Fig. 7.27 Type of target measure

where $\mu = np_0$ is the mean, and σ is the standard deviation in the whole population, which is given by $\sigma = \sqrt{n\,p_0(1 - p_0)}$. A table of the standard normal distribution tells us that the probability of observing a z-value of 2.18 or above is approximately 0.0146. Thus, in $1 - 0.0146 = 98.54\%$ there will be less than 80 observed target flags. A significance level of 0.01 means that we reject the null-hypothesis if this situation occurs only in 1 out of 100 times—actually it may occur in ≈ 1.5 out of 100 times, so the null-hypothesis cannot be rejected (at this significance level), and the subgroup is not reported. As we will perform quite a large number of tests, the significance level must be adjusted appropriately. If we test 1000 subgroups for being interesting and use a significance level of 0.01, we have to expect about 10 subgroups that pass this test (even if there is no substantial deviation).

All interesting subgroups are ranked according to their interestingness. One possibility is to reuse the outcome of the statistical test itself, as it expresses how unusual the finding is. In the above case, this could be the z-score itself:

$$z = \frac{m - n\,p_0}{\sqrt{n\,p_0(1 - p_0)}} = \frac{\sqrt{n}(p - p_0)}{\sqrt{p_0(1 - p_0)}}.$$

The z-score value balances the size of subgroup (factor of $\sqrt{n}$), the deviation (factor of $p - p_0$) and the level of the target share (p_0), which is important because an absolute deviation of $p_0 - p = 0.1$ may represent both, a small increase (from 0.9 to 1.0) or a doubled share (from 0.1 to 0.2). Another aspect that might be considered by an additional factor of $\sqrt{\frac{N}{N-n}}$ is the relative size of the subgroup with respect to the total population size. A frequently used quality function is the **weighted relative accuracy (WRAcc)**, which is typically used for predictive tasks:

$$WRAcc = (p - p_0) \cdot \frac{n}{N},$$

However, the user may want to focus on other (application dependent) aspects such as practical usefulness, novelty, and simplicity and employ other quality function be used to rank the subgroups.

Since subgroups are to be described by a conjunction of features (such as *variable = value*), the number of possible subgroups on the syntactical level grows exponentially with the number of attributes and the number of possible values. Given an attribute A with $\Omega_A = \{high, medium, low, absent\}$, we may form four conditions on identity (e.g., $A = high$), four on imparity (e.g., $A \neq medium$), and as A has ordinal scale, even more features regarding the order (e.g., $A \leq low$ or $A \geq medium$). When constructing 11 features from A alone, another 9 attributes with 4 values would al-

Table 7.12 Beam search for deviation analysis

Algorithm SubgrpDisc($\mathcal{D}$, F)
input: data set $\mathcal{D}$, $|\mathcal{D}| = n$, feature set F parameter: beam width b output: set of b interesting subgroups

1	initialize b subgroups with condition C_i and quality value q_i, $1 \leq i \leq b$
2	**do**
3	**for** all subgroups ($1 \leq i \leq b$)
4	**for** all features $f \in F$
5	create new candidate subgroup with condition $C_i \wedge f$
6	evaluate subgroup and test for significant deviation
7	compute quality q of this new subgroup
8	**if** candidate is evaluated to be better than worst subgroup in beam
9	replace worst subgroup in beam by candidate
10	**while** beam has improved

low us to define $11^{10} \approx 25 \cdot 10^9$ syntactically different subgroups. Examining and testing each of them individually is apparently not an option.

A commonly applied search technique for deviation analysis is **beam search**. The algorithm keeps track of a fixed number b of subgroups (b is called **beam width**) and tries to further specialize all these subgroups, that is, make the subgroups more specific by including another feature in the subgroup description. If these candidate subgroups turn out to be interesting (according to the statistical test), they are evaluated with respect to their quality. In case they evaluate better than one of the b subgroups in the beam, they replace the worst subgroup. This process is iterated until the beam does no longer improve. The algorithm is shown in Table 7.12.

There are other search algorithms besides beam search. For instance, for a highly ranked subgroup, it is not necessarily the case that all subgroups obtained by taking prefixes from the condition list must also be interesting. If that is not the case, the beam search algorithm will not find such a subgroup, because it stops any further specialization of subgroups that were not successfully verified by the significance test. A broad view search would still keep such unverified subgroups for further expansion. Some quality measures, such as the J-measure [49], provide an upper bound for the quality that may be achieved in subsequent specializations. If such information is given, this optimistic estimate might be used to eliminate subgroups only in case there is no chance of becoming one of the k top-ranked subgroups (as in [49]). Various heuristics may be implemented to direct the search to the most promising subgroups and to avoid delving into a large number of very similar subgroups [24, 36].

7.7.3 Variations and Issues

If a high quality subgroup is present, it is very likely that several variants of this subgroup exist, which are also evaluated as good subgroups (e.g., by adding a constraint $A \neq v$ for some attribute A and a rare feature v). Thus, it may happen that many of the subgroups in the beam are variations of a single scheme, which prevents the beam search from focussing on other parts of the dataspace—a small number of diverse subgroups would be preferable. A subset of diverse subsets can be extracted from the beam by selecting successively those subgroups that cover most of the data. Once a subgroup has been selected, the covered data is excluded from the subsequent selection steps [24]. If most of the subgroups in the beam are variations of one core subgroup, only a few diversive subgroups will be selected. A better approach is a sequential covering search, where good subgroups are discovered one after the other. Over several runs, only a few or a single best subgroup is identified, and the data covered by this subgroup is then excluded from subsequent runs. Thereby subgroups cannot be rediscovered, but the method has to focus on different parts of the dataspace. Similar techniques are applied for learning sets of classification rules (see Chap. 8). It is also possible to generate a new sample from the original dataset that no longer exhibits the unusualness that has been discovered by a given subgroup [48]. If this subsampling is applied before any subsequent run, new subgroups rather than known subgroups will be discovered.

Another issue is the efficiency of search, the **scalability** to large datasets. Exhaustive searching is prohibitive unless intelligent pruning techniques are applied that prevent us from losing to much time with redundant, uninteresting subgroups. On the other hand, any kind of heuristic search (like beam search) bears the risk of missing the most interesting subgroups. There are multiple directions how to attack this problem.

For nominal target variables, efficient algorithms from association rule mining can be utilized. If it is possible to transform the database such that association mining (e.g., via the Apriori algorithm) can be applied, the validation and ranking of the patterns found are merely a post-processing step [15]. Missing values require special care in this approach, as the case of missing data is usually not considered in market basket analysis. Furthermore, deviation analysis typically focusses on a target variable rather than associations between any attributes, which offers some optimization potential [6].

If the dataset size becomes an issue, one may use a subsample to test and rank subgroups rather than the full dataset. For a broad range of quality measures, one can derive upper bounds for the size of the sample with guaranteed upper bounds on the subgroup quality estimation error [47]. This speeds up the discovery process considerably, because there is no need for a full database scan.

7.8 Finding Patterns in Practice

As we have seen, finding patterns in data often consists of finding some sort of partitioning of the data: those may be hierarchical clusterings, typical patterns, embeddings of lower dimensional maps or subgroups. Any reasonably modern data analysis tool offers the classic clustering algorithm k-Means and provides association rule mining methods. All the different variants discussed in this chapter will usually only be available in certain flavors, however.

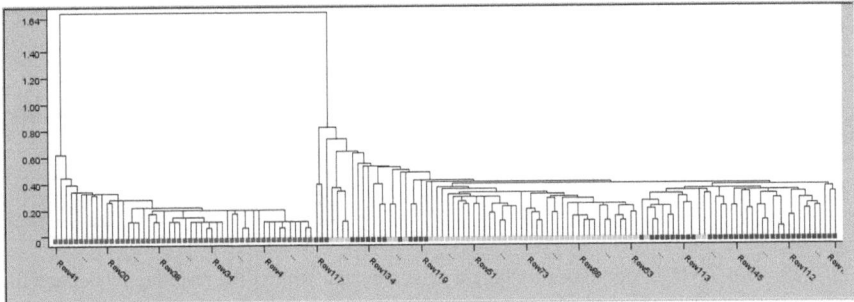

Fig. 7.28 A KNIME dendrogram for the Iris data. The colors indicate the underlying class information of the three different types of iris plants

7.8.1 Finding Patterns with KNIME

KNIME offers a number of nodes for pattern finding. Hierarchical clustering is, of course, among them. Applying the node is fairly straight forward. The user selects the (numerical[4]) columns to be used along with the distance function and the type of linkage. The node will also provide a cluster ID assignment of the input on its outport, and for this, the user can specify the cutoff, i.e., the number of clusters to be created. Figure 7.28 shows a dendrogram for the Iris data described earlier. We have also used the color assigner based on the class labels. However, the Euclidean distance was only computed on the four numerical attributes.

The standard method for prototype based clustering is provided via the *k-Means* node. The options here are again the selection of numerical columns to be used and additionally the number of clusters to be used and the maximum number of iterations. The latter is usually not required, but in the (rare) case of oscillations and hence nontermination of the underlying algorithm, KNIME will force termination after reaching this number of iterations. The node has two outports: one carrying the input data together with cluster labels and a second port holding the cluster model. This model can be exported for use in other tools (using, e.g., the *Model Writer* node), or it can be used to apply the clustering to new data. The *Cluster Assigner* node will compute the closest cluster prototype for each pattern and assign the corresponding label. Figure 7.29 shows two small flows to demonstrate this.

The part on the left reads a file, runs the k-Means algorithm, and writes the resulting cluster model into a file. The flow on the right reads this model and a second file and applies the cluster model to this new data. Of course, in this little example we could have simply connected the clustering node to the node assigning the cluster model, but this demonstrates how we can use one analysis tool to create a clustering and then use a second tool (maybe a nightly batch job clustering customers?) using the very same model. We will discuss this aspect of model deployment in more detail in Chap. 10. KNIME also provides a node which allows one to apply the fuzzy version of k-Means, fuzzy c-Means.

---

[4]We will discuss the use of other types of distance metrics in KNIME later.

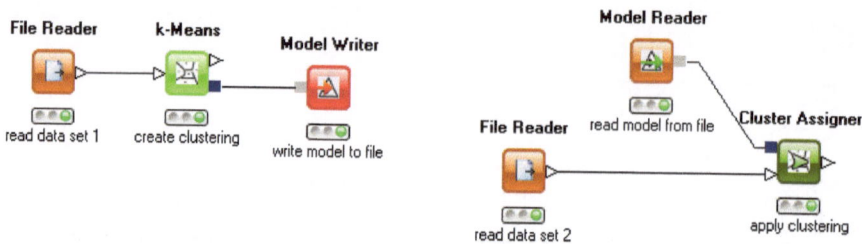

Fig. 7.29 Two KNIME workflows creating a clustering and writing the resulting model to a file (*left*) and reading data and model from file and applying the read clustering model (*right*)

As we have discussed in Sect. 7.2 the underlying notion of (dis)similarity plays a crucial role when using clustering algorithms. The KNIME nodes we have discussed so far assume numerical attributes and then allow to chose a few different distance metrics based on these numerical feature vectors. Obviously this can be rather restrictive. KNIME therefore also allows one to use arbitrary distance matrices for some of the clustering methods which can deal with matrices instead of a well-defined feature space. In particular, KNIME provides nodes to read in distance matrices but also various nodes to compute distance matrices on other than just numerical feature vectors. Using this functionality, one can use (dis)similarities metrics between documents, molecular structures, images, and other objects and the cluster, those using some of the distance-matrix clustering nodes. KNIME, as of version 2.1, provides a k-medoids clustering node and a set of nodes to produce, view, and, apply a hierarchical clustering.

KNIME also offers nodes for association rule mining. Here we will demonstrate a different aspect of itemset mining, however. Through various different extensions, KNIME can be extended to deal with other types of data such as documents, images, and also molecular structures. As we have already shown in Sect. 7.6.3.4, frequent itemset mining algorithms can be extended to find discriminative fragments in molecules. Once the appropriate extension to KNIME is installed (which are available online as well), KNIME can read in different molecular file formats and a set of new nodes is available which can deal with such structures. One of them implements the MoFa algorithm discussed in Sect. 7.6.3.4. The algorithm can be applied very similarly to all other nodes. One or more columns in the input table should hold molecules, which are then also shown in a two-dimensional representation since the chemistry plugin for KNIME also brings along an open-source renderer via the Chemistry Development Kit (CDK [50]). Figure 7.30 shows an example flow with the corresponding output. As one can see, one column in the table holds molecules, and the others are standard types, indicating screening results which are used as classes in this example. The output are parts of molecules which could be responsible for the activity against HIV. More information about the underlying algorithm and application can be found in [12].

Note how the usual KNIME view recognizes the newly registered type and uses the corresponding renderer. The KNIME Hiliting-mechanism allows the user to select a fragment in the view on the right and immediately see in the view on

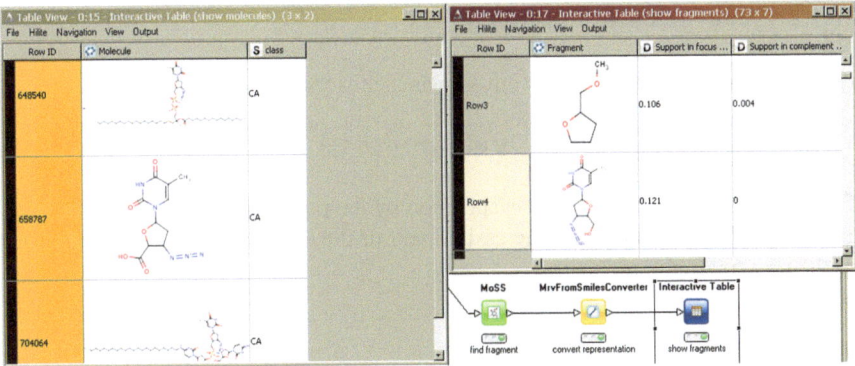

Fig. 7.30 Through additional plugins, KNIME can also process other types of data. In this example, molecular file formats are read and processed in a discriminative fragment mining node

the left (connected to the input data) which rows (and molecules) the fragment is contained in.

7.8.2 Finding Patterns in R

7.8.2.1 Hierarchical Clustering

As an example, we apply hierarchical clustering to the Iris data set, ignoring the categorical attribute *Species*. We use the normalized Iris data set `iris.norm` that is constructed in Sect. 6.6.2.3. We can apply hierarchical clustering after removing the categorical attribute and can plot the dendrogram afterwards:

```
> iris.num <- iris.norm[1:4]
> iris.cl <- hclust(dist(iris.num), method="ward")
> plot(iris.cl)
```

Here, the Ward method for the cluster distance aggregation function as described in Table 7.3 was chosen. For the other cluster distance aggregation functions in the table, one simply has to replace `ward` by `single` (for single linkage), by `complete` (for complete linkage), by `average` (for average linkage), or by `centroid`.

For heatmaps, the library `gplots` is required that needs installing first:

```
> library(gplots)
> rowv <- as.dendrogram(hclust(dist(iris.num),
                               method="ward"))
> colv <- as.dendrogram(hclust(dist(t(iris.num)),
                               method="ward"))
> heatmap.2(as.matrix(iris.num), Rowv=rowv,Colv=colv,
                               trace="none")
```

7.8.2.2 Prototype-Based Clustering

The R-function kmeans carries out k-means clustering.

```
> iris.km <- kmeans(iris.num,centers=3)
```

The desired numbers of clusters is specified by the parameter centers. The location of the cluster centers and the assignment of the data to the clusters is obtained by the print-function:

```
> print(iris.km)
```

For fuzzy c-means clustering, the library cluster is required. The clustering is carried out by the method fanny similar to kmeans:

```
> library(cluster)
> iris.fcm <- fanny(iris.num,3)
> iris.fcm
```

The last line provides the necessary information on the clustering results, especially the membership degrees to the clusters.

Gaussian mixture decomposition automatically determining the number of clusters requires the library mlcust to be installed first:

```
> library(mclust)
> iris.em <- mclustBIC(iris.num[,1:4])
> iris.mod <- mclustModel(iris.num[,1:4],iris.em)
> summary(iris.mod)
```

The last line lists the assignment of the data to the clusters.

Density-based clustering with DBSCAN is implemented in the library fpc which needs installation first:

```
> library(fpc)
> iris.dbscan <- dbscan(iris.num[,1:4],1.0,showplot=T)
> iris.dbscan$cluster
```

The last line will print out the assignment of the data to the clusters. Singletons or outliers are marked by the number zero. The second argument in dbscan (in the above example 1.0) is the parameter ε for DBSCAN. showplot=T will generate a plot of the clustering result projected to the first two dimensions of the data set.

7.8.2.3 Self Organizing Maps

The library som provides methods for self organizing maps. The library som needs to be installed:

```
> library(som)
> iris.som <- som(iris.num,xdim=5,ydim=5)
> plot(iris.som)
```

xdim and ydim define the number of nodes in the mesh in *x*- and *y*-directions, respectively. plot will show, for each node in the mesh, a representation of the values in the form of parallel coordinates.

7.8.2.4 Association Rules

For association rule mining, the library arules is required in which the function apriori is defined. This library does not come along with R directly and needs to be installed first.

Here we use an artificial data set basket that we enter manually. The data set is a list of vectors where each vector contains the items that were bought:

```
> library(arules)
> basket <- list(c("a","b","c"), c("a","d","e"),
                 c("b","c","d"), c("a","b","c","d"),
                 c("b","c"), c("a","b","d"),
                 c("d","e"), c("a","b","c","d"),
                 c("c","d","e"), c("a","b","c"))
> rules <- apriori(baskets,parameter = list(supp=0.1,
                                            conf=0.8,
                                            target="rules"))
> inspect(rules)
```

The last command lists the rules with their support, confidence, and lift.

7.9 Further Reading

There are quite a number of books about cluster analysis. A good reference to start is [22]. A review of subspace clustering methods can be found in [43]. Self-organizing maps are mentioned in almost every book about neural networks, but only a few books are more or less dedicated to them [35, 45]. The usefulness of SOMs for visualization is shown in [51].

There were two workshops on frequent itemset mining implementations in 2003 and 2004 by Bart Goethals, Mohammed J. Zaki, and Roberto Bayardo, and the workshop proceedings are available online http://fimi.cs.helsinki.fi/. The website provides a comparison and the source code of many different implementations.

References

1. Aggarwal, C.C., Wolf, J.L., Yu, P.S., Procopiuc, C., Park, J.S.: Fast algorithms for projected clustering. In: Proc. 1999 ACM SIGMOD Int. Conf. on Management of Data, pp. 61–72. ACM Press, New York (1999)
2. Agrawal, R., Srikant, R.: Fast algorithms for mining association rules. In: Proc. 20th Int. Conf. on very Large Databases (VLDB 1994, Santiago de Chile), pp. 487–499. Morgan Kaufmann, San Mateo (1994)
3. Agrawal, R., Mannila, H., Srikant, R., Toivonen, H., Verkamo, A.: Fast discovery of association rules. In: Fayyad, U.M., Piatetsky-Shapiro, G., Smyth, P., Uthurusamy, R. (eds.): Advances in Knowledge Discovery and Data Mining, pp. 307–328. AAAI Press/MIT Press, Cambridge (1996)
4. Agrawal, R., Gehrke, J., Gunopulos, D., Raghavan, P.: Automatic subspace clustering of high dimensional data. Data Min. Knowl. Discov. 11, 5–33 (2005)
5. Ankerst, M., Breunig, M.M., Kriegel, H.-P., Sander, J.: OPTICS: ordering points to identify the clustering structure. In: ICMD, pp. 49–60, Philadelphia (1999)
6. Atzmueller, M., Puppe, F.: Sd-map: a fast algorithm for exhaustive subgroup discovery. In: Proc. Int. Conf. Knowledge Discovery in Databases (PKDD). Lecture Notes in Computer Science, vol. 4213. Springer, Berlin (2006)
7. Baumgartner, C., Plant, C., Kailing, K., Kriegel, H.-P., Kröger, P.: Subspace selection for clustering high-dimensional data. In: Proc. IEEE Int. Conf. on Data Mining, pp. 11–18. IEEE Press, Piscataway (2003)
8. Bayardo, R., Goethals, B., Zaki, M.J. (eds.): Proc. Workshop Frequent Item Set Mining Implementations (FIMI 2004, Brighton, UK), CEUR Workshop Proceedings 126, Aachen, Germany (2004). http://www.ceur-ws.org/Vol-126/
9. Bellman, R.: Adaptive Control Processes. Princeton University Press, Princeton (1961)
10. Böttcher, M., Spott, M., Nauck, D.: Detecting temporally redundant association rules. In: Proc. 4th Int. Conf. on Machine Learning and Applications (ICMLA 2005, Los Angeles, CA), pp. 397–403. IEEE Press, Piscataway (2005)
11. Böttcher, M., Spott, M., Nauck, D.: A framework for discovering and analyzing changing customer segments. In: Advances in Data Mining—Theoretical Aspects and Applications. Lecture Notes in Computer Science, vol. 4597, pp. 255–268. Springer, Berlin (2007)
12. Borgelt, C., Berthold, M.R.: Mining molecular fragments: finding relevant substructures of molecules. In: Proc. IEEE Int. Conf. on Data Mining (ICDM 2002, Maebashi, Japan), pp. 51–58. IEEE Press, Piscataway (2002)
13. Borgelt, C.: On canonical forms for frequent graph mining. In: Proc. 3rd Int. Workshop on Mining Graphs, Trees and Sequences (MGTS'05, Porto, Portugal), pp. 1–12. ECML/PKDD 2005 Organization Committee, Porto (2005)
14. Borgelt, C., Wang, X.: SaM: a split and merge algorithm for fuzzy frequent item set mining (to appear)
15. Branko, K., Lavrac, N.: Apriori-sd: adapting association rule learning to subgroup discovery. Appl. Artif. Intell. 20(7), 543–583 (2006)
16. Cheng, Y., Fayyad, U., Bradley, P.S.: Efficient discovery of error-tolerant frequent itemsets in high dimensions. In: Proc. 7th Int. Conf. on Knowledge Discovery and Data Mining (KDD'01, San Francisco, CA), pp. 194–203. ACM Press, New York (2001)
17. Cook, D.J., Holder, L.B.: Graph-based data mining. IEEE Trans. Intell. Syst. 15(2), 32–41 (2000)
18. Davé, R.N.: Characterization and detection of noise in clustering. Pattern Recognit. Lett. 12, 657–664 (1991)
19. Ding, C., He, X.: Cluster merging and splitting in hierarchical clustering algorithms. In: Proc. IEEE Int. Conference on Data Mining, p. 139. IEEE Press, Piscataway (2002)
20. Dunn, J.: Well separated clusters and optimal fuzzy partitions. J. Cybern. 4, 95–104 (1974)
21. Ester, M., Kriegel, H.-P., Sander, J., Xiaowei, X.: A density-based algorithm for discovering clusters in large spatial databases with noise. In: Proc. 2nd Int. Conf. on Knowledge Discovery and Data Mining (KDD 96, Portland, Oregon), pp. 226–231. AAAI Press, Menlo Park (1996)

22. Everitt, B.S., Landau, S., Leese, M.: Cluster Analysis. Wiley, Chichester (2001)

23. Finn, P.W., Muggleton, S., Page, D., Srinivasan, A.: Pharmacore discovery using the inductive logic programming system PROGOL. Mach. Learn. **30**(2–3), 241–270 (1998)

24. Gamberger, D., Lavrac, N.: Expert-guided subgroup discovery: methodology and application. J. Artif. Intell. Res. **17**, 501–527 (2007)

25. Goethals, B., Zaki, M.J. (eds.): Proc. Workshop Frequent Item Set Mining Implementations (FIMI 2003, Melbourne, FL, USA), CEUR Workshop Proceedings 90, Aachen, Germany (2003). http://www.ceur-ws.org/Vol-90/

26. Guha, S., Rastogi, R., Shim, K.: ROCK: a robust clustering algorithm for categorical attributes. Inf. Syst. **25**(5), 345–366 (2000)

27. Halkidi, M., Batistakis, Y., Vazirgiannis, M.: On clustering validation techniques. J. Intell. Inf. Syst. **17**(2–3), 107–145 (2001)

28. Han, J., Pei, H., Yin, Y.: Mining frequent patterns without candidate generation. In: Proc. Conf. on the Management of Data (SIGMOD'00, Dallas, TX), pp. 1–12. ACM Press, New York (2000)

29. Hinneburg, A., Keim, D.A.: An efficient approach to clustering in large multimedia satabases with noise. In: Proc. 4th Int. Conf. on Knowledge Discovery and Data Mining (KDD), pp. 224–228. AAAI Press, Menlo Park (1998)

30. Hinneburg, A., Keim, D.A.: Optimal grid-clustering: towards breaking the curse of dimensionality in high-dimensional clustering. In: Proc. 25th Int. Conf. on Very Large Databases, pp. 506–517. Morgan Kaufmann, San Mateo (1999)

31. Höppner, F.: Speeding up Fuzzy C-means: using a hierarchical data organisation to control the precision of membership calculation. Fuzzy Sets Syst. **128**(3), 365–378 (2002)

32. Höppner, F., Klawonn, F.: A contribution to convergence theory of fuzzy C-means and derivatives. IEEE Trans. Fuzzy Syst. **11**(5), 682–694 (2003)

33. Höppner, F., Klawonn, F., Kruse, R., Runkler, T.A.: Fuzzy Cluster Analysis. Wiley, Chichester (1999)

34. Huan, J., Wang, W., Prins, J.: Efficient mining of frequent subgraphs in the presence of isomorphism. In: Proc. 3rd IEEE Int. Conf. on Data Mining (ICDM 2003, Melbourne, FL), pp. 549–552. IEEE Press, Piscataway (2003)

35. Kaski, S., Oja, E., Oja, E.: Kohonen Maps. Elsevier, Amsterdam (1999)

36. Klösgen, W.: Efficient discovery of interesting statements in databases. J. Intell. Inf. Syst. **4**, 53–69 (1995)

37. Klösgen, W.: Explora: a multipattern and multistrategy discovery assistant. In: Advances in Knowledge Discovery and Data Mining. MIT Press, Cambridge (1996). Chap. 10

38. Kohonen, T.: The self-organizing map. Proc. IEEE **78**, 1464–1480 (1990)

39. Kramer, S., de Raedt, L., Helma, C.: Molecular feature mining in HIV data. In: Proc. 7th ACM SIGKDD Int. Conf. on Knowledge Discovery and Data Mining (KDD 2001, San Francisco, CA), pp. 136–143. ACM Press, New York (2001)

40. Kuramochi, M., Karypis, G.: Frequent subgraph discovery. In: Proc. 1st IEEE Int. Conf. on Data Mining (ICDM 2001, San Jose, CA), pp. 313–320. IEEE Press, Piscataway (2001)

41. Leman, D., Feelders, A., Knobbe, A.: Exceptional model mining. In: Proc. Europ. Conf. Machine Learning and Knowledge Discovery in Databases. Lecture Notes in Computer Science, vol. 5212, pp. 1–16. Springer, Berlin (2008)

42. Nijssen, S., Kok, J.N.: A quickstart in frequent structure mining can make a difference. In: Proc. 10th ACM SIGKDD Int. Conf. on Knowledge Discovery and Data Mining (KDD2004, Seattle, WA), pp. 647–652. ACM Press, New York (2004)

43. Parsons, L., Haque, E., Liu, H.: Subspace clustering for high dimensional data: a review. SIGKDD Explor. Newsl. **6**(1), 90–105 (2004)

44. Pei, J., Tung, A.K.H., Han, J.: Fault-tolerant frequent pattern mining: problems and challenges. In: Proc. ACM SIGMOD Workshop on Research Issues in Data Mining and Knowledge Discovery (DMK'01, Santa Babara, CA). ACM Press, New York (2001)

45. Ritter, H., Martinez, T., Schulten, K.: Neural Computation and Self-Organizing Maps: An Introduction. Addison-Wesley, Reading (1992)

46. Rousseeuw, P.J.: Silhouettes: a graphical aid to the interpretation and validation of cluster analysis. J. Comput. Appl. Math. **20**, 53–65 (1987)
47. Scheffer, T., Wrobel, S.: Finding the most interesting patterns in a database quickly by using sequential sampling. J. Mach. Learn. Res. **3**, 833–862 (2003)
48. Scholz, M.: Sampling-based sequential subgroup mining. In: Proc. 11th ACM SIGKDD Int. Conf. on Knowledge Discovery and Data Mining, pp. 265–274. AAAI Press, Menlo Park (2005)
49. Smyth, P., Goodman, R.M.: An information theoretic approach to rule induction from databases. IEEE Trans. Knowl. Discov. Eng. **4**(4), 301–316 (1992)
50. Steinbeck, C., Han, Y., Kuhn, S., Horlacher, O., Luttmann, E., Willighagen, E.: The chemistry development kit (CDK): an open-source Java library for chemo- and bioinformatics. J. Chem. Inf. Comput. Sci. **43**(2), 493–500 (2003)
51. Vesanto, J.: SOM-based data visualization methods. Intell. Data Anal. **3**(2), 111–126 (1999)
52. Wang, X., Borgelt, C., Kruse, R.: Mining fuzzy frequent item sets. In: Proc. 11th Int. Fuzzy Systems Association World Congress (IFSA'05, Beijing, China), pp. 528–533. Tsinghua University Press/Springer, Beijing/Heidelberg (2005)
53. Webb, G.I., Zhang, S.: k-Optimal-rule-discovery. Data Min. Knowl. Discov. **10**(1), 39–79 (2005)
54. Webb, G.I.: Discovering significant patterns. Mach. Learn. **68**(1), 1–33 (2007)
55. Wrobel, S.: An algorithm for multi-relational discovery of subgroups. In: Proc. 1st Europ. Symp. on Principles of Data Mining and Knowledge Discovery. Lecture Notes in Computer Science, vol. 1263, pp. 78–87. Springer, London (1997)
56. Yan, X., Han, J.: gSpan: graph-based substructure pattern mining. In: Proc. 2nd IEEE Int. Conf. on Data Mining (ICDM 2003, Maebashi, Japan), pp. 721–724. IEEE Press, Piscataway (2002)
57. Yan, X., Han, J.: Close-graph: mining closed frequent graph patterns. In: Proc. 9th ACM SIGKDD Int. Conf. on Knowledge Discovery and Data Mining (KDD 2003, Washington, DC), pp. 286–295. ACM Press, New York (2003)
58. Xie, X.L., Beni, G.A.: Validity measure for fuzzy clustering. IEEE Trans. Pattern Anal. Mach. Intell. (PAMI) **3**(8), 841–846 (1991)
59. Zaki, M., Parthasarathy, S., Ogihara, M., Li, W.: New algorithms for fast discovery of association rules. In: Proc. 3rd Int. Conf. on Knowledge Discovery and Data Mining (KDD'97, Newport Beach, CA), pp. 283–296. AAAI Press, Menlo Park (1997)
60. Zhang, T., Ramakrishnan, R., Livny, M.: BIRCH: a new data clustering algorithm and its applications. Data Min. Knowl. Discov. **1**(2), 141–182 (1997)
61. Zhao, Y., Karypis, G., Fayyad, U.: Hierarchical clustering algorithms for document datasets. Data Min. Knowl. Discov. **10**, 141–168 (2005)

Chapter 8
Finding Explanations

In the previous chapter we have discussed methods that find patterns of different shapes in data sets. All these methods needed measures of similarity in order to group similar objects. In this chapter we will discuss methods that address a very different setup: instead of finding structure in a data set, we are now focusing on methods that find explanations for an unknown dependency within the data. Such a search for a dependency usually focuses on a so-called target attribute, that is, we are particularly interested in why one specific attribute has a certain value. In case of the target attribute being a nominal variable, we are talking about a **classification problem**; in case of a numerical value we are referring to a **regression problem**. Examples for such problems would be understanding why a customer belongs to the category of people who cancel their account (e.g., classifying her into a yes/no category) or better understanding the risk factors of customers in general.

We are therefore assuming that, in addition to the object description **x**, we have access to a value for the target attribute Y. In contrast to the methods described in previous chapters, where no target value was available, we now aim to model a dependency towards one particular attribute. Since this can be compared to a teacher model, where the teacher gives us the desired output, this is often called **supervised learning**. Our dataset in this setup consists of tuples $\mathcal{D} = \{(\mathbf{x}_i, Y_i)|i = 1, \ldots, n\}$. Based on this data, we are interested in finding an explanation in the form of some type of interpretable model, which allows us to understand the dependency of the target variable to the input vectors. The focus of these models lies on their interpretability, and in a subsequent chapter we will focus on the quality of the forecasting or predictive ability. Here we are interested in supervised (because we know the desired outcome) and descriptive (because we care about the explanation) data analysis. It is important to emphasize immediately that this model will not necessarily express a causal relationship, that is, the true underlying cause of how the outcome depends on the inputs. All we can extract (automatically, at least) from the given data are relationships expressed by some type of numerical correlations.

This chapter will cover the main representatives of model-based explanation methods: decision trees, Bayes classifiers, regression models, and rule extraction methods. These four types of explanation finding methods nicely cover four rather different flavors.

M.R. Berthold et al., *Guide to Intelligent Data Analysis*,
Texts in Computer Science 42,
DOI 10.1007/978-1-84882-260-3_8, © Springer-Verlag London Limited 2010

Decision trees   Decision trees aim to find a hierarchical structure to explain how different areas in the input space correspond to different outcomes. The hierarchical way to partition this space is particularly useful for applications where we drown in a series of attributes of unknown importance: the final decision tree often only uses a small subset of the available set of attributes. Decision trees are often thrown first at classification problems because they tend to be insensitive to normalization issues and tolerant toward many correlated or noisy attributes. In addition, the structure of the tree also allows for data clean-up. Quite often, the first attempt at generating a decision tree reveals unexpected dependencies in the data which would otherwise be hidden in a more complex model.

Bayes classifiers   Bayes classifiers form a solid baseline for achievable classification accuracy—any other model should at least perform as well as the naive Bayes classifier. In addition they allow quick inspection of possible correlations between any given input attribute and the target value. Before trying to apply more complex models, a quick look at a Bayes classifier can be helpful to get a feeling for realistic accuracy expectations and simple dependencies in the data. Bayes classifiers express their model in terms of simple probabilities (or parameters of simple probability distributions, such as mean and standard deviation) and hence explain nicely how the classifier works.

Regression models   Regression models are the counterpart for numerical approximation problems. Instead of finding a classifier and minimizing the classification error, regression models try to minimize the approximation error, that is, some measure for the average deviation between the expected and predicted numerical output value. Again, many more complex models exist, but the regression coefficients allow easy access to the internals of the model. Each coefficient shows the dependency between any input attribute and the target value.

Rule models   Rule models are the most intuitive, interpretable representation. However, not many efficient or usable algorithms exist to date for complex real-world data sets. As a result, rule methods are generally not the first choice when it comes to finding explanations in complex data sets. We still discuss some of the most prominent methods in this chapter because the knowledge representation is best suited to offer insights into the data set and some of those methods deserve a bit more attention in data analysis setups. Generally, one would only apply rule extraction algorithms only to data set with a reasonably well-understood structure. Many of the algorithms tend to be rather sensitive toward useless or highly correlated attributes and excessive noise in the data.

8.1 Decision Trees

We start this chapter with likely the most prominent and heavily used methods for interpretable model extraction from large data sets: Decision Trees. Decision trees not only allow one to model complex relationships in an easy-to-understand manner, but additionally efficient algorithms exist to build such models from large datasets. Ross Quinlan's c4.5 [18] and CART [5] started the boom, and nowadays practically every data mining toolkit offers at least one algorithm to construct decision trees from data sets.

We will start by explaining what forms of a decision tree exist, then cover the most prominent training algorithm ID3 (which, with various extensions, ends up forming c4.5), continue by outlining some extensions and issues of decision trees, and end this section by showing how decision trees can be build in practice.

8.1.1 Overview

Decision Trees come in two flavors, classification and regression trees. We will first concentrate on the former, since classification problems are more common and most training algorithms and other issues can easily be generalized to regression problems as well. We will discuss issues particularly related to regression trees at the end of this section. Figure 8.1 shows a typical example for a classification tree. The simple tree can be used to classify animals, based on a number of attributes (the ability to swim and fly, in this case). As usual in computer science related areas, trees grow from top to bottom. The tree then builds a hierarchical decision structure which helps to understand the classification process by traversing the tree from the root node until a leaf is reached. At each intermediate node, the relevant attribute is investigated and the branch matching the attribute's value is followed. The leaves then hold the classifications. Obviously, the tree in our example does not work correctly for all types of animals—we have already discussed this issue of generalization and performance on unseen cases in Chap. 5. However, it is a reasonable compact representation of a generic classification mechanism for animals based on easy to observe attributes and—more importantly—it summarizes (most of) our data.

Note how the tree in Fig. 8.1 uses different types of splits. In practice there are really three options:

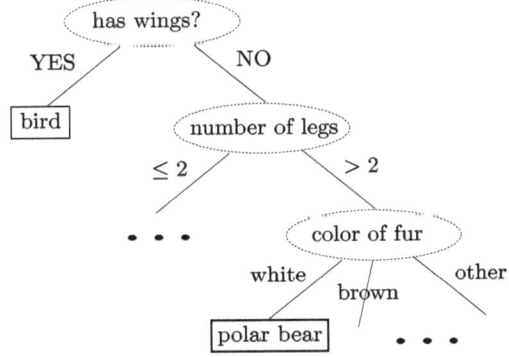

Fig. 8.1 A simple example for a decision tree to classify animals

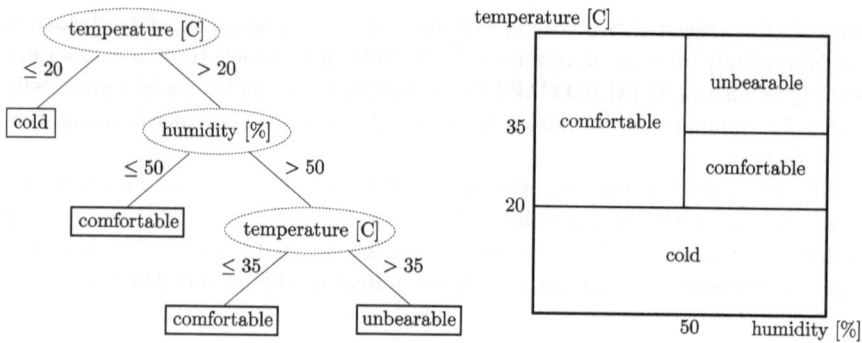

Fig. 8.2 A simple example for a decision tree using two numerical attributes (*left*) and the corresponding partitioning of the underlying feature space (*right*)

1. Boolean splits are considering boolean attributes and have two child nodes. An example for such a split would be an attribute "married" with two leaves "yes" and "no."
2. Nominal splits are based on nominal attributes and can be binary, that is, they split based on two disjoint subsets of the attribute values. Sometimes also splits with more children are considered—in the extreme for each nominal value one. An example for such a split would be connected to the boolean split "married=no" above and could split into "never", "divorced", and "widow(er)". Note that one can model such a split also with a sequence of nodes splitting the same attribute into increasingly smaller subsets. We will see in the section on decision tree construction how such potentially very wide splits can lead to rather useless trees.
3. Splits on continuous attributes finally use a numerical variable and can either split based on one particular value (e.g., "temperature" with two splits ≤80 F and >80 F) or on a series of values defining bins for each child.

Decision tree algorithms are also known as **recursive partitioning** methods since each split of the tree essentially divides the remaining space into two (or more in case of not binary splits) disjoint subpartitions. Figure 8.2 shows an example of a tree operating on two numerical attributes and illustrates the corresponding partitioning of the underlying feature space.

Decision trees are well received by users because they are easy to read and interpret. Additionally, the recursive partitioning approach seems to resemble the human like hierarchical structuring of the description of a classification problem.

8.1.2 Construction

From a data analysis point of view it is now of course interesting to know how we can generate such simple, easy to understand structures from real-world data sets.

Table 8.1 The basic decision tree induction algorithm

Algorithm BuildDecisionTree($\mathcal{D}, \mathcal{A}$)	

input: training data $\mathcal{D}$, set $\mathcal{A}$ of available attributes
output: a decision tree matching $\mathcal{D}$, using all or a subset of $\mathcal{A}$

1	**if** all elements in $\mathcal{D}$ belong to one class
2	**return** node with corresponding class label
3	**elseif** $\mathcal{A} = \emptyset$
4	**return** node with majority class label in $\mathcal{D}$
5	**else**
6	select attribute $A \in \mathcal{A}$ which best classifies $\mathcal{D}$
7	create new node holding decision attribute A
8	**for** each split v_A of A
9	add new branch below with corresponding test for this split
10	create $\mathcal{D}(v_A) \subset \mathcal{D}$ for which split condition holds
11	**if** $\mathcal{D}(v_A) = \emptyset$
12	**return** node with majority class label in $\mathcal{D}$
13	**else**
14	add subtree returned by calling
	BuildDecisionTree($\mathcal{D}(v_A), \mathcal{A}\backslash\{A\}$)
15	**endif**
16	**endfor**
17	**return** node.
18	**endif**

Finding the optimal decision tree for a given set of training examples is nontrivial, but in most cases it is sufficient to find a reasonably small tree, which explains the training data well. Note also that "optimal" is not obvious to define: do we mean the smallest tree or the one with the best accuracy on the training data. And if we are indeed interested in the smallest tree, does this relate to the number of nodes or the depth or width of the tree? At the end of this chapter we will discuss this issue of "learning bias" in more detail.

The most prominent algorithms therefore do not attempt to optimize a global measure but employ a greedy strategy, that is, they focus on building the tree root-first and then add subsequent branches and splits along those branches for the remainder of the training data—they recursively find the best split at each point. Generally, such an algorithm looks like the one shown in Table 8.1, where $\mathcal{D}$ indicates the available training examples, $\mathcal{C}$ the target attribute, and $\mathcal{A}$ the set of available input attributes.

The algorithm recursively splits up the data and constructs the decision tree starting with its root node. The recursion stops at steps 1 or 3 if a subset of patterns of only one class is encountered (the resulting tree is then simply a leaf carrying that class label) or no further attributes for splits are available (the leaf then carries the label of the majority class in the subset). Otherwise we find the best split for the subset at hand (line 6) and create a new split node. We then recursively call the tree construction method on the subsets created by applying the chosen split (lines 8–14).

Looking at this algorithm, two issues remain unclear:

- How do we determine which of the attributes is the "best" one?
- Which kind of splits are possible?

The second question is already answered above: for nominal attributes, we can either carve out one value, resulting in a binary split or split the set of values into two or more subsets ultimately leading to a split into as many branches as there exist attribute values. (We will discuss other types of attributes later.)

If we focus on the last option, we are dealing with ID3, the classic decision tree learning algorithm [17],[1] at developing such an algorithm. ID3 only constructs decision trees for training data with nominal values and considers solely wide splits, that is, splits into all possible values of an attribute in a single node. This makes the choice of possible splits simpler (only one choice exists); however, we still do not know how to pick the attribute resulting in the "best" split.

Intuitively, a best split would be one which helps us create leaves as soon as possible. Leaves are created only if we reach a branch holding a subset with training patterns, all belonging to one class (or, of course, if the subset is empty, but that is an extreme case). So we need to make sure that we quickly create subsets that have mostly patterns of one class. An often used measure for such an **impurity** is the Shannon Entropy, which measures the impurity of a labeled set, based on the column of labels $\mathcal{C}$:

$$H_{\mathcal{D}}(\mathcal{C}) = - \sum_{k \in \mathrm{dom}(\mathcal{C})} \frac{|\mathcal{D}_{\mathcal{C}=k}|}{|\mathcal{D}|} \log \frac{|\mathcal{D}_{\mathcal{C}=k}|}{|\mathcal{D}|}$$

with $0 \log 0 := 0$. The entropy ranges from 0 to 1 and is maximal for the case of two classes and an even $50 : 50$ distribution of patterns of those classes. An entropy of exactly 0 tells us that only patterns of the same class exist. Entropy therefore provides us with a measure of how impure (with respect to the class variable $\mathcal{C}$) a dataset is.

We can now determine what the best attribute at any given point is: it is the attribute A with the biggest reduction in entropy compared to the original set (recall that our aim is to find pure leaves). We call this reduction also **information gain**:

$$I_{\mathcal{D}}(\mathcal{C}, A) = H_{\mathcal{D}}(\mathcal{C}) - H_{\mathcal{D}}(\mathcal{C}, A),$$

where

$$H_{\mathcal{D}}(\mathcal{C}, A) = \sum_{a \in \mathrm{dom}(A)} \frac{|\mathcal{D}_{A=a}|}{|\mathcal{D}|} H_{\mathcal{D}_{A=a}}(\mathcal{C}),$$

and $\mathcal{D}_{A=a}$ indicates the subset of $\mathcal{D}$ for which attribute A has value a. $H_{\mathcal{D}}(\mathcal{C}, A)$ then denotes the entropy that is left in the subsets of the original data after they have been split according to their values of A. It is interesting to note that the information

---

[1] Rumors say that ID3 stands for "Iterative Dichotomiser 3" (from Greek dichotomia: divided), supposedly it was Quinlans' third attempt. Another interpretation of one of the authors is "Induction of Decision 3rees."

gain can never be negative, that is, no matter which split we choose, the entropy spread out over the resulting subsets is not going to increase.

However, what happens if we have an attribute ID with unique values for every single example pattern and we use these kinds of splits and the entropy to measure the information gain? Simple, the attribute holding the unique IDs will be chosen first because it results in a node with n branches for $n = |\mathcal{D}|$ training examples: each example has its individual ID resulting in its own branch. The entropy of the sets passed down to all of these branches goes down to zero (since only one element of one class is contained in each set), and the information gain is maximized. This is clearly not desirable in most applications. The reason for choosing this odd split is that we are reducing the entropy at all costs, completely ignoring the costs of actually making the split. In order to compensate for this we can also compute the **split information**:

$$SI_{\mathcal{D}}(A) = - \sum_{a \in \text{dom}(A)} \frac{|\mathcal{D}_{A=a}|}{|\mathcal{D}|} \log \frac{|\mathcal{D}_{A=a}|}{|\mathcal{D}|},$$

which computes the entropy of distribution of our original set into subsets, and use this to normalize the purely entropy driven information gain:

$$GR_{\mathcal{D}}(\mathcal{C}, A) = \frac{I_{\mathcal{D}}(\mathcal{C}, A)}{SI_{\mathcal{D}}(A)},$$

resulting in GR, the **gain ratio** of a given split. The gain ratio penalizes very wide splits, that is, splits with many branches, and biases the selection toward more narrow splits.

Of course, building decision trees on nominal values only is of limited interest. In the following we will touch upon some of the extensions to address this and other limitations. However, the main underlying algorithm always remains the same— a greedy search which attempts to locally maximize some measure of information gain over the available attributes and possible splits at each stage until a certain criteria is reached.

8.1.3 Variations and Issues

Decision trees are a widely researched topic in machine learning and data analysis. There are numerous extensions and variants around which various shortcomings or specific types of data and applications address. In the following we try to briefly summarize some of the more general variations.

8.1.3.1 Numerical Attributes

As mentioned above, the biggest limitation of ID3 is the focus on nominal attributes only. In order to extend this to numerical values, let us first explore possible splits

Fig. 8.3 A numerical attribute together with class values and a few possible splits

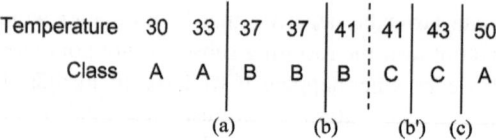

for such variables and worry about how to actually find those splits a tad later. A small example for a numerical attribute (top row) and the corresponding class values (bottom row) is shown in Fig. 8.3.

For convenience, we have sorted the training instances (columns showing values of one input and one class value in this case) according to the numerical attribute. Clearly, splits can divide the range of the numerical attribute "temperature" only into two (or more) ranges. Hence, looking at the sorted series of training instances, we can really only divide this into two pieces, a set of patterns to the left of the "cut" and a set to the right. One could imaging performing more than one split at each branching point in the tree, but this can easily be modeled by a series of binary splits. So in the following we will concentrate on the binary split setup.

In addition, in the case of a decision tree, we need to care about splits that occur within class boundaries only—why is that? An intuitive explanation is simple: if we already have instances of one class (e.g., C) in one branch of our considered split, it really does not make sense to assign any patterns of that class to the other branch, that is, splitting a uniform subset into even smaller subsets never gives us any advantage. This, of course, depends on the chosen measure of information gain, but in most cases, this assumption holds.

Therefore the only split points we need to consider are the splits indicated in Fig. 8.3. Note that even though we would like to split at the dashed lined (i.e., in between the patterns of class B and class C), we cannot do this since the value of temperature is 41 for both instances. So in this particular case we have to consider neighboring splits, e.g., (b) and (b'). This results in four possible splits that we need to investigate further. And suddenly, the problem can be converted back into the issue of looking into finding the one best nominal split: we can simply define four binary variables `temp_leq_35`, `temp_leq_39`, `temp_leq_42`, and `temp_leq_46.5` that take on values of `true` resp. `false` when the value of the original attribute `temperature` lies below resp. above the specific threshold. We can then apply the strategies described in the previous section to determine information gains for each one of these attributes and compare them with other possible splits on nominal or numerical attributes.

Note that using this approach, a numerical attribute may appear several times along the branch of a decision tree, each time being split at a different possible split point. This may make the tree somewhat harder to read. One could circumvent this by adding nominal values that split the numerical attributes in class-wise pure ranges only, but especially for somewhat noisy attributes, this will result in many small bins and subsequently many nominal values for the resulting variable. As we have seen above, such splits are favored by pure entropy-based information gains and discouraged by other measures. Either way, these splits will be hard to interpret as well. So for many practical applications, it is preferable to restrict splits on numerical attributes to binary divisions.

8.1.3.2 Numerical Target Attributes: Regression Trees

In addition to handling numerical attributes for the feature vector, one is often also interested in dealing with numerical target values. The underlying issue then turns from predicting a class (or the probability of an input vector belonging to a certain class) to the problem of predicting a continuous output value. We call this regression, and, in case of decision trees, the resulting tree is called a **regression tree**. At the same time when Quinlan was discussing algorithms for construction of classification tree, Breiman and his colleagues described **CART** (Classification And Regression Trees, see [5]).[2] The structure of such a tree is the same as that of the trees discussed above; the only difference is the leaves: instead of class labels (or distributions of class labels), the leaves now contain numerical constants. The tree output is then simply the constant corresponding to the leaf the input pattern ends up in. So the regression process is straightforward, but what about constructing such trees from data? Instead of an entropy-based measurement which aims to measure class impurity, we now need to find a measure for the quality of fit of a tree, or at least a branch. For this, we can use the sum of squares measurement, discussed earlier. The sum of squares for a specific node n may then simply be defined as the squared mean error of fit:

$$SME(\mathcal{D}_n) = \frac{1}{|\mathcal{D}_n|} \sum_{(\mathbf{X},Y)\in\mathcal{D}_n} (Y - f(x))^2,$$

where $f(n)$ indicates the constant values assigned to the node n, and $\mathcal{D}_n$ are the data points ending up at the node. Using this quality measure, we can again apply our decision tree construction algorithm and continue analogously to the classification case: we greedily investigate all possible splits, determine the error on the respective splits, combine them (weighted by the respective sizes of the subsets, of course), and choose the split which results in the biggest reduction of error.

More sophisticated versions of regression trees, so-called **model trees**, not only allow for constants in the leaves but more complex functions on all or a subset of the input variables. However, these trees are much less commonly used, since normal regression trees are usually sufficient to approximate regression problems. In effect, model trees allow one to move some of the complexity from the tree structure into the functions in each leaf, resulting in smaller trees with leaves that are harder to interpret.

---

[2]Quinlan later also developed methods for regression problems, similar to CART.

Fig. 8.4 A decision tree (*middle*) with one node (red cross) pruned by subtree replacement (*left*) and subtree raising (*right*)

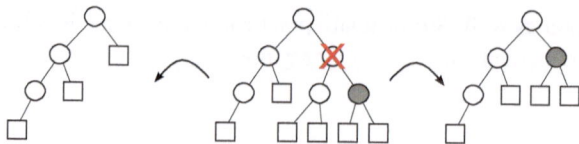

8.1.3.3 Pruning

So far we have not really worried much about when to stop growing a tree. Driven to the extreme, the base algorithm described in Sect. 8.1.2 simply continues until the tree cannot be refined any further, that is, no split attributes remain to be used, or the dataset at a node contains one pattern or only patterns of the same class resp. output value. As we have discussed, this behavior is most likely not desirable since it leads to overfitting and, in this particular case will also make the interpretation of the resulting tree unnecessarily complex.

This is why pruning of decision trees gains importance quickly in real-world applications. Pruning decision trees comes in two variants:

- **prepruning** refers to the process of stopping the construction of the decision tree already during the training process, that is, specific splits will not be added since they do not offer sufficient advantage even though they would produce a (albeit small) numerical improvement;
- **postpruning** refers to the reduction of a decision tree which was build to a clearly overly large size. Here we distinguish two variants; one replaces subtrees with leaves (subtree replacement), and the other one also removes nodes from within the tree itself (subtree raising). The latter is substantially more complex since the instances contained in a node need to be redistributed among the branches of the raised subtree, and its benefits are questionable since in practice trees rarely become large enough to make this approach beneficial.

Figure 8.4 illustrates both approaches.

The important question is, independent of when we prune, how do we determine if a node can be removed (or should not be split in the first place)? Most commonly, one splits the available data into a training and validation data set, where the training set is used to determine possible splits, and the validation data then helps to estimate the split's actual usefulness. Often a sufficient amount of training data is, however, not available, and one needs to rely on other measures. One possibility is statistical tests that help to estimate if expanding a node will likely bring a significant improvement. Quinlan proposes a heuristic which is based on the training data and computes confidence intervals for each node. From these, a standard Bernoulli-based estimate (see Sect. A.3 in the appendix) is used to determine which nodes can be pruned. The statistical foundation of this method is somewhat shaky, but it works well in practice. Other ways for pruning are based on measures for complexity, similar to the split information we discussed above. The Minimum Description Principle (see Sect. 5.5.4.1) can help to weight the expected improvement in accuracy over the required enlargement of the model.

There is also a variant to post pruning which deserves to be mentioned at least briefly: rule post pruning. We can easily convert a decision tree into a set of independent rules by simply creating one rule for each leaf of the final decision tree: the rule will then consist of a conjunction of conditions found along the branch back to the root of the tree. We can then apply rule pruning techniques where all conditions in a rule body which do not affect accuracy of the tree are removed. Note that the final set of rules then needs to be sorted somehow since the conditions of all rules are not necessarily disjunct anymore. One usually sorts them by their estimated accuracy on unseen validation data. We discuss rule generation and processing methods later in this chapter in more detail.

8.1.3.4 Missing Values

Decision trees make it relatively easy to deal with missing values. Instead of simply ignoring rows which contain missing values, we can make use of the remaining information both during training and classification. What all approaches for missing value handling during training have in common is that they use more or less sophisticated ways to estimate the impact of the missing value on the information gain measure. The most basic one simply adds a fraction of each class to each partition if the split attribute's value is missing for that record. During classification, dealing with missing attribute values is straightforward: special treatment is only required if, during tree traversal, a node relies on the value which is missing in the pattern to be classified. Then the remainder of the tree traversal can be simply done in both branches and later on, the results that are encountered in the two (or more, if more than one missing values was encountered) leaves are merged.

8.1.3.5 Additional Notes

Decision trees are one of the most well-known and prominently used examples of more sophisticated data analysis methods. However, one often forgets that they are notoriously unstable. **Stability** means that small changes to the training data, such as removing just one example, can result in drastic changes in the resulting tree. This is mostly due to the greedy nature of the underlying algorithm, which never reconsiders chosen splits but is also part of the general nature of the tree structure: two very similar but not necessarily highly correlated attributes can exchange roles in a decision tree when a few training examples are added or removed, in turn affecting the rest of the tree dramatically.

When better, more stable performance is needed but interpretation is not such an issue, one often refers to **forests of decision trees** or decision stumps, which belong to the family of ensemble methods: instead of building one large tree, a set of differently initialized, much smaller decision trees ("stumps") are created, and the classification (or regression) output is created by committee voting. Forests of trees have the advantage that they are more stable than classical decision trees and often show superior generalization performance. This type of model class can also be used for feature selection wrapper methods, such as backward feature elimination as discussed in Sect. 6.1.1.

There are many other variations of decision trees around: for instance, **fuzzy decision trees** [12] which allow one to process imprecise data and handle degrees of class membership. Other variations allow one to include costs of attributes, that is, they allow one to consider during the coding phase that different attribute values may be obtainable at different costs. The result is a decision tree which attempts to optimize both the quality and the expected cost of processing new instances.

8.2 Bayes Classifiers

Bayes classifiers exploit Bayes' rule and some simplifying assumptions in order to model the conditional probability distributions of different classes given the values of the descriptive attributes. They predict, for each new instance, the most probable class based on this probability model. Bayes classifiers differ mainly in the kind and the structure of the simplifying assumptions that are made.

8.2.1 Overview

Let us consider the question: what is the best possible classifier? The obvious and immediate answer seems to be: a classifier that always predicts the correct class, that is, the class the instance under consideration actually belongs to. Although this is certainly the ideal we strive for, it is rarely possible to achieve it in practice.

A fundamental obstacle is that an instantiation of the available descriptive attributes rarely allows us to single out one class as the obtaining one. Rather, there are usually several possible classes, either because there exist one or more hidden (unobserved) attributes that influence the class an instance belongs to, or because there is some random influence which cannot be predicted in principle (see Sect. 5.4 for a more detailed discussion).

Such a situation may be reflected in the available data by the fact that there are contradictory instances, that is, instances that coincide in their values for all descriptive attributes but differ in the class they belong to. Note, however, that the absence of contradictory instances does not guarantee that a perfect classifier is possible: the fact that no contradictory instances were observed (up to now) does not imply that they are impossible. Even if the training data set does not contain contradictory instances, it may still be that future cases, for which we want to predict the class, contradict each other or contradict instances from the training data set.

In a situation where there exists no unique class that can be predicted with certainty, but we still have to predict a single class (and not only a set of possible classes), the best we can do is to predict the class that has the highest probability (provided, of course, that all misclassification costs are equal—unequal misclassification costs are discussed in Sect. 8.2.3). The reason is that this scheme obviously yields (at least on average) the highest number or rate of correct predictions.

If we try to build a classifier that predicts, for any instantiation of the available descriptive attributes, the most probable class, we face two core problems: (1) how can we properly estimate which is the most probable class for a given instantiation? and (2) how can we store all of these estimates in a feasible way, so that they are easily accessible whenever we have to classify a new case? The second problem is actually a more fundamental one: if we are able to estimate, but cannot store efficiently, obtaining the estimates is obviously pointless.

Unfortunately, the most straightforward approach, namely simply storing the most probable class or a probability distribution over the classes for each possible instance is clearly infeasible: from a theoretical point of view, a single metric attribute would give rise to a supercountably infinite number of possible instances. In addition, even if we concede that in practice metric attributes are measured with finite precision and thus can have only a finite number of values, the number of possible instances grows exponentially with the number of attributes.

As a consequence, we have to introduce simplifying assumptions. The most common is to assume that the descriptive attributes are conditionally independent given the class (see Sect. A.3.2.3 in the appendix for a definition of this notion), thus reducing the number of needed parameters from the product of the domain sizes to their sum times the number of classes. However, the disadvantage is that this assumption is very strong and not particularly realistic. Therefore the result is also known as the **naive** or even the **idiot's Bayes classifier**. Despite this pejorative name, naive Bayes classifiers perform very well in practice and are highly valued in domains in which large numbers of descriptive attributes have to be taken into account (for example, chemical compound and text document classification).

Other simplifying assumptions concern metric (numeric) attributes. They are usually treated by estimating the parameters of one conditional distribution function per class—most commonly a normal distribution. Of course, this approach limits how well the actual distribution can be fitted but has the advantage that it considerably reduces the number of needed parameters. As a consequence, it may even become possible to abandon the naive conditional independence assumption: if we model all metric attributes jointly with one (multivariate) normal distribution per class, the number of parameters is merely quadratic in the number of (metric) attributes. The result is known as the **full Bayes classifier**. Note, however, that this applies only to metric (numeric) attributes. If categorical attributes are present, assuming conditional independence or something similar may still be necessary.

Extensions of the basic approach deal with mitigating the objectionable conditional independence assumption, selecting an appropriate subset of descriptive attributes to simplify the classifier, or incorporating misclassification costs.

8.2.2 Construction

As outlined in the preceding section, we want to construct a classifier that predicts the most probable class, that is, a classifier that computes its prediction as

$$\text{pred}(\mathbf{x}) = \arg\max_{y \in \text{dom}(Y)} P(y \mid \mathbf{x}).$$

Here $\mathbf{x}$ is an instantiation of the descriptive attributes, and Y is the class attribute. Exploiting **Bayes' rule**, we can rewrite this formula as[3]

$$\text{pred}(\mathbf{x}) = \arg\max_{y \in \text{dom}(Y)} \frac{P(\mathbf{x} \mid y)P(y)}{P(\mathbf{x})}.$$

Since $P(\mathbf{x})$ is independent of the class y, this formula is clearly equivalent to

$$\text{pred}(\mathbf{x}) = \arg\max_{y \in \text{dom}(Y)} P(\mathbf{x} \mid y)P(y).$$

Note that this simplification is possible even if we desire not only to predict the most probable class, but want to report its probability as well. The reason is that by exploiting the **law of total probability** we can compute $P(\mathbf{x})$ for any $\mathbf{x}$ as

$$P(\mathbf{x}) = \sum_{y \in \text{dom}(Y)} P(\mathbf{x} \mid y)P(y).$$

This leads to what is also known as the **extended Bayes' rule**:

$$P(y \mid \mathbf{x}) = \frac{P(\mathbf{x} \mid y)P(y)}{\sum_{z \in \text{dom}(Y)} P(\mathbf{x} \mid z)P(z)}.$$

It shows that in order to find the class probabilities, it suffices to compute $P(\mathbf{x} \mid y)P(y)$ for all $y \in \text{dom}(Y)$, and we have to do this anyway in order to find the maximum.

In the formula $\text{pred}(\mathbf{x}) = \arg\max_{y \in \text{dom}(Y)} P(\mathbf{x} \mid y)P(y)$ the factor $P(y)$ is clearly manageable: since the number of classes y is limited, we can store their probabilities explicitly. The problem resides with the other factor, that is, with $P(\mathbf{x} \mid y)$: even if all attributes are categorical, storing the $|\text{dom}(Y)| \cdot \prod_{X \in \mathbf{X}} |\text{dom}(X)|$ needed probabilities cannot be accomplished unless the number of attributes is *very* small.

In order to reach a manageable number of parameters, we make the crucial (but naive) assumption that all descriptive attributes are conditionally independent given the class (see Sect. A.3.3.4 in the appendix for a more detailed treatment of conditional independence). This allows us to rewrite the classification formula as

$$\text{pred}(\mathbf{x}) = \arg\max_{y \in \text{dom}(Y)} P(y) \prod_{X \in \mathbf{X}} P(x \mid y),$$

---

[3] Since one or more of them may be metric, we may have to use a probability density function f to refer to descriptive attributes: $f(\mathbf{x} \mid y)$. However, we ignore such notational subtleties here.

where x is the element of the data tuple $\mathbf{x}$ that refers to the descriptive attribute X. This is the core classification formula of the **naive Bayes classifier**.

If an attribute X is categorical, the corresponding factor $P(x \mid y)$ is fairly easily manageable: it takes only $|\operatorname{dom}(X)| \cdot |\operatorname{dom}(Y)|$ parameters to store it explicitly. However, if X is metric, a distribution assumption is needed to store the corresponding conditional probability density $f(x \mid y)$. The most common choice is a normal distribution $f(x \mid y) = N(\mu_{X|y}, \sigma^2_{X|y})$ with parameters $\mu_{X|y}$ (mean) and $\sigma^2_{X|y}$ (variance).

Once we have the above classification formula, estimating the model parameters becomes very simple. Given a data set $\mathbf{D} = \{(\mathbf{x}_1, y_1), \ldots, (\mathbf{x}_n, y_n)\}$, we use

$$\forall y \in \operatorname{dom}(Y): \quad \hat{P}(y) = \frac{\gamma + n_y}{\gamma |\operatorname{dom}(Y)| + n}$$

as an estimate for the prior class probability[4] and

$$\forall y \in \operatorname{dom}(Y): \forall x \in \operatorname{dom}(X): \quad \hat{P}(x|y) = \frac{\gamma + n_{xy}}{\gamma |\operatorname{dom}(X)| + n_y}$$

as an estimate for the conditional probabilities of the values of a categorical descriptive attribute X given the class. The n_y and n_{xy} in these formulas are defined as

$$n_y = \sum_{i=1}^{n} \tau(y_i = y) \quad \text{and} \quad n_{xy} = \sum_{i=1}^{n} \tau(\mathbf{x}_i[X] = x \wedge y_i = y),$$

where $\mathbf{x}_i[X]$ is the value that attribute X has in the ith tuple, and y_i is the class of the i-tuple (that is, the value of the class attribute Y). The function τ is a kind of truth function, that is, $\tau(\varphi) = 1$ if φ is true and 0 otherwise. Hence, n_y is simply the number of sample cases belonging to class y, and n_{xy} is the number of sample cases in class y for which the attribute X has the value x.

Finally, γ is a constant that is known as **Laplace correction**. It may be chosen as $\gamma = 0$, thus reducing the procedure to simple maximum likelihood estimation (see Sect. A.4.2.3 in the appendix for more details on maximum likelihood estimation). However, in order to appropriately treat categorical attribute values that do not occur with some class in the given data set but may nevertheless be possible, it is advisable to choose $\gamma > 0$. This also renders the estimation more robust, especially for a small sample size (small data set or small number of cases for a given class). Clearly, the larger the value of γ, the stronger the tendency toward a uniform distribution. Common choices are $\gamma = 1$, $\gamma = \frac{1}{2}$, or $\gamma = \frac{\gamma_0}{|\operatorname{dom}(X)| \cdot |\operatorname{dom}(Y)|}$, where γ_0 is a user-specified constant that is known as the **equivalent sample size** (since it can be justified with an argument from Bayesian statistics, where it represents the weight of the prior distribution, measured as the size of a sample having the same effect).

---

[4]This is a *prior* probability, because it describes the class probability *before* observing the values of any descriptive attributes.

For a metric (numeric) attribute X, a (conditional) normal distribution is commonly assumed (see above), that is, it is assumed that

$$f(x \mid y) = N\left(\mu_{X|y}, \sigma^2_{X|y}\right) = \frac{1}{\sqrt{2\pi}\,\sigma_{X|y}} \exp\left(-\frac{(x - \mu_{X|y})^2}{2\sigma^2_{X|y}}\right).$$

The parameters $\mu_{X|y}$ and $\sigma^2_{X|y}$ of this conditional density function are estimated as

$$\hat{\mu}_{X|y} = \frac{1}{n_y} \sum_{i=1}^{n} \tau(y_i = y) \cdot x_i[X] \quad \text{and}$$

$$\hat{\sigma}^2_{X|y} = \frac{1}{n'_y} \sum_{i=1}^{n} \tau(y_i = y) \cdot \left(x_i[X] - \hat{\mu}_{X|y}\right)^2,$$

where either $n'_y = n_y$ (maximum likelihood estimator) or $n'_y = n_y - 1$ (unbiased estimator for the variance) is chosen. (See Sect. A.4.2 in the appendix for more details about estimators and their desirable properties.)

As already briefly mentioned above, the normal distribution assumption also permits us to treat all metric attributes jointly if we assume that their distribution can be modeled with a multivariate normal distribution, that is,

$$f(\mathbf{x}_M \mid y) = \frac{1}{(2\pi)^{\frac{|\mathbf{X}_M|}{2}} \sqrt{|\Sigma_{\mathbf{X}_M|y}|}} \exp\left(-\frac{1}{2} (\mathbf{x}_M - \mu_{\mathbf{X}_M|y})^\top \Sigma^{-1}_{\mathbf{X}_M|y} (\mathbf{x}_M - \mu_{\mathbf{X}_M|y})\right),$$

where $\mathbf{X}_M$ is the set of metric attributes, and $\mathbf{x}_M$ is the vector of values of these attributes. The (class-conditional) mean vectors $\mu_{\mathbf{X}_M|y}$ and the corresponding covariance matrices $\Sigma_{\mathbf{X}_M|y}$ can be estimated from a given data set $\mathbf{X}$ as

$$\forall y \in \text{dom}(Y): \quad \hat{\mu}_{\mathbf{X}_M|y} = \frac{1}{n_y} \sum_{i=1}^{n} \tau(y_i = y) \cdot x_i[\mathbf{X}_M]$$

$$\text{and} \quad \hat{\Sigma}_{\mathbf{X}_M|y} = \frac{1}{n'_y} \sum_{i=1}^{n} \tau(y_i = y)$$

$$\times \left(x_i[\mathbf{X}_M] - \mu_{\mathbf{X}_M|y}\right)\left(x_i[\mathbf{X}_M] - \mu_{\mathbf{X}_M|y}\right)^\top,$$

with either $n'_y = n_y$ or $n'_y = n_y - 1$ (as above). If all attributes are metric, the result is

$$\text{pred}(\mathbf{x}) = \arg\max_{y} P(y) f(\mathbf{x} \mid y),$$

that is, the core classification formula of the **full Bayes classifier**.

As a consequence, the classification formula for a **mixed Bayes classifier** is

$$\text{pred}(\mathbf{x}) = \arg\max_{y} P(y) f(\mathbf{x}_M \mid y) \prod_{X \in \mathbf{X}_C} P(x \mid y),$$

Table 8.2 A naive Bayes classifier for the iris data. The class-conditional normal distributions are described by $\hat\mu \pm \hat\sigma$ (that is, expected value $\pm$ standard deviation)

Iris type	Iris setosa	Iris versicolor	Iris virginica
Prior probability	0.333	0.333	0.333
Petal length	1.46 ± 0.17	4.26 ± 0.46	5.55 ± 0.55
Petal width	0.24 ± 0.11	1.33 ± 0.20	2.03 ± 0.27

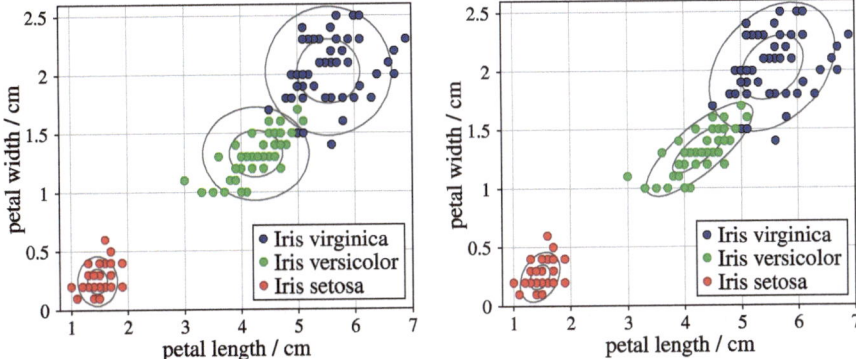

Fig. 8.5 Naive Bayes density functions for the Iris data (axes-parallel ellipses, *left*) and density functions that take the covariance of the two measures into account (general ellipses, *right*). The ellipses are the $1\hat\sigma$- and $2\hat\sigma$-boundaries (lines of equal probability density)

where $\mathbf{X}_C$ is the set of categorical attributes. In a mixed Bayes classifier the metric attributes are modeled with a multivariate normal distribution (as in a full Bayes classifier), while the categorical attributes are treated with the help of the conditional independence assumption (as in a naive Bayes classifier).

In order to illustrate the effect of the conditional independence assumption, we consider a naive Bayes classifier for the well-known iris data [2, 8]. The goal is to predict the iris type (Iris setosa, Iris versicolor, or Iris virginica) from measurements of the length and width of the petals and sepals. Here we confine ourselves to the measurements of the petals, which are most informative w.r.t. the class. The parameters of a naive Bayes classifier, derived with a normal distribution assumption, are shown in Table 8.2, and its graphical illustration in Fig. 8.5 on the left. In addition, Fig. 8.5 contains, on the right, an illustration of a full Bayes classifier, in which the two descriptive attributes (petal length and width) are modeled with class-conditional bivariate normal distributions.

These diagrams show that the conditional independence assumption is expressed by the orientation of the ellipses, which are lines of equal probability density: in a naive Bayes classifier, the major axes of these ellipses are always parallel to the co-ordinate axes, while a full Bayes classifier allows them to have arbitrary directions. As a consequence, the dependence of the attributes petal length and width can be modeled better, especially for Iris versicolor.

8.2.3 Variations and Issues

Bayes classifiers have been extended in several ways, for example, by mitigating the conditional independence assumption (Sect. 8.2.3.3) and by taking misclassification costs into account (Sect. 8.2.3.5). In addition, we argue in this section why naive Bayes classifiers often perform very well despite the strong independence assumptions they make (Sect. 8.2.3.1), study how full Bayes classifiers are related to linear discriminant analysis (Sect. 8.2.3.2), and consider how missing values are handled in Bayes classifiers (Sect. 8.2.3.4).

8.2.3.1 Performance

The naive assumption that the descriptive attributes are conditionally independent given the class attribute is clearly objectionable: it is rarely (exactly) satisfied in practice. Even in the simple iris classification example studied above it does not hold. Nevertheless, naive Bayes classifiers usually perform surprisingly well and are often not much worse than much more sophisticated and complicated classifiers.

The reasons for this behavior are investigated in detail in [7], where it is revealed that the good performance of naive Bayes classifiers is less surprising than one may think at first sight. In a nutshell, this behavior is due to the fact that a classifier is usually evaluated with **accuracy** or **0–1 loss**, that is, it is simply counted how often it makes a correct prediction (on test data). However, in order to make a correct prediction, it is not necessary that the class probabilities are predicted with high accuracy. It suffices that the most probable class receives the highest probability assignment.

For example, in a classification problem with only two classes, the true probability of class 1 given some instantiation of the descriptive attributes may be 94%. If a naive Bayes classifier predicts it to be 51% instead, it wildly misestimates this probability but still yields the correct classification result.

In a large fraction of all cases in which the conditional independence assumption is not satisfied, the effect of this assumption is mainly the distortion of the probability distribution, while the most probable class remains the same for most instantiations. Therefore the classification performance can still be very good, and this is what one actually observes in practice on many (though not all) data sets.

Note also that, even though it sounds highly plausible, it is not the case that a (naive) Bayes classifier can only profit from additional attributes. There are cases where additional attributes actually worsen the prediction performance, even if these additional attributes carry information about the class. An example is shown in Fig. 8.6: a naive Bayes classifier that uses only the horizontal axis to discriminate between cases belonging to class ● from cases belonging to class ○ yields a perfect result (solid line). However, if the vertical axis is also used, the classification boundary changes because of the diagonal location of the mass of the data points (dashed line). As a consequence, two instances are misclassified, namely the data point marked with ● at the top and the data point marked with ○ at the bottom.

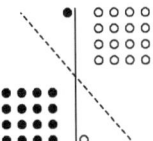

Fig. 8.6 The classification performance of a (naive) Bayes classifier need not get better if additional attributes are used. In this example adding the vertical axis, despite its being informative, yields a worse classification accuracy than using the horizontal axis alone, which allows for a perfect classification of the data points

8.2.3.2 Linear Discriminant Analysis

Linear discriminant analysis, as developed by Fisher [8], can be seen as a special case of a full Bayes classifier, which results from additional simplifying assumptions. In the first place, this is a method for discriminating between two classes, while more classes have to be handled by ensembles of classifiers (see Sect. 9.4). In addition, linear discriminant analysis assumes that the class-conditional multivariate normal distributions have the same shape. Formally, this means that they have the same covariance matrix and differ only in their mean vectors. As a consequence, the class boundary becomes a straight line (in two dimensions) or generally a hyperplane, thus explaining the name *linear* discriminant analysis.

Since there are only two classes, it is common to use a **decision function**, which is simply the difference of the class probability estimates:

$$
\mathrm{dec}(\mathbf{x}) = \frac{1}{(2\pi)^{\frac{|X|}{2}} \sqrt{|\Sigma|}} \left(\exp\left(-\frac{1}{2}(\mathbf{x} - \mu_1)^{\top} \Sigma^{-1}(\mathbf{x} - \mu_1) \right) \right.
$$
$$
\left. - \exp\left(-\frac{1}{2}(\mathbf{x} - \mu_0)^{\top} \Sigma^{-1}(\mathbf{x} - \mu_0) \right) \right).
$$

Class 1 is predicted if the decision function is positive, class 0 if it is negative:

$$
\mathrm{pred}(\mathbf{x}) = \frac{1}{2}\mathrm{sgn}(\mathrm{dec}(\mathbf{x})) + \frac{1}{2}.
$$

In the decision function, μ_0 and μ_1 are the mean vectors for the two classes that are estimated in the same way as for a full Bayes classifier. In order to estimate the covariance matrix, however, the data points from the two classes are pooled:

$$
\hat{\Sigma} = \frac{1}{n'} \sum_{i=1}^{n} \left(\mathbf{x}_i[\mathbf{X}] - \hat{\mu}_{y_i} \right) \left(\mathbf{x}_i[\mathbf{X}] - \hat{\mu}_{y_i} \right)^{\top},
$$

where either $n' = n$ or $n' = n - 1$ (as usually for variance estimation, see Sect. A.4.2.1 in the appendix). That is, the covariance matrix is computed from the difference vectors of the data points to their respective class means.

This pooling of data points is possible because of the assumption that the class-conditional density functions have the same shape. It has the advantage that the covariance estimate becomes more robust (since it is computed from more data points) but also the disadvantage that it misrepresents the density functions if the assumption of equal shape is not valid.

8.2.3.3 Augmented Naive Bayes Classifiers

Bayes classifiers can be seen as a special case of Bayes networks. Although we cannot provide here a comprehensive treatment of Bayes network (an interested reader is referred to, for instance, [4, 13] for an in-depth treatment), we try to capture some basics, because otherwise the name of the most common augmented naive Bayes classifier remains mysterious. The core idea underlying **Bayes networks** is to exploit conditional independence statements that hold in a multidimensional probability distribution, to decompose the distribution function. This already shows the connection to Bayes classifiers, in which we also factorize the joint probability distribution of the descriptive attributes based on the assumption that they are conditionally independent given the class.

The other core ingredient of Bayes networks is that, generally, the set of conditional independence statements that hold in a probability distribution has properties that are highly analogous to those of certain separation statements in graphs or networks. As a consequence, the idea suggests itself to use graphs or networks to express these conditional independence statements in a concise and intuitive form. Although we cannot go into the full formal details here, we would like to mention that a naive Bayes classifier is a Bayes network with a star-like structure, see Fig. 8.7 on the left: all edges are directed from the class attribute Y at the center to the descriptive attributes $X_1, \ldots, X_m$. This graph structure expresses the conditional independence assumption by a graph-theoretic criterion that is known as d-separation: any path from a descriptive attribute X_i to another attribute X_j, $i \neq j$, is blocked by the class attribute Y if it is known (details can be found, for instance, in [4]).

Drawing on the theory of representing conditional independencies by graphs or networks, these conditional independence assumptions can be mitigated by adding edges between descriptive attributes that are not conditionally independent given the class, see Fig. 8.7 on the right: the direct connections between these attributes express that even if the class is known, and thus the path through the class attribute

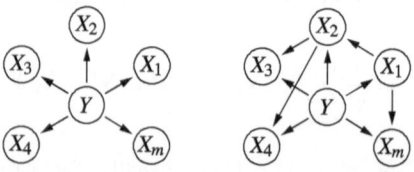

Fig. 8.7 A naive Bayes classifier is a Bayes network with a star-like structure (with the class attribute at the center). Additional edges can mitigate the conditional independence assumptions

is blocked, the two attributes are still dependent on each other. In this case the classification formula of a Bayes classifier is generalized with the standard factorization formula for a Bayes network, namely to

$$\text{pred}(\mathbf{x}) = \arg\max_{y \in Y} P(y) \prod_{X \in \mathbf{X}} P(x \mid \text{parents}(X)),$$

where parents(X) denotes the set of parents of the attribute X in the graph structure. Clearly, with a purely star-like structure (as in Fig. 8.7 on the left), this formula coincides with the standard classification formula of a naive Bayes classifier, because in a purely star-like structure the class attribute Y is the only parent of any descriptive attribute. If, however, there are additional edges, a descriptive attribute may have additional parents (other than the class attribute Y) on which it depends and which are needed in order to compute the correct conditional probability.

The most common form of a Bayes classifier that is extended in this way is the **tree-augmented naive Bayes classifier** (TAN) [9, 10]. In this classifier the number of parents of each descriptive attribute is restricted to two: the class attribute Y and at most one descriptive attribute. Since Bayes networks are required to be acyclic directed graphs (that is, there must not be a directed path connecting a node to itself), this constraint allows at most edges forming a tree to be added to the star structure of a naive Bayes classifier (and hence the name *tree-augmented* naive Bayes classifier).

The standard way to choose the additional edges is to construct a maximum weight spanning tree for the descriptive attributes with conditional mutual information providing the edge weights. This measure is defined as

$$I(X_i, X_j \mid Y) = H(X_i \mid Y) + H(X_j \mid Y) - H(X_i, X_j \mid Y)$$

$$= \sum_{x_i \in \text{dom}(X_i)} \sum_{x_j \in \text{dom}(X_j)} \sum_{y \in \text{dom}(Y)} P(x_i, x_j, y)$$

$$\times \log_2 \frac{P(x_i, x_j \mid y)}{P(x_i \mid y) \cdot P(x_j \mid y)}$$

for two descriptive attributes X_i and X_j, conditional on the class attribute Y, where H denotes the Shannon entropy. This procedure has the advantage that it selects those edges that, under the tree constraint, produce a Bayes network that yields the smallest Kullback–Leibler information divergence between the actual multivariate probability distribution and the probability distribution represented by the Bayes network/classifier. The edges may then be directed by choosing an arbitrary descriptive attribute as the root of the added tree and directing all added edges away from it.

More general augmented Bayes classifiers, which allow for more edges than just a directed tree to be added, together with methods for selecting the edges resulting in a good classifier, have been studied, for example, in [14, 20]. They are generally Bayes networks in which one variable representing the class attribute plays a special role, so that all computations are directed toward it.

8.2.3.4 Missing Values

Handling missing values is particularly easy in naive Bayes classifiers. In the construction step, it is convenient that the (conditional) distributions refer to at most two attributes, namely the class and at most one conditioned attribute. Therefore data points that are only partially known (that is, miss some attribute values) can still be used for the estimation: they are useless only for estimating those distributions that involve one of the attributes for which the value is missing. For all other attributes, the data point can be exploited. The only exception is the case for which the class is unknown, since the class is involved in all relevant (conditional) distributions.

In the execution step—that is, when a naive Bayes classifier is employed to compute a prediction for a new case—the factors of the classification formula, which refer to attributes the values of which are unknown, are simply dropped. The prediction computed in this way is the same as that of a naive Bayes classifier which has been constructed on the subspace of those attributes that are known for the sample case under consideration. Hence a naive Bayes classifier can flexibly adapt to whatever subset of the descriptive attributes is known.

The situation is less convenient for a full Bayes classifier: since all descriptive attributes are treated jointly (no conditional independence assumption), a single missing value can already make it impossible to use the sample case in the estimation of the class-conditional distributions. With a multivariate normal distribution assumption, however, the case may still be used to estimate the covariances of known attribute pairs. One merely has to keep track of the different values of n', which may be different for each element of the covariance matrix. In the execution step integrating the class-conditional multivariate normal distributions over unknown attributes provides a means of computing a prediction despite the fact that values are missing.

8.2.3.5 Misclassification Costs

Up to now we assumed that the costs of misclassifying an instance do not depend on the true class of the instance. In practice, however, these costs usually differ. Consider, for example, medical diagnosis: if a sick patient is misclassified as healthy, the consequences are usually much more severe than if a healthy patient is misclassified as sick. In the former case, the disease is left untreated and thus may develop into a severe state that may cause considerable harm to the patient. In the latter case the actually healthy status of the patient will likely be discovered by future tests.

Another example is the task to select customers for a special mail advertisement. In this case one would like to send the mail only to those customers that will actually buy the advertised product, that is, the class to predict is whether the customer buys or not. In this case misclassifying a nonbuying customer as buying incurs only a relatively small loss, namely, the cost of a useless mailing. However, if a buying customer is misclassified as nonbuying and thus does not receive the mailing (and therefore does not buy, because he/she never gets to know about the offer), the costs are high, because the selling company loses the revenue it could have made.

Since Bayes classifiers predict a class by estimating the probability of the different possible classes and then selecting the most probable one, misclassification costs can fairly easily be incorporated into the process. Suppose that we model the misclassification costs with a cost function $c(y)$, which assigns to each class y the costs of wrongly classifying an instance of this class as belonging to some other class y'. Then the classification formula should be modified to

$$\text{pred}(\mathbf{x}) = \arg\min_{y \in \text{dom}(Y)} c(y)(1 - P(y \mid \mathbf{x})).$$

In this way a class with nonmaximal probability may be predicted if the costs for misclassifying an instance of it as belonging to some other class are high.

A more complicated situation is where we are given misclassification costs in the form of a matrix $\mathbf{C}$ of size $\text{dom}(Y) \times \text{dom}(Y)$. Each element c_{yz} of this matrix states the costs of misclassifying an instance of class y as belonging to class z. Naturally, the diagonal elements c_{yy} are all zero (as these refer to a correct classification). In this case the classification formula is modified to

$$\text{pred}(\mathbf{x}) = \arg\min_{y \in \text{dom}(Y)} \sum_{z \in \text{dom}(Y)} c_{zy} P(z \mid \mathbf{x}).$$

With this prediction procedure, the expected costs (under the induced probability model) that result from possible misclassifications are minimized.

Note that both this and the previous modified classification formula reduce to the standard case (that is, predicting the most probable class) where all costs are equal:

$$\text{pred}(\mathbf{x}) = \arg\min_{y \in \text{dom}(Y)} \sum_{z \in \text{dom}(Y)-\{y\}} c \cdot P(z \mid \mathbf{x})$$

$$= \arg\min_{y \in \text{dom}(Y)} c \cdot (1 - P(y \mid \mathbf{x}))$$

$$= \arg\max_{y \in \text{dom}(Y)} P(y \mid \mathbf{x}),$$

where c denotes the (equal) misclassification costs.

8.3 Regression

Generally, regression is the process of fitting a real-valued function of a given class to a given data set by minimizing some cost functional, most often the sum of squared errors. Most commonly, the fitting function is linear, which allows for a direct solution. Actually, for any polynomial, the solution that minimizes the sum of squared errors can be computed with methods from linear algebra. Transformations, like logistic regression, allow us to use more general function classes and still to obtain a direct solution. In other cases, gradient descent methods can be employed.

8.3.1 Overview

Up to now we considered classification, that is, the task to predict a class from a finite set of possible classes. However, in many applications the quantity to predict is not categorical, but rather metric (numerical), whether it is the price of the shares of a given company, the electrical power consumption in a given area, the demand for a given product, etc. In such cases classification techniques may, in principle, still be applied, but doing so requires us to discretize the quantity to predict into a finite set of intervals, which are then considered as the classes. Clearly, however, a better approach is to use a prediction method that can yield a numeric output.

The main problem of such an approach is that (physical) measurement values rarely show the exact relationship between the considered quantities, because they are inevitably afflicted by errors. If one wants to determine the relationship between the considered quantities nevertheless, at least approximately, one faces the task to find a function that fits the given data as well as possible, so that the measurement errors are "neutralized." Naturally, for such an approach, one should possess at least a conjecture about how the target attribute (in statistics also called the **response variable**) depends on the descriptive attributes (in statistics also called the **explanatory** or **regressor variables**), so that one can choose a (parameterized) function class and thus can reduce the problem to a parameter estimation task. This choice is a critical issue: if a chosen function class does not fit the data (for example, if one tries to fit a linear function to nonlinear data), the result can be completely useless, because the function cannot, in principle, be made to fit the data.[5]

Generally, one has to be particularly careful not to choose a function class with too many degrees of freedom (too many free parameters), as this invites overfitting. For example, any set of n data points with one explanatory and one response variable can be fitted perfectly with a polynomial of degree $n - 1$ (and thus n free parameters). An example with eight data points is shown in Fig. 8.8 (blue curve: polynomial

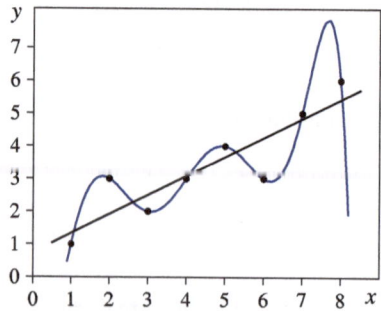

Fig. 8.8 The function class for regression has to be chosen with care. If a very complex function is chosen, a perfect fit can be achieved, but this fit does not allow for reliable inter- or extrapolations. A simpler function, although it fits the training data less well, is usually a much better predictor for new function values

---

[5]For more details see also Sect. 5.4.

of degree 7). Clearly, even though the blue curve fits the data points perfectly, it is completely useless for interpolation or extrapolation purposes: it is not likely that a data point with a value for the explanatory variable (x-axis) other than those of the given data points actually has a value for the response variable (y-axis) so that it lies on the blue curve. This is particularly obvious for points with an x-value beyond 8. A better fit of this data set can be obtained with a simple straight line (shown in black). Intuitively, interpolating or extrapolating the values of the response variable based on this straight line are much more likely to yield a useful result.

In order to deal with the unavoidable measurement errors and to achieve a good fit to the data within the limitations imposed by the choice of the function class, we have to choose a cost function that penalizes (at least large) deviations from the actual values. The most common cost function is the **sum of squared errors** (SSE), with which the approach is also known as the **method of least squares** (OLS for "ordinary least squares"). It has the advantage that for a large family of parameterized function classes—in particular, any polynomial of the explanatory variables— the solution (that is, the set of parameters yielding the least sum of squared errors) can be obtained by taking derivatives and applying simple methods from linear algebra. However, it has the disadvantage that outliers, either in the explanatory or the response variable, can have a distorting effect due to the fact that the errors are squared, and thus outliers strongly influence the estimation. As a consequence, other cost functions—like the **sum of absolute errors** or functions that limit the contribution of a single data point to the total cost to a certain maximum—are also used.

Once the (parameterized) function class and the cost function have been chosen, the model construction process is straightforward: try to find the set of parameters that identifies the function from the chosen class that minimizes the costs. Whether the result is unique or not and whether it can be computed directly or has to be found by an iterative improvement procedure (like gradient descent) depends mainly on the cost function used. As we will see below, a direct solution can be obtained for any polynomial as the function class and the sum of squared errors as the cost function.

8.3.2 Construction

We start our description of the model construction procedure with the simplest case, finding a linear regression function for a single explanatory variable. In this case one has to determine the parameters a and b of a straight line $y = f(x) = a + bx$. However, due to the unavoidable measurement errors, it will usually not be possible to find a straight line such that all n data points (x_i, y_i), $1 \le i \le n$, lie exactly on this straight line. Rather we have to find a straight line from which the given data points deviate as little as possible. Hence it is plausible to determine the parameters a and b in such a way that the sum of squared deviations, that is,

$$F(a, b) = \sum_{i=1}^{n} \left(f(x_i) - y_i \right)^2 = \sum_{i=1}^{n} (a + bx_i - y_i)^2,$$

is minimal. In other words, the y-values that are computed with the linear equation should (in total) deviate as little as possible from the measured values. The reasons for choosing squared deviations are twofold: (1) by using squares the error function becomes continuously differentiable everywhere (whereas the derivative of the sum of absolute errors is discontinuous/does not exist at zero), and (2) the squares weight larger deviations higher, so that individual large deviations are avoided.[6]

A necessary condition for a minimum of the error function $F(a, b)$ is that the partial derivatives of this function w.r.t. the parameters a and b vanish, that is,

$$\frac{\partial F}{\partial a} = \sum_{i=1}^{n} 2(a + bx_i - y_i) = 0 \quad \text{and} \quad \frac{\partial F}{\partial b} = \sum_{i=1}^{n} 2(a + bx_i - y_i)x_i = 0.$$

As a consequence, we obtain (after a few simple steps) the so-called **normal equations**

$$na + \left(\sum_{i=1}^{n} x_i\right)b = \sum_{i=1}^{n} y_i \quad \text{and} \quad \left(\sum_{i=1}^{n} x_i\right)a + \left(\sum_{i=1}^{n} x_i^2\right)b = \sum_{i=1}^{n} x_i y_i,$$

that is, a linear two-equation system with two unknowns a and b. It can be shown that this equation system has a unique solution unless all x-values are identical and that this solution specifies a minimum of the function F. The straight line determined in this way is called **regression line** for the data set $(x_1, y_1), \ldots, (x_n, y_n)$.

Note that finding a regression line can also be seen as a **maximum likelihood estimation** (see Sect. A.4.2.3 in the appendix) of the parameters of the linear model

$$Y = aX + b + \xi,$$

where X is a random variable with arbitrary distribution, and ξ is a normally distributed random variable with mean 0 and arbitrary variance (but independent of the value of X). The parameters that minimize the sum of squared deviations (in y-direction) from the data points maximize the probability of the data given this model class.

In order to illustrate the procedure, we consider a simple example. Let the following data set be given, which consists of eight data points $(x_1, y_1), \ldots, (x_8, y_8)$:

x	1	2	3	4	5	6	7	8
y	1	3	2	3	4	3	5	6

The regression line computed with the above procedure is

$$y = \frac{3}{4} + \frac{7}{12}x.$$

---

[6]Note, however, that this second property can also be a disadvantage, as it can make outliers have an overly strong influence on the regression result.

This line is shown in Fig. 8.8 on page 230 in black. Obviously, it provides a reasonable fit to the data and may be used to interpolate between or to extrapolate the data points.

The least squares method is not limited to straight lines but can be extended (at least) to **regression polynomials**. In this case one tries to find a polynomial

$$y = p(x) = a_0 + a_1 x + \cdots + a_m x^m,$$

with a given fixed degree m, which best fits the n data points $(x_1, y_1), \ldots, (x_n, y_n)$. Consequently, we have to minimize the error function

$$F(a_0, a_1, \ldots, a_m) = \sum_{i=1}^{n} (p(x_i) - y_i)^2 = \sum_{i=1}^{n} (a_0 + a_1 x_i + \cdots + a_m x_i^m - y_i)^2.$$

In analogy to the linear case, we form the partial derivatives of this function w.r.t. the parameters a_k, $0 \leq k \leq m$, and equate them to zero (as this is a necessary condition for a minimum). The resulting system of linear equations ($m + 1$ unknowns and $m + 1$ equations) can be solved with one of the standard methods of linear algebra (elimination method according to Gauß, inverting the coefficient matrix, etc.).

Note that finding a regression polynomial can also be interpreted as a maximum likelihood estimation of the parameters (as it was possible for linear regression) if one assumes a normally distributed error term ξ that is independent of X and Y.

Furthermore, the least squares method can be used for functions with more than one argument. This case is known as **multivariate regression**. We consider only **multilinear regression**, that is, we are given a data set $((\mathbf{x}_1, y_1), \ldots, (\mathbf{x}_n, y_n))$ with input vectors $\mathbf{x}_i$ and the corresponding responses y_i, $1 \leq i \leq n$, for which we want to determine the linear regression function

$$y = f(x_1, \ldots, x_m) = a_0 + \sum_{k=1}^{m} a_k x_k.$$

In order to derive the normal equations, it is convenient to write the functional to minimize in matrix form, that is, as

$$F(\mathbf{a}) - (\mathbf{X}\mathbf{a} - \mathbf{y})^\top (\mathbf{X}\mathbf{a} - \mathbf{y}), \tag{8.1}$$

where

$$\mathbf{X} = \begin{pmatrix} 1 & x_{11} & \cdots & x_{1m} \\ \vdots & \vdots & \ddots & \vdots \\ 1 & x_{n1} & \cdots & x_{nm} \end{pmatrix} \quad \text{and} \quad \mathbf{y} = \begin{pmatrix} y_1 \\ \vdots \\ y_n \end{pmatrix} \tag{8.2}$$

represent the data set, and $\mathbf{a} = (a_0, a_1, \ldots, a_m)^\top$ is the vector of coefficients we have to determine. (Note that the ones in the matrix $\mathbf{X}$ refer to the coefficient a_0.) Again a necessary condition for a minimum is that the partial derivatives of this function w.r.t. the coefficients a_k, $0 \leq k \leq m$, vanish. Using the differential operator ∇, we can write these conditions as

$$\nabla_\mathbf{a} F(\mathbf{a}) = \frac{\mathrm{d}}{\mathrm{d}\mathbf{a}} F(\mathbf{a}) = \left(\frac{\partial}{\partial a_0} F(\mathbf{a}), \frac{\partial}{\partial a_1} F(\mathbf{a}), \ldots, \frac{\partial}{\partial a_m} F(\mathbf{a}) \right) = \mathbf{0}.$$

The derivative can most easily be computed by realizing that formally the differential operator

$$\nabla_\mathbf{a} = \left(\frac{\partial}{\partial a_0}, \frac{\partial}{\partial a_1}, \ldots, \frac{\partial}{\partial a_m} \right)$$

behaves like a vector (as one can easily confirm by writing out the computations). As a consequence, we obtain

$$
\begin{aligned}
\mathbf{0} &= \nabla_\mathbf{a}(\mathbf{Xa} - \mathbf{y})^\top (\mathbf{Xa} - \mathbf{y}) \\
&= \left(\nabla_\mathbf{a}(\mathbf{Xa} - \mathbf{y}) \right)^\top (\mathbf{Xa} - \mathbf{y}) + \left((\mathbf{Xa} - \mathbf{y})^\top \left(\nabla_\mathbf{a}(\mathbf{Xa} - \mathbf{y}) \right) \right)^\top \\
&= \left(\nabla_\mathbf{a}(\mathbf{Xa} - \mathbf{y}) \right)^\top (\mathbf{Xa} - \mathbf{y}) + \left(\nabla_\mathbf{a}(\mathbf{Xa} - \mathbf{y}) \right)^\top (\mathbf{Xa} - \mathbf{y}) \\
&= 2\mathbf{X}^\top (\mathbf{Xa} - \mathbf{y}) \\
&= 2\mathbf{X}^\top \mathbf{Xa} - 2\mathbf{X}^\top \mathbf{y},
\end{aligned}
$$

from which we immediately obtain the system of normal equations

$$\mathbf{X}^\top \mathbf{Xa} = \mathbf{X}^\top \mathbf{y}.$$

This system is (uniquely) solvable iff $\mathbf{X}^\top \mathbf{X}$ is invertible. In this case we have

$$\mathbf{a} = \left(\mathbf{X}^\top \mathbf{X} \right)^{-1} \mathbf{X}^\top \mathbf{y}.$$

The expression $\left(\mathbf{X}^\top \mathbf{X} \right)^{-1} \mathbf{X}^\top$ is also known as the (Moore–Penrose) **pseudo-inverse** of the (usually nonsquare) matrix $\mathbf{X}$ [1]. With this pseudo-inverse matrix, the solution of the regression task can be written down immediately.

It should be clear that the least squares method can also be extended to polynomials in multiple explanatory variables. In this case one starts most conveniently from the above matrix representation but also enters into the matrix $\mathbf{X}$ all monomials (that is, products of powers of the regressors, up to a given degree) that can be formed from the explanatory variables. Formally, these can be seen as additional inputs. In this way the derivation of the normal equations remains unchanged and can be used directly to find the solution.

8.3.3 Variations and Issues

As we have seen above, an analytical solution of the least squares problem can easily be obtained for polynomials. However, by exploiting transformations (for example, the logit-transformation), the regression approach can also be applied for other functions classes (Sect. 8.3.3.1). As least squares regression is sensitive to outliers, several more robust methods have been developed (Sect. 8.3.3.3). Finally, methods to automatically select the function class, at least from a wider family of functions of different complexity, may be worth considering (Sect. 8.3.3.4).

8.3.3.1 Transformations

In certain special cases the procedure described above can also be used to find other regression functions. In order for this to be possible, one has to find a suitable transformation which reduces the problem to the problem of finding a regression line or regression polynomial. For example, regression functions of the form

$$y = ax^b$$

can be found by finding a regression line. By simply taking the (natural) logarithm of this equation, we arrive at

$$\ln y = \ln a + b \cdot \ln x.$$

This equation can be handled by finding a regression line if we take the (natural) logarithms of the data points (x_i, y_i), $1 \leq i \leq n$, set $a' = \ln a$, and then carry out all computation with these transformed values.

It should be noted, though, that with such an approach, only the sum of squared errors in the transformed space (coordinates $x' = \ln x$ and $y' = \ln y$) is minimized, but not necessarily also the sum of squared errors in the original space (coordinates x and y). Nevertheless, this approach often yields good results. In addition, one may use the result only as a good starting point for a subsequent gradient descent (see generally Sect. 5.3, also Sect. 9.2 on training artificial neural network, and Sect. 8.3.3.2 below) in the original space, with which the solution can be improved, and the true minimum in the original space may be obtained.

For practical purposes, it is important that one can transform the **logistic function**,

$$y = \frac{y_{max}}{1 + e^{a+bx}},$$

where y_{max}, a, and b are constants, to the linear or polynomial case (so-called **logistic regression**). The logistic function is relevant for many applications, because it describes growth processes with a limit, for example, the growth of an animal population with a habitat of limited size or the sales of a (new) product with a limited market. In addition, it is popular in artificial neural networks (especially multilayer perceptrons, see Sect. 9.2), where it is often used as the activation function of the neurons.

In order to linearize the logistic function, we first take the reciprocal values

$$\frac{1}{y} = \frac{1 + e^{a+bx}}{y_{max}}.$$

As a consequence, we have

$$\frac{y_{max} - y}{y} = e^{a+bx}.$$

Taking the (natural) logarithm of this equation yields

$$\ln\left(\frac{y_{max} - y}{y}\right) = a + bx.$$

Obviously, this equation can be handled by finding a regression line if we transform the y-values of all data points according to the left-hand side of the above equation. (Note that the value of y_{max}, which mainly scales the values, must be known.) This transformation is known under the name **logit transformation** and can be seen as an inverse of the logistic function. By finding a regression line for the transformed data points, we obtain a logistic regression curve for the original data points.

However, note again that this procedure minimizes the sum of squared errors in the transformed space (coordinates x and $z = \ln(\frac{y_{max}-y}{y})$), but not necessarily also the sum of squared errors in the original space (coordinates x and y). If one needs to know parameter values that actually describe the minimum of the sum of squared errors in the original space, one may (as above) use the obtained solution as a starting point, which is then improved with a gradient descent (see Sect. 8.3.3.2 below).

In order to illustrate the procedure, we consider a simple example. Let the following simple data set be given that consists of five data points $(x_1, y_1), \ldots, (x_5, y_5)$:

x	1	2	3	4	5
y	0.4	1.0	3.0	5.0	5.6

By transforming these data points with the logit transformation using $y_{max} = 6$ and finding a regression line for the result, we obtain

$$z \approx 4.133 - 1.3775x,$$

and thus, for the original data, the logistic regression curve is

$$y \approx \frac{6}{1 + e^{4.133-1.3775x}}.$$

These two regression functions are shown together with the (transformed or original) data points in Fig. 8.9.

8.3.3.2 Gradient Descent

If a transformation was applied in order to reduce a more complex functional relationship to the polynomial case, the sum of squared errors is, as emphasized above, minimized only in the transformed space, but not necessarily in the original space.

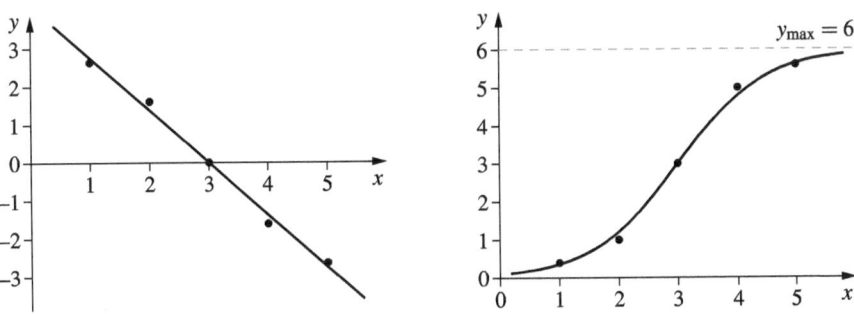

Fig. 8.9 Transformed data (*left*) and original data (*right*) as well as the regression line (*left*) and the corresponding logistic regression curve (*right*) computed with the method of least squares

Hence there is room for improvement, which may be achieved by using the solution that is obtained by solving the system of normal equations in the transformed space only as an initial point in the parameter space of the regression model. This initial solution is then iteratively improved by gradient descent (see Sect. 5.3 for a general treatment), that is, by repeatedly computing the gradient of the objective function (here, the sum of squared errors) at the current point in the parameter space and then making a small step in this direction.

If the functional relationship is logistic (as in the example in the previous section), this procedure is actually equivalent to training a multilayer perceptron without a hidden layer (so actually a two-layer perceptron, with only an input layer and one output neuron) with standard error backpropagation, provided, of course, that the activation function of the output neuron is logistic. Details for this special case can be found in Sect. 9.2.

8.3.3.3 Robust Regression

As pointed out above, solutions of (ordinary) least squares regression can be strongly affected by outliers. More robust results can usually be obtained by minimizing the **sum of absolute deviations** (least absolute deviations, LAD). However, this approach has the disadvantage of not being analytically solvable (like least squares) and thus has to be addressed with iterative methods right from the start. In addition, least absolute deviation solutions can be unstable in the sense that small changes in the data can lead to "jumps" (discontinuous changes) of the solution parameters, while least squares solutions always changes "smoothly" (continuously). Finally, severe outliers can still have a distorting effect on the solution.

As a consequence, more sophisticated regression methods, categorized as **robust regression,** have been developed in order to improve robustness. Among these are **M-estimation** and **S-estimation** for regression and **least trimmed squares** (LTS), the latter of which simply uses a subset of at least half the size of the data set (with the exact size to be specified by a user) that yields the smallest sum of squared errors.

Table 8.3 Error measures ρ for different approaches

Method	$\rho(e)$
Least squares	e^2
Huber	$\begin{cases} \frac{1}{2}e^2 & \text{if } \|e\| \leq k, \\ k\|e\| - \frac{1}{2}k^2 & \text{if } \|e\| > k. \end{cases}$
Tukey's bisquare	$\begin{cases} \frac{k^2}{6}(1 - (1 - (\frac{e}{k})^2)^3) & \text{if } \|e\| \leq k, \\ \frac{k^2}{6} & \text{if } \|e\| > k. \end{cases}$

As an example for robust regression, we take a closer look at M-estimators. Let us revisit the linear regression problem as it was formulated in (8.1) and (8.2). We rewrite the error function (8.1), i.e., the sum of squared errors, to be minimized in the form

$$\sum_{i=1}^{n} \rho(e_i) = \sum_{i=1}^{n} \rho(\mathbf{x}_i^\top \mathbf{a} - y_i), \tag{8.3}$$

where $\rho(e_i) = e_i^2$, and e_i is the (signed) error of the regression function at the ith point. Is this the only reasonable choice for the function ρ? The answer is definitely no. However, ρ should satisfy at least some reasonable restrictions. ρ should always be positive, except for the case $e_i = 0$. Then we should have $\rho(e_i) = 0$. The sign of the error e_i should not matter for ρ, and ρ should be increasing when the absolute value of the error increases. These requirements can formalized in the following way:

$$\rho(e) \geq 0, \tag{8.4}$$

$$\rho(0) = 0, \tag{8.5}$$

$$\rho(e) = \rho(-e), \tag{8.6}$$

$$\rho(e_i) \geq \rho(e_j) \quad \text{if } |e_i| \geq |e_j|. \tag{8.7}$$

Parameter estimation (here the estimation of the parameter vector b) based on an objective function of the form (8.3) and an error measure satisfying (8.4)–(8.7) is called an M-estimator. The classical least squares approach is based on the quadratic error, i.e. $\rho(e) = e^2$. Table 8.3 provides the error measure ρ for the classical least squares method and for two approaches from robust statistics.

In order to understand the more general setting of an error measure ρ satisfying (8.4)–(8.7), it is useful to consider the derivative of the error measure $\psi = \rho'$.

Taking the derivatives of the objective function (8.3) with respect to the parameters a_i, we obtain the system of $(m + 1)$ linear equations

$$\sum_{i=1}^{n} \psi_i (\mathbf{x}_i^\top \mathbf{a} - y_i) \mathbf{x}_i^\top = 0. \tag{8.8}$$

Table 8.4 The computation of the weights for the corresponding approaches

Method	$w(e)$
Least squares	1
Huber	$\begin{cases} 1 & \text{if } \|e\| \leq k, \\ k/\|e\| & \text{if } \|e\| > k. \end{cases}$
Tukey's bisquare	$\begin{cases} (1-(\frac{e}{k})^2)^2 & \text{if } \|e\| \leq k, \\ 0 & \text{if } \|e\| > k. \end{cases}$

Defining $w(e) = \psi(e)/e$ and $w_i = w(e_i)$, Equation (8.8) can be rewritten in the form

$$\sum_{i=1}^{n} \frac{\psi_i(\mathbf{x}_i^\top \mathbf{a} - y_i)}{e_i} \cdot e_i \cdot \mathbf{x}_i^\top = \sum_{i=1}^{n} w_i \cdot \left(y_i - x_i^\top b\right) \cdot x_i^\top = 0. \qquad (8.9)$$

Solving this system of linear equations corresponds to solving a standard least squares problem with (nonfixed) weights in the form

$$\sum_{i=1}^{n} w_i e_i^2. \qquad (8.10)$$

However, the weights w_i depend on the residuals e_i, the residuals depend on the coefficients a_i, and the coefficients depend on the weights. Therefore, it is in general not possible to provide an explicit solution to the system of equations. Instead, the following iteration scheme is applied.

1. Choose an initial solution $\mathbf{a}^{(0)}$, for instance, the standard least squares solution setting all weights to $w_i = 1$.
2. In each iteration step t, calculate the residuals $e^{(t-1)}$ and the corresponding weights $w^{(t-1)} = w(e^{(t-1)})$ determined by the previous step.
3. Compute the solution of the weighted least squares problem $\sum_{i=1}^{n} w_i e_i^2$ which leads to

$$\mathbf{a}^{(t)} = \left(\mathbf{X}^\top \mathbf{W}^{(t-1)} X\right)^{-1} \mathbf{X}^\top \mathbf{W}^{(t-1)} \mathbf{y}, \qquad (8.11)$$

where $\mathbf{W}$ stands for the diagonal matrix with weights w_i on the diagonal.

Table 8.4 lists the formulae for the weights in the regression scheme based on the error measures listed in Table 8.3.

Figure 8.10 shows the graph of the error measure ρ and the weighting function for the standard least squares approach. The error measure ρ increases in a quadratic manner with increasing distance. The weights are always constant. This means that extreme outliers will have full influence on the regression coefficients and can corrupt the result completely.

In the more robust approach by Huber the change of the error measure ρ switches from a quadratic increase for small errors to a linear increase for larger errors. As a

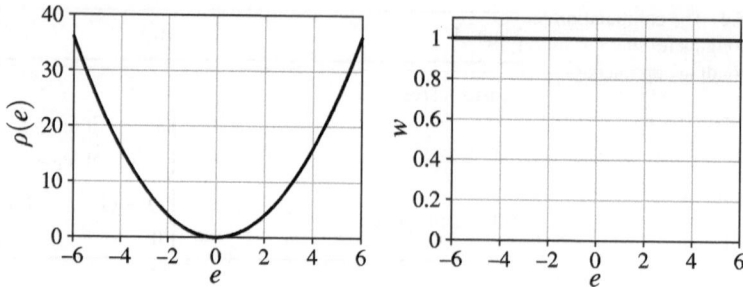

Fig. 8.10 The error measure ρ and the weight w for the standard least squares approach

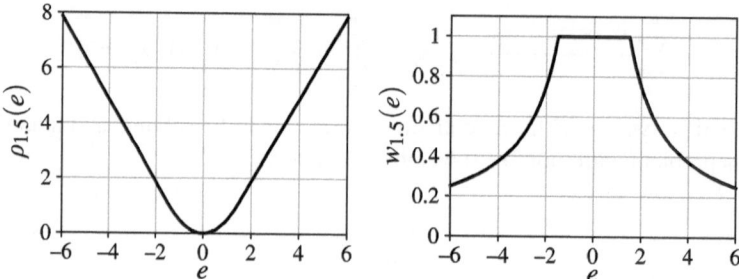

Fig. 8.11 The error measure ρ and the weight w for Huber's approach

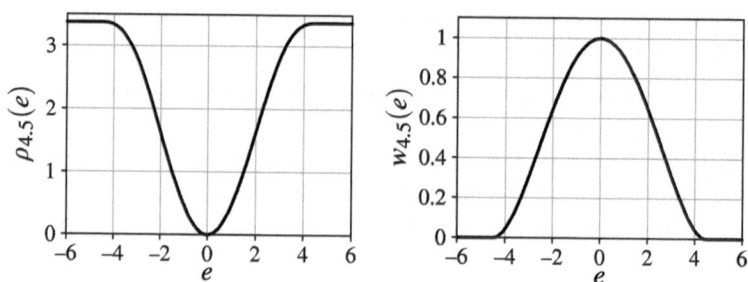

Fig. 8.12 The error measure ρ and the weight w for the bisquare approach

result, only data points with small errors will have the full influence on the regression coefficients. For extreme outliers, the weights tend to zero. This is illustrated by the corresponding graphs in Fig. 8.11.

Tukey's bisquare approach is even more drastic than Huber's approach. For larger errors, the error measure ρ does not increase at all but remains constant. As a consequence, the weights for outliers drop to zero when they are too far away from the regression curve. This means that extreme outliers have no influence on the regression curve at all. The corresponding graphs for the error measure and the weights are shown in Fig. 8.12.

Fig. 8.13 Least squares (*red*) vs. robust regression (*blue*)

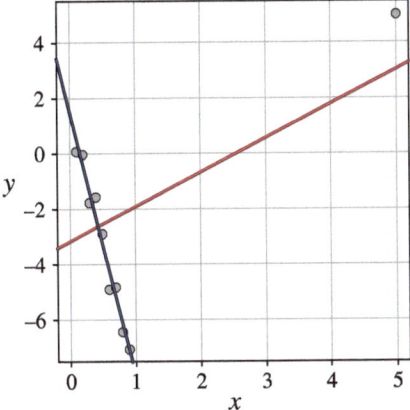

Fig. 8.14 The weights for the robust regression function in Fig. 8.13

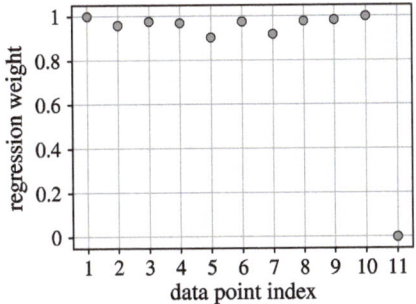

To illustrate how robust regression works, consider the simple regression problem in Fig. 8.13. There is one outlier that leads to the red regression line that neither fits the outlier nor the other points. With robust regression, for instance, based on Tukey's ρ-function, we obtain the blue regression line that simply ignores the outlier.

An additional result in robust regression is the computed weights for the data points. The weights for the regression problem in Fig. 8.13 are plotted in Fig. 8.14. All weights, except one, have a value close to 1. The right-most weight with the value close to 0 is the weight for the outlier. In this way, outlier can be identified by robust regression. This applies also to the case of multivariate regression where we cannot simply plot the regression function as in Fig. 8.13. But the weights can still be computed and plotted. We can also take a closer look at the data points with low weights for the regression function. They might be exceptional cases or even erroneous measurements.

8.3.3.4 Function Selection

In their basic form, regression approaches require a conjecture about the form of the functional relationship between the involved variables, so that the task only consists in estimating its parameters. However, sometimes several different function classes present themselves as reasonable hypotheses. For example, one may be reasonably sure that the relationship can be described by a polynomial but does not know the degree. In such a case one can try to determine the function class automatically with an approach that is based on the minimum description length principle.

The core idea of this principle is to imagine that the training data has to be transmitted as efficiently as possible (that is, using a message of minimal length) from a sender to a receiver, possibly using a model to code the data. In this way it becomes possible to measure the complexity of the model (which, if used, needs to be specified in the message, so that the receiver can decode the data) and the data on the same scale, namely as the length of a message describing it. Therefore one goes through the different functions in the hypothesized family, builds the best model for each function, and then determines which model, if used to code the data, yields the shortest message (model description plus data description).

Applied to regression, this general principle takes the form that one considers functions that sum different numbers of monomials (products of powers of the regressors). The coefficients of these monomials are determined to build the model (using the techniques described above). The model description consists of a specification of the monomials that are used in the function and an encoding of the corresponding coefficients (for which a finite precision has to be fixed) plus a variance for the deviations of the response values from the function. The data is described by specifying how far the value of the response variable differs from the value predicted by the model, assuming a normal distribution with the specified variance for coding.

Another approach to select an appropriate function is based on bias/variance decomposition (see Sect. 5.4.5): a more complex function has a lower bias, because it can adapt better to the actual function due to its larger number of degrees of freedom (larger number of free parameters). However, it leads to a higher variance, because the more parameters have to be estimated, the less reliable these estimates become. As a consequence, we would like to minimize the total, that is, lower the bias only as far as the gain is not eaten up by an increased variance.

8.3.4 Two Class Problems

Even though mainly directed at numeric prediction problems, regression, especially logistic regression, is also a popular method for solving classification problems with two classes. That is, each given data point is assigned to one of two classes, and we desire to find a function (a so-called **discrimination function**) that maps the values of the other, descriptive attributes to a class. More precisely, we desire to find a function that models the probability of one of the classes for the different points in the data space. In this way the classification problem is conveniently mapped to a numeric prediction problem, and, in addition, the classifier yields not only a class but also a confidence value, namely the class probability. This approach is particularly popular in the finance industry to model the credit-worthiness of customers.

Formally the problem to solve can be described as follows: let Y be a class attribute with the domain $dom(Y) = \{c_1, c_2\}$, and $\mathbf{X} = (X_1, \ldots, X_m)$ an m-dimensional random vector. Furthermore, let

$$P(Y = c_1 \mid \mathbf{X} = \mathbf{x}) = p(\mathbf{x}) \quad \text{and thus}$$

$$P(Y = c_2 \mid \mathbf{X} = \mathbf{x}) = 1 - p(\mathbf{x}).$$

Finally, let a data set $\mathcal{D} = \{(\mathbf{x}_1, y_1), \ldots, (\mathbf{x}_n, y_n)\}$ be given, the elements of which are assigned to one of the classes c_1 or c_2 (that is, $y_i = c_1$ or $y_i = c_2$ for $i = 1, \ldots, n$).

We desire to find a reasonably simple description of the function $p(\mathbf{x})$, the parameters of which have to be estimated from the data set $\mathbf{X}$. A common approach is to model $p(\mathbf{x})$ as a logistic function, that is, as

$$p(\mathbf{x}) = \frac{1}{1 + e^{a_0 + \mathbf{ax}}} = \frac{1}{1 + \exp(a_0 + \sum_{i=1}^{r} a_i x_i)}.$$

By applying the logit-transformation we discussed in Sect. 8.3.3.1, we obtain

$$\ln\left(\frac{1 - p(\mathbf{x})}{p(\mathbf{x})}\right) = a_0 + \mathbf{ax} = a_0 + \sum_{i=1}^{r} a_i x_i,$$

that is, a multilinear regression problem, which can easily be solved with the techniques introduced above.

What remains to be clarified is how we determine the values $p(\mathbf{x})$ that enter the above equation. If the data space is small enough, so that there are sufficiently many realizations for every possible point (that is, for every possible instantiation of the random variables $X_1, \ldots, X_m$), we may estimate the class probabilities simply as the relative frequencies of the classes (see Sect. A.4.2 in the appendix).

If this is not the case, we may rely on an approach known as **kernel estimation** in order to determine the class probabilities at the data points. The basic idea of such an estimation is to define a kernel function K which describes how strongly a data point influences the estimation of the probability (density) at a neighboring point (see Sect. 9.1.3.1 for a related approach in connection with k-nearest-neighbor classifiers and Sect. 9.3 on support vector machines). The most common choice is a Gaussian function, that is,

$$K(\mathbf{x}, \mathbf{y}) = \frac{1}{(2\pi\sigma^2)^{\frac{m}{2}}} \exp\left(-\frac{(\mathbf{x} - \mathbf{y})^\top (\mathbf{x} - \mathbf{y})}{2\sigma^2}\right),$$

where the variance σ^2 has to be chosen by a user. With the help of this kernel function the probability density at a point $\mathbf{x}$ is estimated from an (unclassified) data set $\mathcal{D} = \{\mathbf{x}_1, \ldots, \mathbf{x}_n\}$ as

$$\hat{f}(\mathbf{x}) = \frac{1}{n} \sum_{i=1}^{n} K(\mathbf{x}, \mathbf{x}_i).$$

If we deal with a two-class problem, one estimates the class probabilities by relating the probability density resulting from datapoints of one of the classes to the total probability density. That is, we estimate

$$\hat{p}(\mathbf{x}) = \frac{\sum_{i=1}^{n} c(\mathbf{x}_i) K(\mathbf{x}, \mathbf{x}_i)}{\sum_{i=1}^{n} K(\mathbf{x}, \mathbf{x}_i)},$$

where

$$c(\mathbf{x}_i) = \begin{cases} 1 & \text{if } x_i \text{ belongs to class } c_1, \\ 0 & \text{if } x_i \text{ belongs to class } c_2. \end{cases}$$

Solving the resulting regression problem yields a (multidimensional) logistic function, which describes the probability of one of the two classes for the points of the data space. For this function, one has to choose a threshold value: if the function value exceeds the threshold, the class the function refers to is predicted; otherwise the other class is predicted. Note that with such a threshold a linear separation of the input space is described (see also Sect. 9.3).

If this method is used in finance to assess the credit-worthiness of a customer, one of the classes means that the loan applied for is granted, while the other means that the application is declined. As a consequence, several threshold values are chosen, which refer to different loan conditions (interest rate, liquidation rate, required securities, loan duration, etc.)

8.4 Rule learning

The last type of explanation-based methods that we are discussing in this chapter are rule learning methods. In contrast to association rule learners as discussed before, we are now concentrating on algorithms that generate sets of rules which explain all of the training data. Hence we are interested in a global rule system vs. association rules that are a collection of local models. These local models can be highly redundant (and they usually are), and we expect a global rule system to explain the data reasonably free of overlaps. Rule systems are one of the easiest ways (if not the easiest way) to express knowledge in a human-readable form. By the end of this section we hope to have given a better understanding what these types of methods can achieve and why they are still relatively unknown to data analysis practitioners.

Rule Learning algorithms can be roughly divided into two types of methods. Simple, "if this fact is true, then that fact holds"-type rules and more complex rules which allow one to include variables, something along the lines of "if x has wings then x is a bird".[7] The first type of rules are called propositional rules and will be discussed first. Afterwards we will then discuss first-order rules or the field of inductive logic programming as this area of rule learning is often also called.

---

[7] Note that we are not saying much about the truthfulness or precision of rules at this stage.

8.4.1 Propositional Rules

Propositional rules are rules consisting of atomic facts and combinations of those using logical operators. In contrast to first-order logic (see next section), no variables are allowed to be part of those rules. A simple propositional classification rule could look like this:

$$\text{IF } x_1 \leq 10 \text{ AND } x_3 = red \text{ THEN class A.}$$

Note that we have an **antecedent** part of the rule (to the left of THEN) indicating the conditions to be fulfilled in order for the **consequent** part (to the right) to be true. As with typical implications, we do not know anything about the truth value of the consequent if the antecedent does not hold. The atomic facts of propositional rules more commonly occurring in data analysis tasks are constraints on individual attributes:

- constraints on numerical attributes, such as greater/smaller (-or-equal) than a constant, equality (usually for integer values) or containment in given intervals;
- constraints on nominal attributes, such as checks for equality or containment in a set of possible values;
- constraints on ordinal attributes, which add range checks to the list of possible constraints on numeric attributes.

The rule above shows examples for a numerical constraint and a constraint on a nominal attribute. How can we now find such rules given a set of training instances?

8.4.1.1 Extracting Rules from Decision Trees

One very straight forward way to find propositional rules was already presented earlier in this chapter: we can simply train a decision tree and extract the rules from the resulting representation. The tree shown in Fig. 8.2 can also be interpreted as four (mutually exclusive—we will get back to this in a second) rules:

R_a: IF temperature ≤ 20 THEN class "cold"

R_b: IF temperature > 20 AND humidity ≤ 50 THEN class "comfortable"

R_c: IF temperature $\in (20, 35]$ AND humidity > 50 THEN class "comfortable"

R_d: IF temperature > 35 AND humidity > 50 THEN class "unbearable"

Those rules are simply generated by traversing the tree from each leaf up to the root and collecting the conditions along the way. Note that in rule R_c we collapsed the two tests on temperature into one. Since the rules are disjunctive and disjunction is commutative, this does not alter the antecedent.

Rules extracted from decision trees have two interesting properties:

- mutual exclusivity: this means that a pattern will be explained by one rule only. This stems from the origin of the rules: they were generated from a hierarchical structure where each branch partitions the feature space into two disjoint partitions.
- unordered: the rules are extracted in arbitrary order from the tree, that is, no rule has preference over any other one. This is not a problem since only one (and exactly one!) rule will apply to a training pattern. However, as we will see later, rules can also overlap, and then a conflict avoidance strategy may be needed: one such strategy requires rules to be ordered, and the first on that matches is the one creating the response.

Creating rules from decision trees is straightforward and uses well-established and efficient training algorithms. However, we inherit all of the disadvantages of decision tree learning algorithms (most notably with respect to their notorious instability!), and the rules can also be quite redundant. We can sometimes avoid this redundancy by transforming the rule set into a set of ordered rules:

R_1: IF temperature ≤ 20 THEN class "cold"

R_2: IF humidity ≤ 50 THEN class "comfortable"

R_3: IF temperature ≤ 25 THEN class "comfortable"

R_4: IF true THEN class "unbearable"

Now we really need to apply these rules in the correct order, but then they still represent exactly the same function as the original decision tree and the unordered rule set shown before. This type of conversion only works because we have an unbalanced tree—no branch going to the left actually carries a subtree. So we could simply assign labels for those branches and carry the rest forward to the next rule describing the branch to the right. In general, simplifying rule sets extracting from decision trees are considerably more complex.

8.4.1.2 Extracting Propositional Rules

Real-world propositional Rule Learners can essentially be categorized by looking at two properties, supported attribute types and learning strategy. There are a number of algorithms that learn rules on nominal attributes only and a second class which concentrates on learning rules for numerical attributes. The second categorization looks at the strategy: most if not all rule learning methods are either specializing or generalizing, that is, they start with very specific rules and make them more general until they fit the training data or they start with one very general rule and iteratively divide and/or specialize the rules until no further conflicts with the training data are encountered. Specializing and generalizing rules are concepts which are more formally defined in [15], however, for the purpose of this chapter, it suffices to understand that we can increase (or decrease) a rule's support and thereby increase (or decrease) the set of training patterns that fall within the influence of that rule's constraint.

The first type of rule learners generally starts with a set of extremely small special rules. In the extreme, these will be rules centered on one or more training examples having constraints that limit numerical values to the exact value of that instance and a precise match of the nominal value. So, for a training instance $(\mathbf{v}, k)$ with

$$\mathbf{v} = (12, 3.5, \text{red}),$$

such a rule would look like

$$\text{IF } x_1 = 12 \text{ AND } x_2 = 3.5 \text{ AND } x_3 = \text{'red' THEN class } k.$$

There are two typical generalization operators that can now be applied iteratively. If we start with one rule, we will attempt to make this rule cover more training instances by finding one (usually the closest) example that can be included in this rule without including any other example. A second training instance $(\mathbf{v}_2, k)$ with

$$\mathbf{v}_2 = (12.3, 3.5, \text{blue})$$

would result in a revised rule

$$\text{IF } x_1 \in [12, 12.3] \text{ AND } x_2 = 3.5 \text{AND } x_3 = \text{'red' OR 'blue' THEN class } k.$$

Alternatively, we can also combine two rules into a new one, paying attention that we are not accidentally including a third rule of different class. The first approach (extending rules by covering one additional example) can be seen as a special case of the second approach (merging two rules) since we can always model a single training example as a (very) specific rule.

More generally, we can regard these training algorithms as a heuristic greedy search which iteratively tries to make rules more general by either merging two rules of the original set or by enlarging one rule to cover an additional training example.[8] The biggest question is then which two rules (or one rule and training example) to pick in the next step and how to do the generalization in order to merge the rules (or cover the additional example). As with all greedy algorithms, no matter how these two questions are answered, the resulting algorithm will usually not return the optimal set of rules, but in most cases the result will be relatively close to the optimum.

Specializing Rule Learners operate exactly the opposite way. They start with very general rules, in the extreme with one rule of the following form:

$$\text{IF true THEN class } k$$

---

[8]Note that this is a substantial deviation from the abstract concepts of rule learners in Mitchell's version space setup: real-world rule learners usually do not investigate *all* more general (or more specific) rules but only a subset of those chosen by the employed heuristic(s).

Table 8.5 The basic algorithm to learn a set of rules

Algorithm BuildRuleSet($\mathcal{D}$, p_{min})

---

input: training data $\mathcal{D}$
parameter: performance threshold p_{min}
output: a rule set R matching $\mathcal{D}$ with performance $\geq p_{min}$

---

1      $R = \emptyset$
2      $\mathcal{D}_{rest} = \mathcal{D}$
3   **while** (Performance(R, $\mathcal{D}_{rest}$) $< p_{min}$)
4         $r = $ FindOneGoodRule($\mathcal{D}_{rest}$)
5         $R = R \cup \{r\}$
6         $\mathcal{D}_{rest} = \mathcal{D}_{rest} - $ covered(r, $\mathcal{D}_{rest}$)
7   **endwhile**
8   **return** R

---

for each class k. Then they iteratively attempt to avoid misclassifications by specializing these rules, i.e., adding new or narrowing existing constraints.

We have now, of course, not yet really talked about learning more than one rule; all we know is how to generate *one* rule for a data set. However, it is quite optimistic to assume that one simple rule will be sufficient to explain a complex real-world data set. Most rule learning algorithms hence wrap this "one rule learning" approach into an outer loop, which tries to construct an entire set of rules. This outer loop often employs a set covering strategy, also known as **sequential covering** and generically looks as shown in Table 8.5.

The initialization (steps 1 and 2) creates an empty set of rules R and sets the training instances which are still to be explained to the entire dataset. The while loop runs into the performance of the rule set R, reaches a certain threshold p_{min} and iteratively creates a new rule r, adds it to the rule base R, and removes the instances it covers from the "still to be explained" instances. Once the threshold is reached, the resulting rule set is returned.

The biggest variations in existing implementations of this base algorithm are the chosen error measure to measure the performance of a given rule (or sets of rules) and the strategy to find one good rule which optimizes this performance criterion. One of the earlier and still very prominent methods is called CN2 [6] and uses a simple generalizing rule searching heuristic as shown in Table 8.6.

This routine essentially performs a search for all hypothesis starting with a general one (line 1 and 2) and iteratively specializing them (line 4). During each iteration, the so far best hypothesis is remembered (line 5). There are two heuristics involved controlling the specialization (line 4) and the termination criteria; the latter is done in line 6, where all newly generated hypotheses which do not fulfill a validity criteria are eliminated. For this, CN2 uses a significance test on the dataset. The specialization in line 4 returns only consistent, maximally specific hypotheses. In line 8, finally, a rule is assembled, assigning to the chosen antecedent the majority class of the patterns covered by the best hypothesis.

CN2 therefore not only returns 100% correct hypotheses but also rules which do make some errors on the training data—the amount of this error is controlled by the statistical significance test in the update-routing used in line 6. The CN2 algorithm

Table 8.6 The algorithm to learn one rule as used in CN2

Algorithm FindOneGoodRule($\mathcal{D}_{\text{rest}}$)

input: (subset of) training data $\mathcal{D}_{\text{rest}}$
output: one good rule r explaining some instance of the training data

1	$h_{\text{best}} = \texttt{true}$
2	$H_{\text{candidates}} = \{h_{\text{best}}\}$
3	**while** $H_{\text{candidates}} \neq \emptyset$
4	$H_{\text{candidates}} = \text{specialize}(H_{\text{candidates}})$
5	$h_{\text{best}} = \arg\max_{h \in H_{\text{candidates}} \cup \{h_{\text{best}}\}} \{\text{Performance}(h, \mathcal{D}_{\text{rest}})\}$
6	update($H_{\text{candidates}}$)
7	**endwhile**
8	**return** $'$IF h_{best} THEN $\arg\max_k \{|\text{covered}_k(h_{\text{best}}, \mathcal{D}_{\text{rest}})|\}'$

essentially performs a beam search with variable beam width (controlled by the significance test) and evaluates all hypotheses generated within the beam based on the performance criteria used also in the main algorithm.

We have not yet specified which attribute types we are dealing with, the "specialized" routine just assumes that we know how to specialize (or generalize in other types of algorithms) our rules. For nominal attributes it is pretty straightforward: specialization removes individual values from the set of allowed ones, and generalization adds some. In the extreme we have either only one value left (leaving none does not make much sense as it results in a rule that is never valid) or all possible values, resulting in a constraint that is always true. But what about numerical values? The most specific case is easy as it translates an equality to one exact numerical value. Also the most generic case is simple: it results in an interval from minus to plus infinity. But also generalizing a numerical constraint so that it contains a value it previously did not contain is simple: we only have to expand the interval so that it uses this new value either as new lower or upper bound, depending on which side of the interval the value lied. Specializing is a bit more complicated as we have two options. In order to move a specific value out of a given numerical interval, we can either move the lower or upper bound to be just above (or below) the given value. This is one point where heuristics start playing a role. However, the much bigger impact of heuristics happens during multidimensional specialization operations. Figure 8.15 illustrates this for the case of two numerical attributes. On the left we show how we can generalize a given rule (solid line) to cover an additional training instance (cross). The new rule (dashed lines) is given without any ambiguity. On the right we show a given rule (solid line) and four possibilities to avoid a given conflict (cross). Note that the different rectangles are not drawn right on top of each other as they should be but slightly apart for better visibility. It is obvious that we have a number of choices and this number will not decrease with an increase in dimensionality.

However, there is one big problem with all of the current propositional rule induction approaches following the above schema: for real-world data sets, they tend to generate an enormous amount of rules. Each one of these rules is, of course, interpretable and hence fulfills the requirements of the methods described in this chapter.

Fig. 8.15 A simple example
for two-dimensional rule
generalization (*left*) and
generalization (*right*)

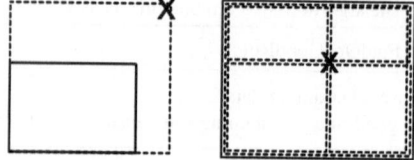

However, the overall set of rules is often a way too large to be even remotely comprehensible. Some approaches have been proposed to learn hierarchical rule models
or find rules with outlier models, but those have not yet gained great prominence,
in part also because they rely on a set of heuristics which are hard to control. In the
following section we discuss a few approaches that try to address this problem.

8.4.1.3 Other Types of Propositional Rule Learners

In order to handle the larger number of rules, various approaches have been proposed. Some simply prune rules by their importance, which is usually measured by
the number of patterns they cover. Others include the minimum description length
principle to balance the complexity (e.g., length) of a rule against the amount of
data it covers. Similarly to decision tree learning, we can of course here also employ
pruning strategies to reduce collapse several rules into one or completely eliminate
them. Other approaches attempt to further reduce the number of rules by looking
at their discriminative power between classes or other measures of interestingness.
The most prominent such measure is the **J-measure** [11], which essentially estimates how dissimilar the a priori and a posteriori beliefs about the rule's consequent
are. Only if these two likelihoods are substantially different, a rule is potentially interesting. In the J-measure this difference is additionally weighted by the generality
of the rule (probability that the rule's conditional part holds), because the rule becomes the more interesting the more often it applies. See Chap. 5 for further details
on some of these techniques.

Note that in the toy setup, as used in the Version Spaces, another very elegant
way to dramatically reduce the number of matching rules was presented: by only
reporting the most general and most specific rules which cover the training data, the
entire part of the lattice in between these two boundaries is reported as well. Unfortunately this cannot as easily be applied to rules stemming from real-world data
since coverage will hardly ever be exactly 100%. However, the setup described in
Sect. 7.6.3.3, where we describe a propositional rule learning system finding association rules, uses a related strategy. Also, here it is hard to dig through the resulting
list of association rules. However, looking at closed or maximum itemsets (and the
resulting association rules) drastically reduces the set of rules to consider. So we,
in effect, report only the most specific rules describing a local aspect of the training
data.

One of the reasons why propositional rule learners tend to produce excessive
numbers of rules, especially in numerical feature spaces, is the sharp boundaries
they attempt to impose. If training instances of different class are not separable by

axes parallel lines in some reasonably large local area, the rule learning algorithms are forced to introduce many small rules to model these relationships. Often the precise nature of these boundaries is not interesting or, worse, caused only by noisy data in the first place. To be able to incorporate this imprecision into the learning process, a large number of fuzzy rule learners have been introduced. They are all using the notion of fuzzy sets which allows one to model degrees of membership (to rules in this case) and hence results in the ability to model gray zones where areas of different class overlap. Fuzzy rule systems have the interesting side effect that it is possible to express a fuzzy rule system as a system of differentiable equations of the norm and membership functions are used accordingly. Then other adaptation methods can be applied, often motivated by work in the neural network community. The resulting Neuro-Fuzzy systems allow one, for instance, to use expert knowledge to initialize a set of fuzzy rules and then employ a gradient descent training algorithm to update those rules to better fit training data. In addition Takagi–Sugeno–Kang-type fuzzy rules also allow one to build regression systems based on fuzzy antecedents and (local) regression consequents. We refer to [3] for more details on fuzzy rules and the corresponding training methods and [16] for details on Neuro-Fuzzy systems.

8.4.2 Inductive Logic Programming or First-Order Rules

Propositional rules are quite limited in their expressive power. If we want to express a rule of the form

IF x is Father of y AND y is female THEN y is Daughter of x

using propositional rules, we would need to enumerate all possible values for x and y and put those pairs into individual rules. In order to express such types of rules, we need to introduce variables. First Order rules allow one to do just that and are based on only a few base constructs:

- constants, such as Bob, Luise, red, green,
- variables, such as x and y in the example above,
- predicates, such as is_father(x, y), which produce truth values as a result, and
- functions, such as age(x), which produce constants.

From this we can construct

- terms which are constants, variables, or functions applied to a term,
- literals, which are predicates (or negations of predicates) applied to any set of terms,
- ground literals, which are literals that do not contain a variable,
- clauses, which are disjunctions of literals whose variables are universally quantified, and
- horn clause, which are clauses with at most one positive literal.

The latter is especially interesting because any disjunction with at most one positive literal can be written as

$$H \vee \neg L_1 \vee \cdots \vee \neg L_n$$

$$\hat{=} H \Rightarrow (L_1 \wedge \cdots \wedge L_n)$$

$$\hat{=} \text{IF } L_1 \text{ AND} \ldots \text{AND } L_2 \text{ THEN } H.$$

So horn clauses express rules and H, the *head* of the rule, is the consequent, and the L_i together form the *body* or consequent.

Horn clauses are also used to express Prolog programs, which is also the reason why learning first-order rules is often referred to as **inductive logic programming** (ILP), because we can see this as learning (Prolog) programs, not only sets of rules.

We talk about *substitutions* which denote any replacement, called *binding* of variables in a literal with appropriate constants. A rule body is satisfied if at least one binding exists that satisfies the literals. A few example rules are:

- simple rules:
  IF x is Parent of y AND y is male THEN x is Father of y
- existentially qualified variables (z in this case):
  IF y is Parent of z AND z is Parent of x THEN x is Granddaughter of y
- Recursive rules:
  IF x is Parent of z AND z is Anchestor of y THEN x is Anchestor of y
  IF x is Parent of y THEN x is Anchestor of y

Inductive Logic Programming allows one to learn concepts also from data spread out over several multirelation databases—without the need for previous integration as discussed in Sect. 6.5. But how do we learn these types of rules from data? A number of learning methods have been proposed recently, and we will explain one of earlier algorithms in more detail in the following section.

FOIL, the First Order Inductive Learning method developed by Quinlan [19], operates very similar to the rule learners we have discussed in the previous section. It also follows the strategy of a sequential covering approach—the only difference being the inner FindOneGoodRule() routine. It creates not quite Horn clauses but something very similar besides two differences:

- the rules learned by FOIL are more restrictive in that they do not allow literals to contain function symbols (this reduces the hypothesis space dramatically!), and
- FOIL rules are more expressive than Horn clauses because they allow literals in the body to be negated.

The FindOneGoodRule() routine of FOIL again specializes candidate rules greedily. It does so by adding literals one by one where these new literals can be one of the following:

- P(v1,..., vr), where P is any predicate name occurring in the set of available predicates. At least one of the variables must already be present in the original rule, the others can be either new or existing;

- Equal(x, y), where *x* and *y* are variables already present in the rule; or
- the negation of either of the above forms of literals.

How does FOIL now pick the "best" specialization to continue? For this, Quinlan introduced a measure (FoilGain) which essentially estimates the utility of the new added literal by comparing the number of positive and negative bindings of the original and the extended rules.

This is, of course, a system that does not scale well with large, real-world data sets. ILP systems are in general not heavily used for real data analysis. They are more interesting for concept learning in data sets with purely nominal values and/or structured data sets. However, some of the ideas are used in other applications such as molecular fragment mining.

8.5 Finding Explanations in Practice

Finding explanations is, similar to finding patterns, a central piece of any analytics software. The main difference is the way the extracted model is presented to the user: earlier tools produced long lists of rules or other ASCII representations, but nowadays tools increasingly offer interactive views which allow one to select individual parts of an explanation, e.g., a leaf in a decision tree, and propagate this selection to other views on the underlying data—see Sect. 4.8 for a more detailed discussion of this type of visual brushing. KNIME is, of course, inherently better suited for such type of visual explorations, but also R offers quite a number of graphical representations of the discovered explanations.

8.5.1 Finding Explanations with KNIME

When constructing decision trees in data mining software, a number of options are available. KNIME allows one to adjust the information gain criteria and a number of other options. Figure 8.16 (left) shows the dialog of the native KNIME decision tree learner. We need to first and foremost select the target attribute (class—it has to be a nominal attribute, a string in this case). Afterwards, we can choose between two different ways to compute the information gain (Gini index and Gain ratio) and if a pruning of the tree is to be performed (KNIME offers to either skip pruning or performs a minimum description length (MDL)-based pruning). Most of the remaining options are self-explanatory or related to the hiliting support of KNIME (number of records to be stored for the interaction). Noteworthy is the last option "number threads," which allows one to control how many threads KNIME can use to execute the learning method in parallel on, e.g., a multicore machine. Once the node is run, we can display the resulting decision tree. Figure 8.16 (right) shows the resulting tree for the training part of the iris data, which, hopefully, shows interesting insights into the structure of the data. The KNIME view shows the color

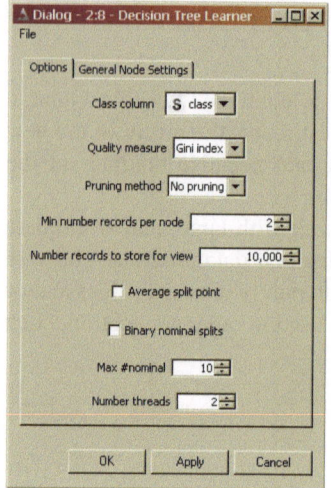
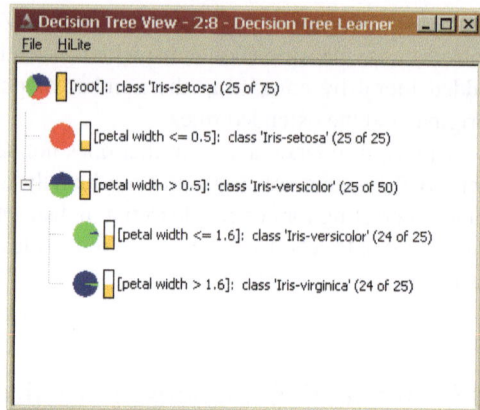

Fig. 8.16 The KNIME dialog of the decision tree construction node (*left*) along with the tree view (*right*) after running on the training part of the iris data

Fig. 8.17 Through the Weka integration, KNIME also offers access to the vast variety of algorithm and views contained in the open-source data mining library Weka

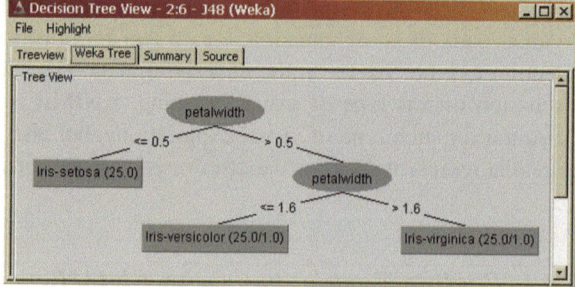

distribution (if available) for each node of the tree, along with the split attribute and the majority class. The numbers in brackets show the number of training patterns classified correctly together with the overall number of training patterns falling into this branch. Additionally, the fraction of patterns falling into splits are displayed by the vertical orange bar charts. The Weka decision tree learning node works similarly. Although the Weka j4.8 (an implementation following closely the original c4.5, revision 8 version of Quinlan's algorithm) offers more options to fine tune the training algorithm, most of those options will hardly be used in practice. After installing the Weka integration, all learning and clustering algorithms are available as individual nodes in KNIME as well. Additionally, also the Weka views are accessible. Figure 8.17 shows the view of the Weka decision tree nodes. The Weka integration also offers access to a regression tree implementation, which is—as of version 2.1—not yet available in KNIME.

Many other explanation finding methods are available in KNIME, among them, Naive Bayes, as described in Sect. 8.2. The corresponding node in KNIME produces

Fig. 8.18 KNIME's Naive
Bayes Learner shows the
distribution parameters in the
node view

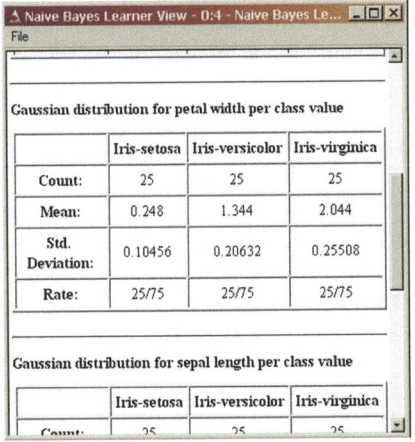

the class-conditional probabilities or normal distributions which can be inspected in
the node's view. Figure 8.18 shows this for the Iris data. This corresponds to the
distribution parameters listed in Table 8.2, fortunately. KNIME also provides nodes
for Linear, Polynomial, and Logistic Regression. For rule learning, a constructive
fuzzy rule learning method is available, and through the Weka integration, also other
rule learners can be accessed. Those methods are all accessible via individual nodes,
and their dialogs (and inline documentation) should explain the available options
sufficiently.

8.5.2 Using Explanations with R

8.5.2.1 Decision Trees

R also allows for much finer control of the decision tree construction. The script
below demonstrates how to create a simple tree for the Iris data set using a training
set of 100 records. Then the tree is displayed, and a confusion matrix for the test
set—the remaining 50 records of the Iris data set—is printed. The libraries rpart,
which comes along with the standard installation of R, and rattle, that needs to
be installed, are required:

```
> library(rpart)
> iris.train <- c(sample(1:150,75))
> iris.dtree <- rpart(Species~.,data=iris,
                                  subset=iris.train)
> library(rattle)
> drawTreeNodes(iris.dtree)
> table(predict(iris.dtree,iris[-iris.train,],
                                  type="class"),
              iris[-iris.train,"Species"])
```

In addition to many options related to tree construction, R also offers many ways to beautify the graphical representation. We refer to R manuals for more details.

8.5.2.2 Naive Bayes Classifers

Naive Bayes classifiers use normal distributions by default for numerical attributes. The package e1071 must be installed first:

```
> library(e1071)
> iris.train <- c(sample(1:150,75))
> iris.nbayes <- naiveBayes(Species~.,data=iris,
                             subset=iris.train)
> table(predict(iris.nbayes,iris[-iris.train,],
                             type="class"),
                 iris[-iris.train,"Species"])
```

As in the example of the decision tree, the Iris data set is split into a training and a test data set, and the confusion matrix is printed. The parameters for the normal distributions of the classes can be obtained in the following way:

```
> print(iris.nbayes)
```

If Laplace correction should be applied for categorical attribute, this can be achieved by setting the parameter laplace to the desired value when calling the function naiveBayes.

8.5.2.3 Regression

Least squares linear regression is implemented by the function lm (linear model). As an example, we construct a linear regression function to predict the petal width of the Iris data set based on the other numerical attributes:

```
> iris.lm <- lm(iris$Petal.Width ~ iris$Sepal.Length
                 + iris$Sepal.Width + iris$Petal.Length)
> summary(iris.lm)
```

The summary provides the necessary information about the regression result, including the coefficient of the regression function.

If we want to use a polynomial as the regression function, we need to protect the evaluation of the corresponding power by the function I inhibiting interpretation. As an example, we compute a regression function to predict the petal width based on a quadratic function in the petal length:

```
> iris.lm <- lm(iris$Petal.Width ~ iris$Petal.Length +
                 I(iris$Petal.Length^2))
```

Robust regression requires the library MASS, which needs installation. Otherwise it is handled in the same way as least squares regression, using the function rlm instead of lm:

```
> iris.rlm <- rlm(iris$Petal.Width ~ iris$Sepal.Length
                + iris$Sepal.Width + iris$Petal.Length)
```

The default method is based on Huber's error function. If Tukey's biweight should be used, the parameter method should be changed in the following way:

```
> iris.rlm <- rlm(iris$Petal.Width ~ iris$Sepal.Length
                + iris$Sepal.Width + iris$Petal.Length,
                method="MM")
```

A plot of the computed weights can be obtained by the following command:

```
> plot(iris.rlm$w)
```

8.6 Further Reading

Explanation Finding Methods are almost always the center piece of data mining books, so any book on this topic will likely cover most, if not all, of what we discussed in this chapter. For Decision trees, there are two main directions:

- Classification and Regression Trees (CART), which are covered in much more detail in the following book: Hastie, T., Tibshirani, R. and Friedman, J.H.: *Elements of Statistical Learning* (Springer, 2001).
- Quinlan's c4.5 algorithm and subsequent improvements, with the original text book describing c4.5r8: J.R. Quinlan: *Induction of Decision Trees* (Springer, 1986).

For Bayes and Regression, any more statistically oriented book offers more background; these (and much more) with a more statistical view are also described in the book by Hastie, Tibshirani, and Friedman.

Tom Mitchell's excellent *Machine Learning* book (McGraw Hill, 1997) introduces the concept of version space learning and also describes decision tree induction, among others. In order to get a better feeling for how learning systems work in general, this is highly recommended reading, especially the first half. For fuzzy rule learning systems, we recommend the chapter on "Fuzzy Logic" in *Intelligent Data Analysis: An Introduction*, M.R. Berthold and D.J. Hand (Eds), published by Springer Verlag (2003) and the book on *Foundations of Neuro-Fuzzy Systems* by D. Nauck, F. Klawonn, and R. Kruse (Wiley, 1997). For inductive logic programming, Peter Flach and Nada Lavrac contributed a nice chapter in the *Intelligent Data Analysis* book edited by Berthold and Hand.

References

1. Albert, A.: Regression and the Moore–Penrose Pseudoinverse. Academic Press, New York (1972)
2. Anderson, E.: The irises of the Gaspe Peninsula. Bull. Am. Iris Soc. **59**, 2–5 (1935)
3. Berthold, M.R.: Fuzzy logic. In: Berthold, M.R., Hand, D.J. (eds.) Intelligent Data Analysis: An Introduction, 2nd edn. Springer, Berlin (2003)
4. Borgelt, C., Steinbrecher, M., Kruse, R.: Graphical Models—Representations for Learning, Reasoning and Data Mining, 2nd edn. Wiley, Chichester (2009)
5. Breiman, L., Friedman, J.H., Olshen, R.A., Stone, C.J.: CART: Classification and Regression Trees. Wadsworth, Belmont (1983)
6. Clark, P., Niblett, T.: The CN2 induction algorithm. Mach. Learn. **3**(4), 261–283 (1989)
7. Domingos, P., Pazzani, M.: On the optimality of the simple Bayesian classifier under zero-one loss. Mach. Learn. **29**, 103–137 (1997)
8. Fisher, R.A.: The use of multiple measurements in taxonomic problems. Ann. Eugen. **7**(2), 179–188 (1936)
9. Friedman, N., Goldszmidt, M.: Building classifiers using Bayesian networks. In: Proc. 13th Nat. Conf. on Artificial Intelligence (AAAI'96, Portland, OR, USA), pp. 1277–1284. AAAI Press, Menlo Park (1996)
10. Geiger, D.: An entropy-based learning algorithm of Bayesian conditional trees. In: Proc. 8th Conf. on Uncertainty in Artificial Intelligence (UAI'92, Stanford, CA, USA), pp. 92–97. Morgan Kaufmann, San Mateo (1992)
11. Goodman, R.M., Smyth, P.: An information-theoretic model for rule-based expert systems. In: Int. Symposium in Information Theory. Kobe, Japan (1988)
12. Janikow, C.Z.: Fuzzy decision trees: issues and methods. IEEE Trans. Syst. Man, Cybern., Part B **28**(1), 1–14 (1998)
13. Jensen, F.V., Nielsen, T.D.: Bayesian Networks and Decision Graphs, 2nd edn. Springer, London (2007)
14. Larrañaga, P., Poza, M., Yurramendi, Y., Murga, R., Kuijpers, C.: Structural learning of Bayesian networks by genetic algorithms: a performance analysis of control parameters. IEEE Trans. Pattern Anal. Mach. Intell. **18**, 912–926 (1996)
15. Mitchell, T.: Machine Learning. McGraw-Hill, New York (1997)
16. Nauck, D., Klawonn, F., Kruse, R.: Neuro-Fuzzy Systems. Wiley, Chichester (1997)
17. Quinlan, J.R.: Induction of decision trees. Mach. Learn. **1**(1), 81–106 (1986)
18. Quinlan, J.R.: C4.5: Programs for Machine Learning. Morgan Kaufmann, San Mateo (1993)
19. Quinlan, J.R., Cameron-Jones, R.M.: FOIL: a midterm report. In: Proc. European Conference on Machine Learning. Lecture Notes in Computer Science, vol. 667, pp. 3–20. Springer, Berlin (1993)
20. Sahami, M.: Learning limited dependence Bayesian classifiers. In: Proc. 2nd Int. Conf. on Knowledge Discovery and Data Mining (KDD'96, Portland, OR, USA), pp. 335–338. AAAI Press, Menlo Park (1996)

Chapter 9
Finding Predictors

In this chapter we consider methods of constructing predictors for class labels or numeric target attributes. However, in contrast to Chap. 8, where we discussed methods for basically the same purpose, the methods in this chapter yield models that do not help much to explain the data or even dispense with models altogether. Nevertheless, they can be useful, namely if the main goal is good prediction accuracy rather than an intuitive and interpretable model. Especially artificial neural networks and support vector machines, which we study in Sects. 9.2 and 9.3, are known to outperform other methods w.r.t. accuracy in many tasks. However, due to the abstract mathematical structure of the prediction procedure, which is usually difficult to map to the application domain, the models they yield are basically "black boxes" and almost impossible to interpret in terms of the application domain. Hence they should be considered only if a comprehensible model that can easily be checked for plausibility is not required, and high accuracy is the main concern.

Nearest-Neighbor Predictors   A very natural approach to predict a class or a numeric target value is to look for historic cases that are similar to the new case at hand and to simply transfer the class or target value of these historic cases. This principle, which we apply in all sorts of situations in daily life, is the foundation of nearest-neighbor predictors. They abstain from computing an explicit model of the domain or the prediction task but simply store a database of historic cases to draw on if a new case has to be processed. In addition, they only need a formalization of what "similar" means in the application context, usually in the form of a distance measure (which explains the term "nearest" in the name). The advantage of such an approach is that it is usually easy to convince people that this is a reasonable scheme to follow, simply because it is so common in daily life. It also yields good predictive performance in many instances. However, its disadvantage is that it does not yield any insights into the mechanisms that determine the class or the value of the target attribute, because it is not even tried to model the relationships between the target attribute and other attributes; they are left implicit in the database. An indication of why a certain prediction was made can only be obtained by looking at the historic cases that were chosen as the nearest/most similar.

M.R. Berthold et al., *Guide to Intelligent Data Analysis*,
Texts in Computer Science 42,
DOI 10.1007/978-1-84882-260-3_9, © Springer-Verlag London Limited 2010

Artificial Neural Networks Among methods that try to endow machines with learning ability, artificial neural networks are among the oldest and most intensely studied approaches. They take their inspiration from biological neural networks and try to mimic the processes that make animals and human beings able to learn and adapt to new situations. However, the used model of biological processes is very coarse, and several improvements to the basic approach have even abandoned the biological analogy. The most common form of artificial neural networks, *multilayer perceptrons*, can be described as a staged or hierarchical logistic regression (see Sect. 8.3), which is trained with a gradient descent scheme, because an analytical solution is no longer possible due to the staged/hierarchical structure. The advantage of artificial neural networks is that they are very flexible and thus often achieve very good accuracy. However, the resulting models, due to their involved mathematical structure (complex prediction function), are basically impossible to interpret in terms of the application domain. In addition, choosing an appropriate network structure and conducting the training in such a way that overfitting is avoided can be tricky.

Support Vector Machines Though closely related to specific types of artificial neural networks, support vector machines approach the problem of finding a predictor in a way that is more strongly based on statistical considerations, especially on risk minimization approaches. After they were introduced about 15 years ago, they quickly became highly popular, because they usually offer shorter training times than artificial neural networks, while achieving comparable or even better accuracy. The basic principle underlying support vector machines can be described as follows: when it comes to classification (with two classes), linear separation is enough; when it comes to numeric prediction, linear regression is enough—one only has to properly map the given data to a different feature space. However, this mapping is never explicitly computed, but only a certain property of it (to be precise, the scalar product in the image space) is described implicitly by a so-called *kernel function*, which intuitively can be seen as a similarity measure for the data points. As a consequence, the choice of the kernel function is crucial for constructing a support vector machine, and therefore a large number of specialized kernels for different application domains have been developed. The advantage of support vector machines is, as already mentioned, that they usually achieve very good accuracy. However, the support vectors they base their prediction on usually do not provide much insight into the relationships that are exploited.

Ensemble Methods If we are sick and seek help from a physician but have some doubts about the diagnosis we are given, it is standard practice to seek a second or even a third opinion. Ensemble methods follow the same principle: for a difficult prediction task, do not rely on a single classifier or numeric predictor but generate several different predictors and aggregate their individual predictions. Provided that the predictors are sufficiently different (that is, exploit different properties of the data to avoid what is expressed in the saying, usually ascribed to Ludwig Wittgenstein: "If I do not believe the news in todays paper, I buy 100 copies of the paper. Then I believe."), this can often lead to a considerably improved prediction accuracy. The main tasks to consider when constructing such ensemble predictors is how to select or generate a set of predictors that exhibits sufficient variation to offer good chances of improving the prediction quality and how to combine the individual predictions.

9.1 Nearest-Neighbor Predictors

The nearest-neighbor algorithm [13] is one of the simplest and most natural classification and numeric prediction methods: it derives the class labels or the (numeric) target values of new input objects from the most similar training examples, where similarity is measured by distance in the feature space. The prediction is computed by a majority vote of the nearest neighbors or by averaging their (numeric) target values. The number k of neighbors to be taken into account is a parameter of the algorithm, the best choice of which depends on the data and the prediction task.

9.1.1 Overview

Constructing nearest-neighbor classifiers and numeric predictors is a special case of **instance-based learning** [1]. As such, it is a **lazy learning** method in the sense that it is not tried to construct a model that generalizes beyond the training data (as **eager learning** methods do), but the training examples are merely stored. Predictions for new cases are derived directly from these stored examples and their (known) classes or target values, usually without any intermediate model construction.

In a basic nearest-neighbor approach only one neighbor object, namely the closest one, is considered, and its class or target value is directly transferred to the query object. This is illustrated in Fig. 9.1, with classification in a two-dimensional feature space on the left and regression in a one-dimensional input space on the right.

In the left diagram the training examples are drawn as small dots. For illustration purposes, a **Voronoi diagram** [2, 34], which is a special kind of decomposition of a

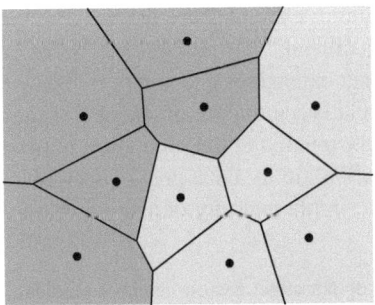
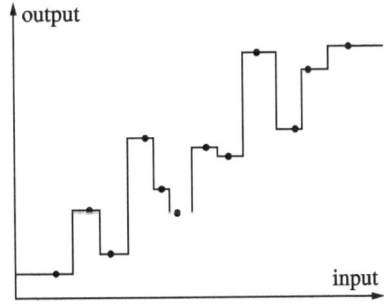

Fig. 9.1 Illustrations of nearest neighbor classification (*left*) and numeric prediction (*right*)

metric space, has been derived from them. In such a Voronoi diagram each training example s is a Voronoi site, which defines a Voronoi cell consisting of all points that are closer to s than to any other site. The shown line segments comprise those points that are equidistant to their two nearest sites, the nodes (i.e., points where line segments meet) are the points that are equidistant to three (or more) sites. A nearest-neighbor classifier transfers the class of the site s of a Voronoi cell to all points in the cell of s, illustrated here by two different colorings (shades of grey) of the cells, which indicates two classes and thus the regions where these classes are predicted. For points on the line segments, if they border Voronoi cells with different classes, some (arbitrary) tie-breaking rule has to be applied.

In the right diagram, the horizontal axis represents the input space, and the vertical axis the output (or target) space of a regression task (i.e., a numeric prediction task). The training examples are again drawn as small dots. The (now metric) target value is transferred from a training example s to all query values that are closer to (the input value of) s than to (the input value of) any other training example. In effect, this yields a piecewise constant regression function as the predictor.

Note, however, that neither the Voronoi tessellation nor the piecewise linear function is actually computed in the learning process, and thus no model is built at training time. The prediction is determined only in response to a query for the class or target value of a new input object, namely by finding the closest neighbor of the query object and then transferring its class or target value. Hence the diagrams should be seen as illustrations that summarize the results of all possible queries within the range of the diagrams rather than depictions of learned models.

A straightforward generalization of the nearest-neighbor approach is to use not just the one closest, but the k **nearest neighbors** (usually abbreviated as k-NN). If the task is classification, the prediction is then determined by a majority vote among these k neighbors (breaking ties arbitrarily); if the task is numeric prediction, the average of the target values of these k neighbors is computed. Not surprisingly, using more than one neighbor improves the robustness of the algorithm, since it is not so easily fooled by individual training instances that are labeled incorrectly or outliers for a class (that is, data points that have an unusual location for the class assigned to them).[1] However, using too many neighbors can reduce the capability of the algorithm as it may smooth the classification boundaries or the interpolation too much to yield good results. As a consequence, apart from the core choice of the distance function that determines which training examples are the nearest, the choice of the number of neighbors to consider is crucial.

Once multiple neighbors are considered, further extensions become possible (see the next sections for details). For example, the (relative) influence of a neighbor may be made dependent on its distance from the query point, and the prediction may be computed from a local model that is constructed on the fly for a given query point (i.e., from its nearest neighbors) rather than by a simple majority or averaging rule.

---

[1]Outliers for the complete data set, on the other hand, do not affect nearest-neighbor predictors much, because they can only change the prediction for data points that should not occur or should occur only very rarely (provided that the rest of the data is representative).

9.1.2 *Construction*

Generally, any nearest-neighbor algorithm has four main ingredients:

- **distance metric**
  The distance metric, together with a possible task-specific scaling or weighting of the attributes, determines which of the training examples are nearest to a query data point and thus selects the training example(s) used to compute a prediction.
- **number of neighbors**
  The number of neighbors of the query point that are considered can range from only one (the basic nearest-neighbor approach) through a few (like k-nearest-neighbor approaches) to, in principle, all data points as an extreme case.
- **weighting function for the neighbors**
  If multiple neighbors are considered, it is plausible that closer (and thus more similar) neighbors should have a stronger influence on the prediction result. This can be expressed by a weighting function defined on the distance of a neighbor from the query point, which yields higher values for smaller distances.
- **prediction function**
  If multiple neighbors are considered, one needs a procedure to compute the prediction from the classes or target values of these neighbors, since they may differ and thus may not yield a unique prediction directly.

Necessary fundamentals about similarity and distance metrics have already been discussed in the context of clustering in Sect. 7.2 and are thus not repeated here. Note that the choice of the distance function is crucial for the behavior of a nearest neighbor algorithm. Therefore it should be chosen with a lot of care and not just be set to the Euclidean distance of the data points in their original representation.

As already mentioned in the preceding section, the main effect of the number k of considered neighbors is how much the class boundaries or the numeric prediction is smoothed. If only one neighbor is considered, the prediction is constant in the Voronoi cells of the training data set. This makes the prediction highly susceptible to the deteriorating effects of incorrectly labeled instances or outliers w.r.t. their class, because a single data point with the wrong class spoils the prediction in its whole Voronoi cell. Considering several neighbors ($k > 1$) mitigates this problem, since neighbors having the correct class can override the influence of an outlier. However, choosing a very large k is also not generally advisable, because it can prevent the classifier from being able to properly approximate narrow class regions or narrow peaks or valleys in a numeric target function. An example of 3-nearest-neighbor prediction (using a simple averaging of the target values of these nearest neighbors), which already shows the smoothing effect (especially at the borders of the input range), is shown in Fig. 9.2 on the left.

A common method to automatically determine an appropriate value for the number k of neighbors is **cross-validation** (see Sect. 5.5.2 for a general treatment): the training data set is divided into r folds of (approximately) equal size.[2] Then r clas-

---

[2]The fold sizes may differ by one data point, to account for the fact that the total number of training examples may not be divisible by r, the number of folds.

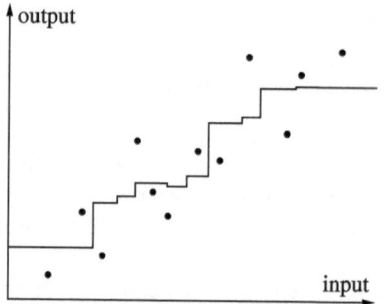
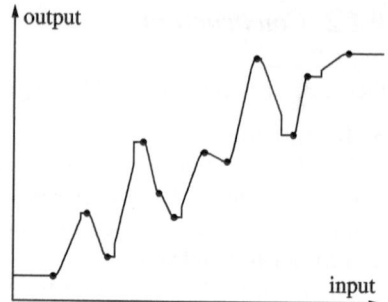

Fig. 9.2 Illustrations of standard 3-nearest neighbor prediction (average of three nearest neighbors, *left*) and distance-weighted 2-nearest neighbor prediction (*right*) in one dimension

sification or prediction experiments are performed: each combination of $r - 1$ folds is once chosen as the training set, with which the remaining fold is classified or with which the (numeric) target value is predicted, using all numbers k of neighbors from a user-specified range. The classification accuracy or the prediction error is aggregated, for the same value k, over these experiments. Finally, the number k of neighbors that yields the lowest aggregated error is chosen.

Approaches that weight the considered neighbors differently based on their distance to the query point are known as **distance-weighted k-nearest neighbor** or (for numeric targets) **locally weighted regression** or **locally weighted scatterplot smoothing** (abbreviated LOWESS or LOESS) [11, 12]. Such weighting is mandatory in the extreme case in which all n training examples are used as neighbors, because otherwise only the majority class or the global average of the target values can be predicted. However, it is also recommended for $k < n$, since it can, at least to some degree, counteract the smoothing effect of a large k, because the excess neighbors are likely to be farther away and thus will influence the prediction less. It should be noted, though, that distance-weighted k-NN is not a way of avoiding the need to find a good value for the number k of neighbors.

A typical example of a weighting function for the nearest neighbors is the so-called tricubic weighting function [11, 12] defined as

$$w(s_i, q, k) = \left(1 - \left(\frac{d(s_i, q)}{d_{\max}(q, k)} \right)^3 \right)^3.$$

Here q is the query point, s_i is (the input vector of) the ith nearest neighbor of q in the training data set, k is the number of considered neighbors, d is the employed distance function, and $d_{\max}(q, k)$ is the maximum distance between any two points from the set $\{q, s_1, \ldots, s_k\}$, that is, $d_{\max}(q, k) = \max_{a, b \in \{q, s_1, \ldots, s_k\}} d(a, b)$. As an illustration, Fig. 9.2 shows on the right the distance-weighted 2-nearest-neighbor prediction for a one-dimensional input. Note that the interpolation is mainly linear (because *two* nearest neighbors are used), except for some small plateaus close to the data points, which result from the weighting, and certain jumps at points where the two closest neighbors are on the same side of the query point.

The most straightforward choices for the prediction function are, as already mentioned, a (weighted) majority vote for classification or a simple average for numeric prediction. However, especially for numeric prediction, one may also consider more complex prediction functions, like building a local regression model from the neighbors (usually with a low-degree polynomial), thus arriving at locally weighted polynomial regression (see Sect. 9.1.3 for some details).

A core issue of implementing nearest-neighbor prediction is the data structure used to store the training examples. In a naive implementation they are simply stored as a list, which requires merely $O(n)$ time, where n is the number of training examples. However, though fast at training time, this approach has the serious drawback of being very slow at execution time, because a linear traversal of all training examples is needed to find the nearest neighbor(s), requiring $O(nm)$ time, where m is the dimensionality of the data. As a consequence, this approach becomes quickly infeasible with a growing number of training examples or for high-dimensional data.

Better approaches rely on data structures like a kd-tree (short for k-dimensional tree[3]) [4, 20], an R- or R*-tree [3, 24], a UB-tree [33], etc. With such data structures, the query time can be reduced to $O(\log n)$ per query data point. The time to store the training examples—that is, the time to construct an efficient access structure for them—is, of course, worse than for storing them in a simple list. However, it is usually acceptably longer. For example, a kd-tree is constructed by iterative bisections in different dimensions that split the set of data points (roughly) equally. As a consequence, constructing it from n training examples takes $O(n \log n)$ time if a linear time algorithm for finding the median in a dimension [5, 14] is employed.

9.1.3 Variations and Issues

The basic k-nearest-neighbor scheme can be varied and extended in several ways. By using kernel functions to weight the neighbors (Sect. 9.1.3.1), all data points or a variable number of neighbors, depending on the query point, may be used. The simple averaging of the target values for numeric prediction may be replaced by building a (simple) local regression model (Sect. 9.1.3.2). Feature weights may be used to adapt the employed distance function to the needs of the prediction task (Sect. 9.1.3.3). In order to mitigate the problem to extract the nearest neighbors from the training data, one may try to form prototypes in a preprocessing step (Sect. 9.1.3.4).

9.1.3.1 Weighting with Kernel Functions

An alternative approach to distance-weighted k-NN consists in abandoning the requirement of a predetermined number of nearest neighbors. Rather a data point is

---

[3]Note that this k is independent of and not to be confused with the k denoting the number of neighbors. This equivocation is an unfortunate accident, which, however, cannot be avoided without deviating from standard nomenclature.

weighted with a kernel function K that is defined on its distance d to the query point and that satisfies the following properties: (1) $K(d) \geq 0$, (2) $K(0) = 1$ (or at least that K has its mode at 0), and (3) $K(d)$ decreases monotonously for $d \to \infty$. In this case all training examples for which the kernel function yields a nonvanishing value w.r.t. a given query point are used for the prediction. Since the density of training examples may, of course, differ for different regions of the feature space, this may lead to a different number of neighbors being considered, depending on the query point. If the kernel function has an infinite support (that is, does not vanish for any finite argument value), *all* data points are considered for any query point. By using such a kernel function, we try to mitigate the problem of choosing a good value for the number K of neighbors, which is now taken care of by the fact that instances that are farther away have a smaller influence on the prediction result. On the other hand, we now face the problem of having to decide how quickly the influence of a data point should decline with increasing distance, which is analogous to choosing the right number of neighbors and can be equally difficult to solve.

Examples of kernel functions with a finite support, given as a radius σ around the query point within which training examples are considered, are

$$K_{\text{rect}}(d) = \tau(d \leq \sigma),$$

$$K_{\text{triangle}}(d) = \tau(d \leq \sigma) \cdot (1 - d/\sigma),$$

$$K_{\text{tricubic}}(d) = \tau(d \leq \sigma) \cdot \left(1 - d^3/\sigma^3\right)^3,$$

where $\tau(\phi)$ is 1 if ϕ is true and 0 otherwise. A typical kernel function with infinite support is the Gaussian function

$$K_{\text{gauss}}(d) = \exp\left(-\frac{d^2}{2\sigma^2}\right),$$

where d is the distance of the training example to the query point, and σ^2 is a parameter that determines the spread of the Gaussian function. The advantage of a kernel with infinite support is that the prediction function is smooth (has no jumps) if the kernel is smooth, because then a training case does not suddenly enter the prediction if a query point is moved by an infinitesimal amount, but its influence rises smoothly in line with the kernel function. One also does not have to choose a number of neighbors. However, the disadvantage is, as already pointed out, that one has to choose an appropriate radius σ for the kernel function, which can be more difficult to choose than an appropriate number of neighbors.

An example of kernel regression with a Gaussian kernel function is shown in Fig. 9.3 on the left. Note that the regression function is smooth, because the kernel function is smooth and always refers to all data points as neighbors, so that no jumps occur due to a change in the set of nearest neighbors. The price one has to pay for this advantage is an increased computational cost, since the kernel function has to be evaluated for all data points, not only for the nearest neighbors.

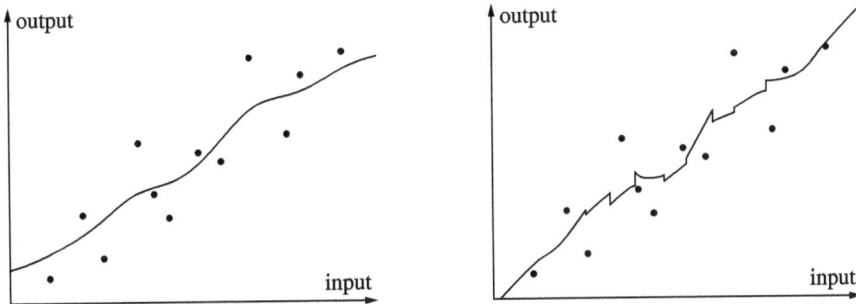

Fig. 9.3 Illustrations of kernel-weighted regression (*left*) and distance-weighted 4-nearest-neighbor regression (tricubic weighting function, *right*) in one dimension

9.1.3.2 Locally Weighted Polynomial Regression

Up to now we considered only a simple averaging of the numeric target values of the nearest neighbors as the prediction function. However, as a straightforward improvement, computing a (simple) local regression function suggests itself. That is, the k nearest neighbors are used to estimate the parameters of, for example, a linear regression function (see Sect. 8.3) in the vicinity of the query point, with which the prediction is then computed.

Not surprisingly, distance weighting (as we discussed it above, both with a fixed number of neighbors or with kernel functions) may also be used in such a setting, leading, for example, to the 4-nearest-neighbor distance-weighted locally linear regression shown in Fig. 9.3 on the right (using a tricubic weighting function, see Sect. 9.1.2). Note how the distance weighting leads to deviations from straight lines between the data points. Note also the somewhat erratic behavior of the resulting regression function (jumps at points where the set of nearest neighbors changes). This indicates that such an approach should be employed with a larger number of neighbors, so that a change of the set of nearest neighbors leads to less severe changes of the local regression line (although there will still be jumps of the predicted value in this case; they are just less high).

Locally weighted regression is usually applied with very simple regression polynomials: most of the time linear, rarely quadratic, basically never any higher order. The reason is that the local character of the regression is supposed to take care of the global shape of the function, so that the regression function is not needed to model it. As a consequence, the advantage of locally weighted polynomial regression is that no global regression function, derived from some data generation model, needs to be found, thus making the method applicable to a broad range of prediction problems. Its disadvantages are that its prediction can be less reliable in sparsely sampled regions of the feature space, where the locally employed regression function is necessarily stretched to a larger area and thus may be a bad fit to the actual target function.

9.1.3.3 Feature Weights

As already mentioned in Sect. 9.1.2, it is crucial for the success of a nearest-neighbor approach that a proper distance function is chosen. A very simple and natural way of adapting a distance function to the needs of the prediction problem is to use distance weights, thus giving certain features a greater influence than others. If prior information is available about which features are most informative w.r.t. the target value, this information can be incorporated directly into the distance function.

However, one may also try to determine appropriate feature weights automatically. The simplest approach is to start with equal feature weights and to modify them iteratively in a hill climbing fashion (see Sect. 5.3.2 for a general description): apply a (small) random modification to the feature weights and check with cross validation whether the new set of feature weights improves the prediction quality. If it does, accept the new weights, otherwise keep the old. Repeat the generation of random changes for a user-specified number of steps or until a user-specified number of trials were unsuccessful (did not yield an improvement of the prediction quality). Such a hill climbing approach has, of course, the disadvantage that it is prone to getting stuck in local optima. To avoid this, more sophisticated search approaches like simulated annealing and genetic algorithms (see Sect. 5.3.4 for a general description) may be employed.

9.1.3.4 Data Set Reduction and Prototype Building

As already pointed out in Sect. 9.1.2, a core problem of nearest-neighbor approaches is to quickly find the nearest neighbors of a given query point. This becomes an important practical problem if the training data set is large and predictions must be computed (very) quickly. In such a case one may try to reduce the set of training examples in a preprocessing step, so that a set of relevant or prototypical data points is found, which yields basically the same prediction quality. Note that this set may or may not be a subset of the training examples, depending on whether the algorithm used to construct this set merely samples from the training examples or constructs new data points if necessary. Note also that there are usually no or only few actually redundant data points, which can be removed without affecting the prediction at all. This is obvious for the numerical case and a 1-nearest-neighbor classifier but also holds for a k-nearest-neighbor classifier with $k > 1$, because any removal of data points may change the vote at some point and potentially the classification.

A straightforward approach is based on a simple iterative merge scheme [9]: at the beginning each training example is considered as a prototype. Then successively two nearest prototypes are merged as long as the prediction quality on some hold-out test data set is not reduced. Prototypes can be merged, for example, by simply computing a weighted sum, with the relative weights determined by how many original training examples a prototype represents.

More sophisticated approaches may employ, for example, genetic algorithms or any other method for solving a combinatorial optimization problem. This is possible, because the task of finding prototypes can be viewed as the task to find a subset of the training examples that yields the best prediction quality (on a given test data set): finding best subsets is a standard combinatorial optimization problem.

9.2 Artifical Neural Networks

Artificial neural networks are a family of models and accompanying training methods that try to mimic the learning ability of higher organisms by taking inspiration from (certain aspects of) biological neural networks. Here we consider the special types of multilayer perceptrons and radial basis function networks, both of which are trained with gradient descent schemes. Artificial neural networks usually achieve high quality in both classification and numerical prediction tasks but suffer from the fact that they are "black boxes," that is, they are difficult to interpret in terms of the application domain because of the abstract mathematical prediction procedure.

9.2.1 Overview

At least many higher organisms exhibit the ability to adapt to new situations and to learn from experience. Since these capabilities clearly derive from the fact that these organisms are endowed with a brain or at least a central nervous system, it is a plausible approach to try to achieve similar capabilities in an artificial system by mimicking the functionality of (biological) neurons and their interaction.

Depending on which aspects of (theories of) biological neurons are emphasized and which plausible modifications are introduced, different types of artificial neural networks can be distinguished. The most popular type is the **multilayer perceptron**, which is based on a "threshold logic" view of neuronal activity, but can also be seen—from a statistical perspective—as a staged or hierarchical logistic regression system. The basic idea underlying multilayer perceptrons relies on the following operation principle of biological neurons: a neuron receives, through "extension lines" of other neurons, electrical input, which either increases (excitatory input) or decreases (inhibitory input) the electrical potential of the neural cell relative to its environment. If the total input exceeds a certain threshold, the neuron is activated and "fires," that is, emits an electrical signal to other neurons it is connected to. The type (excitatory or inhibitory) and the strength of the influence of an input is determined by the chemical conditions at the connection point (so-called *synapse*) between the transmission lines (biologically: *axons*) and the neuron's receptors (biologically, *dendrites*). By modifying these chemical conditions, the behavior of the network of neurons can be changed, and thus adaptation and learning can be achieved.

By heavily simplifying the actually fairly complex mechanisms, this operation principle gives rise to a neuron model as it is displayed in Fig. 9.4 on the left: the influence of the incoming signals from other neurons is encoded as simple weights (real-valued numbers), which are multiplied with the (strength of) the incoming

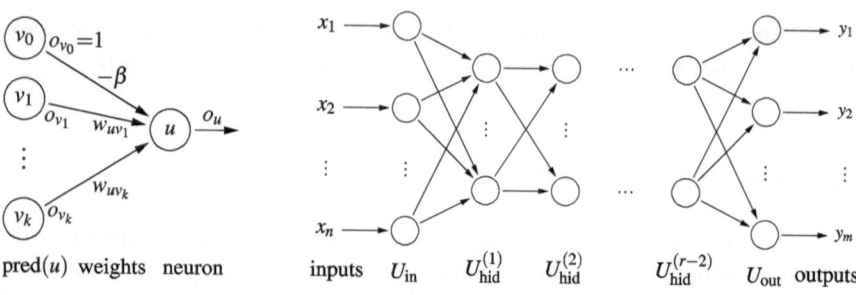

Fig. 9.4 A prototypical artificial neuron with connections to its predecessor neurons, including a dummy neuron v_0 (*left*) and general structure of a multilayer perceptron with r layers (*right*)

signal (also a real-valued number). The weighted signals (or signal strengths) are then summed and submitted to an activation function, which is a kind of threshold function[4] and which determines whether the neuron fires (emits a signal) or not. This signal is then transmitted to other neurons.

A standard connection structure for the neurons is a layered pattern, as it is depicted schematically in Fig. 9.4 on the right. This connection structure and the fact that a single neuron in this structure is called a **perceptron** (for historical reasons [26, 27]) explains the name "multilayer perceptron." The first (leftmost) layer is called the **input layer**, because it receives input (represented by the x_i) from the environment, while the last (rightmost) layer is called the **output layer**, because it emits output (represented by the y_i) to the environment. All intermediate layers are called **hidden layers**, because they do not interact with the environment and thus are "hidden" from it. In principle, a multilayer perceptron may have any number of hidden layers, but it is most common to use only a single hidden layer, based on certain theoretical results about the capabilities of such neural networks [16].

A multilayer perceptron works by executing the neurons in the layers from left to right, computing the output of each neuron based on its weighted inputs and the activation function. This procedure is called **forward propagation** and gives rise to the general name **feed-forward network** for a neural network that operates in this manner. Note that in this fashion the neural network implements, in a staged or layered form, the computation of a (possibly fairly complex) function of the input signals, which is emitted by the output layer at the end of the propagation process.

A multilayer perceptron is trained to implement a desired function with the help of a data set of sample cases, which are pairs of input values and associated output values. The input values are fed into the multilayer perceptron, and its output is computed. This output is then compared to the desired output (as specified by the second part of the sample case). If the two differ, a process called **error backpropagation** is executed, which adapts the connection weights and possibly other parameters of

---

[4]Note that for technical reasons, the threshold or bias value β of the neuronal activation function is turned into a connection weight by adding a connection to a dummy neuron emitting a permanent signal of 1, while the actual activation function has a threshold of zero.

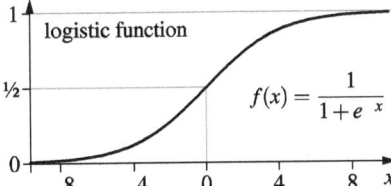
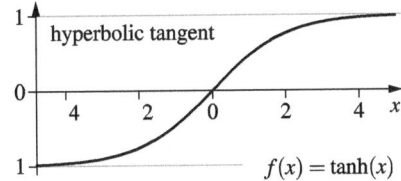

Fig. 9.5 The logistic function and the hyperbolic tangent, two common sigmoid activation functions for multilayer perceptrons, both of which describe a "soft" threshold behavior

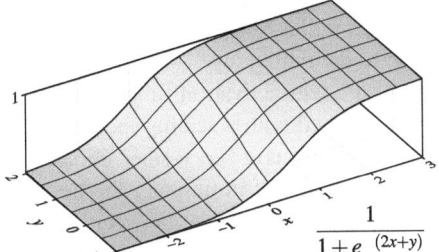
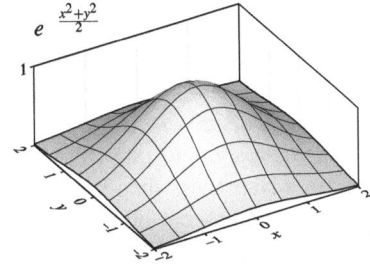

Fig. 9.6 Logistic function and Gaussian radial basis function for a two-dimensional input space

the activation functions in such a way that the output error (that is, the deviation of the actual output from the desired output) is reduced.

For error backpropagation to be feasible, it is mandatory that the activation functions of the neurons are not crisp threshold functions (as suggested in the above description), but differentiable sigmoid (s-shaped) functions. That is, the activation of the neuron should not jump at the threshold from completely inactive to fully active but should rise smoothly, over some input range, from inactive (0 or -1) to active ($+1$). With such functions, the output of a multilayer perceptron is a differentiable function of the inputs, which also depends on the connection weights and the function parameters (basically the threshold or bias value β). As a consequence, the weights and parameters can be adapted with a gradient descent scheme, which minimizes the sum of the squared output errors for the training data set.

The most commonly used activation functions are the **logistic function** (unipolar) and the **hyperbolic tangent** (bipolar). Illustrations for a one-dimensional input space are shown in Fig. 9.5, and a logistic function for a two-dimensional input space with threshold $\beta = 0$ in Fig. 9.6 on the left. Note that the weight vector can be interpreted as the direction in which the logistic function rises.

Like multilayer perceptrons, **radial basis function networks** are feed-forward networks. However, they always have three layers (one hidden layer), while this is only the most common choice for multilayer perceptrons. In addition, they do not employ a sigmoid activation function, at least not in the hidden layer. Rather, a distance from a reference point (or *center*) in the data space, which is represented by the neuron weights, is computed.[5] This distance is transformed with a **radial (basis) function**, the name of which derives from the fact that is defined on a distance from a given point and thus on a radius from this point. Note that *radial basis function* is actually just another name for a *kernel function*, as we considered it in Sect. 9.1.3.1. Such a function is 0 for infinite distance, increases monotonously for decreasing distance, and is 1 at zero distance. Thus it is, in a way, similar to a sigmoid function, which increases monotonously from $-\infty$, where it is 0 (or -1), to $+\infty$, where it is 1. A radial basis function is parameterized with a reference radius σ, which takes the place of the bias value β of the sigmoid activation functions of multilayer perceptrons. An illustration for a two-dimensional input space is shown in Fig. 9.6 on the right, which shows a Gaussian radial basis function centered at $(0, 0)$ and having a reference radius of 1. The value of the radial function for the computed distance from the reference point is passed on to the output layer.

The neurons in the output layer either have a linear activation function (that is, the weighted sum of the inputs is merely transformed with a scaling factor and an offset) or a sigmoid one (like in multilayer perceptrons). The former choice has certain advantages when it comes to the technical task of initializing the network before training,[6] but otherwise no fundamental difference exists between the two choices. Like multilayer perceptrons, radial basis function networks are trained with error backpropagation, which differs only due to the different activation functions.

The term "basis" in the name radial basis function derives from the fact that these functions are the basis, in the sense of the basis of a vector space, for constructing the function, the neural network is desired to compute; especially with a linear activation function in the output layer, the outputs are linear combinations of the basis functions and thus vector representations w.r.t. the basis functions. To some degree one may also say that a multilayer perceptron behaves in this way, even though it uses logistic activation functions for the output neurons, because close to the bias value, a logistic function is almost linear and thus models a linear combination of the activation functions of the hidden layer. The difference is that the basis functions of multilayer perceptrons are not radial functions, but rather logistic functions, which are sigmoid functions along a direction in the data space (see Fig. 9.6).

9.2.2 Construction

The first step of the construction of a neural network model consists in choosing the network structure. Since the number of input and output neurons is fixed by the data

---

[5]Note that with respect to the employed distance function all considerations of Sect. 7.2 can be transferred, although the Euclidean distance is the most common choice.

[6]A linear activation function also explains the term "basis" in "radial basis function," as it allows one to interpret the output as an approximation of the desired function in the vector space spanned by the radial functions computed by the hidden neurons: the connection weights from the hidden layer to the output are the coordinates of the approximating function w.r.t. this vector space.

analysis task (or, for the inputs, by a feature selection method, see Sect. 6.1.1), the only choices left for a multilayer perceptron are the number of hidden layers and the number of neurons in these layers. Since single-hidden-layer networks are most common (because of certain theoretical properties [16]), the choice is usually even reduced to the number of hidden neurons. The same applies to radial basis function networks, which have only one hidden layer by definition.

A simple rule of thumb, which often leads to acceptable training results, is to use $\frac{1}{2}\lfloor\#\text{inputs} + \#\text{outputs}\rfloor$ hidden neurons, where #inputs and #outputs are the numbers of input and output attributes, respectively. There also exist approaches to optimize the number of hidden neurons during the training process [21], or one may employ a wrapper scheme as for feature selection (see Sect. 6.1.1) in order to find the best number of hidden neurons for a given task. However, even though a wrong number of hidden neurons, especially if it is chosen too small, can lead to bad results, one has to concede that other factors, especially the choice and scaling of the input attributes, are much more important for the success of neural network model building.

Once the network structure is fixed, the connection weights and the activation function parameters are initialized randomly. For multilayer perceptrons, the weights and the bias values are usually chosen uniformly from a small interval centered around 0. For a radial basis function network, the reference or center points (the coordinates of which are the weights of the neurons in the hidden layer) may be chosen by randomly selecting data points or by sampling randomly from some distribution (Gaussian or rectangular) centered at the center of the data space. The reference radii are usually initialized to equal values, which are derived from the size of the data space and the number of hidden neurons, for example, as l_d/k, where l_d is the length of the data space diagonal, and k is the number of hidden neurons. If a linear activation function is chosen for the output layer, the connection weights from the hidden to the output layer can be initialized by solving a linear optimization problem. Alternatively, the weights can be initialized randomly.

After all network parameters (connection weights and activation function parameters) have been initialized, the neural network already implements a function of the input attributes. However, unless a radial basis function network with linear activation functions in the output layer has been initialized by solving the corresponding linear optimization problem, this function is not likely to be anywhere close to the desired function as represented by the training samples. Rather it will produce (significant) errors and thus needs to be adapted or trained.

The rationale of **neural network training** is as follows: the deviation of the function implemented by the neural network and the desired function as represented by the given training data is measured by the **sum of squared errors**,

$$e(\mathcal{D}) = \sum_{(\mathbf{x},\mathbf{y})\in\mathcal{D}} \sum_{u\in U_{\text{out}}} (o_u(\mathbf{x}) - y_u)^2,$$

where $\mathcal{D} = \{(\mathbf{x}_1,\mathbf{y}_1),\ldots,(\mathbf{x}_n,\mathbf{y}_n)\}$ is the given training data set, U_{out} the set of output neurons, y_u is the desired output of neuron u for the training case $(\mathbf{x},\mathbf{y})$, and $o_u(\mathbf{x})$ the computed output of neuron u for training case $(\mathbf{x},\mathbf{y})$. Furthermore, this

Fig. 9.7 Illustration of the gradient of a two-dimensional function

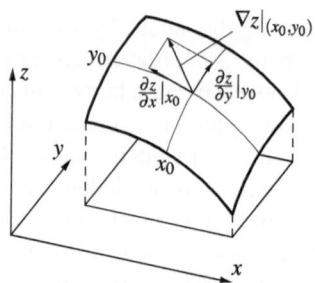

error of the neural network for a given training sample can be seen as a function of the network parameters (connection weights and activation function parameters like bias value and reference radius), since $o_u(\mathbf{x})$ depends on these parameters, while everything else is fixed by the network structure and the given training data:

$$e(\mathcal{D}) = e(\mathcal{D}; \theta),$$

where θ is the total of all network parameters. Provided that all activation functions are (continuously) differentiable (which is the case for the common choices of activation functions, see above, and also one of the reasons why the error is squared[7]), we can carry out a **gradient descent** on the error function in order to minimize the error. Intuitively, the principle of gradient descent (see also Chap. 5 for a general description) is the same that scouts follow in order to find water: always go downhill. Formally, we compute the gradient of the error function, that is, we consider

$$\nabla_\theta e(\mathcal{D}; \theta) = \left(\frac{\partial}{\partial \theta_1}, \dots, \frac{\partial}{\partial \theta_r} \right) e(\mathcal{D}; \theta),$$

where θ_k, $1 \le k \le r$, are the network parameters, $\frac{\partial}{\partial \theta_k}$ denotes the partial derivative w.r.t. θ_k, and ∇_θ (pronounced "nabla") is the gradient operator. Intuitively, the gradient describes the direction of steepest ascent of the error function (see Fig. 9.7 for a sketch). In order to carry out a gradient *descent*, it is negated and multiplied by a factor η, which is called the **learning rate** and which determines the size of the step in the parameter space that is carried out. Formally, we thus have

$$\theta^{(\text{new})} = \theta^{(\text{old})} - \eta \nabla_\theta e\left(\mathcal{D}; \theta^{(\text{old})}\right) \quad \text{or} \quad \theta_k^{(\text{new})} = \theta_k^{(\text{old})} - \eta \frac{\partial}{\partial \theta_k} e\left(\mathcal{D}; \theta^{(\text{old})}\right)$$

if written for an individual parameter θ_k. The exact form of this expression depends on the activation functions that are used in the neural network. A particularly simple case results for the standard case of the logistic activation function

$$f_{\text{act}}(z) = \frac{1}{1 + e^{-z}},$$

which is applied to $z = \mathbf{wx} = \sum_{i=1}^{p} w_i x_i$, that is, the weighted sum of the inputs, written as a scalar product. Here $\mathbf{x} = (x_0 = 1, x_1, \dots, x_p)$ is the input vector of the neuron, extended by a fixed input $x_0 = 1$ (see above), and

---

[7]The other reason is that without squaring the error, positive and negative errors could cancel out.

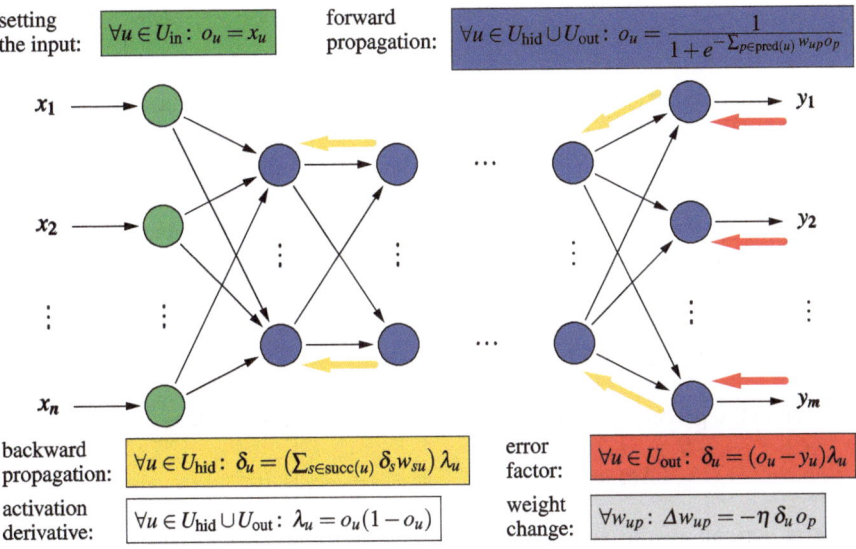

setting the input: $\forall u \in U_{\mathrm{in}}: o_u = x_u$

forward propagation: $\forall u \in U_{\mathrm{hid}} \cup U_{\mathrm{out}}: o_u = \dfrac{1}{1 + e^{-\sum_{p \in \mathrm{pred}(u)} w_{up} o_p}}$

backward propagation: $\forall u \in U_{\mathrm{hid}}: \delta_u = \left(\sum_{s \in \mathrm{succ}(u)} \delta_s w_{su}\right) \lambda_u$

error factor: $\forall u \in U_{\mathrm{out}}: \delta_u = (o_u - y_u)\lambda_u$

activation derivative: $\forall u \in U_{\mathrm{hid}} \cup U_{\mathrm{out}}: \lambda_u = o_u(1 - o_u)$

weight change: $\forall w_{up}: \Delta w_{up} = -\eta\, \delta_u o_p$

Fig. 9.8 Cookbook recipe for the execution (forward propagation) and training (error backpropagation) of a multilayer perceptron for the standard choice of a logistic activation function

$\mathbf{w} = (w_0 = -\beta, w_1, \ldots, w_p)$ is the weight vector of the neuron, extended by the negated bias value $w_0 = -\beta$. The derivative of this function has a particularly simple form, namely

$$f'_{\mathrm{act}}(z) = f_{\mathrm{act}}(z)\left(1 - f_{\mathrm{act}}(z)\right).$$

In addition, the formulas for the different layers are connected by a recursive scheme, which makes it possible to propagate an error term, usually denoted by δ, from a given layer to its preceding layer. We skip the detailed derivation here, which does not pose special mathematical problems (see, for example, [21, 23]).

Rather, we present the final result in the form of a cookbook recipe in Fig. 9.8, which contains all relevant formulas. The computations start at the input layer (green), where the external inputs are simply copied to the outputs of the input neurons. For all hidden and output neurons, forward propagation yields their output values (blue). The output of the output neurons is then compared to the desired output and a first error factor (for the output neurons) is computed (red), into which the derivative of the activation function enters (here, the logistic function, for which the derivative is shown in white). This error factor can be propagated back to the preceding layer with a simple recursive formula (yellow), using the connection weights and again the derivative of the activation function. From the error factors the weight changes are computed, using a user-specified learning rate η and the output of the neuron the connection, the weight is associated with, leads to.

9.2.3 Variations and Issues

Since standard (also called "vanilla"[8]) error backpropagation can be fairly slow and it may also be difficult to choose an appropriate learning rate, several variants have been suggested (see Sect. 9.2.3.1). An important technique to achieve robust learning results with neural networks is weight decay (see Sect. 9.2.3.2). The relative importance of the different inputs for predicting the target variable can be determined with sensitivity analysis (see Sect. 9.2.3.3).

9.2.3.1 Backpropagation Variants

As seen in the preceding section, the standard weight update rule is

$$\Delta\theta_k = \theta_k^{(new)} - \theta_k^{(old)} = -\eta \frac{\partial}{\partial\theta_k} e\left(\mathcal{D}; \theta^{(old)}\right).$$

Several variants of this rule have been proposed, mainly to speed up the training process. In the first place, one may exploit that the error is basically a sum of the errors for the individual training examples, and thus one may consider **online training**, where the weights are updated each time the error for one training example is computed. This can speed up training considerably, because less computations are needed before an update can be carried out. However, especially if the training examples are not properly shuffled, online training can be unstable. This is especially the case for some of the more complex modifications of backpropagation (see below), which more or less require **batch training**—that is, the weight change is computed from the aggregated error over the whole data set—to be stable.

A popular modification of backpropagation is the introduction of a **momentum term**, which means that a certain fraction (to be specified by a user) of the previous weight change is added to the current one:

$$\Delta\theta_k^{(new)} = -\eta \frac{\partial}{\partial\theta_k} e(T; \theta) + m\Delta\theta_k^{(old)}.$$

The underlying intuition is that of a ball rolling down the error surface, which gains momentum when rolling down a slope on which the gradient direction does not change much. Other modifications include **super self-adaptive backpropagation**, which introduces a separate learning rate for each weight and adapts these weights according to the directions of the two preceding learning steps: if they have the same sign, the learning rate is increased to cover a slope faster; if they have opposite signs, the learning rate is decreased, because the error minimum was jumped over. The growth and shrink factors for the learning rate are parameters that have to be specified by a user. A similar approach is employed by **resilient backpropagation**, which, however, directly adapts the (previous) weight change, thus eliminating a learning rate and using only the sign of the gradient for the update. Finally,

---

[8] Because vanilla is the standard ice cream flavor.

quick backpropagation approximates the error function locally and per weight by a parabola (derived from the gradient of the current and the previous step) and then sets the weight directly to the value of the apex of this parabola.

9.2.3.2 Weight Decay

It is usually disadvantageous if the (absolute values of the) weights of a neural network become too large. In such a case the gradient becomes very small, since the activation function argument lies in the saturation region of the sigmoid function, thus impeding learning. In addition, this increases the likelihood of overfitting. In order to counteract the general tendency of the (late) learning process to drive the weights to extreme values, **weight decay** is recommend, which means that prior to any update, a weight is multiplied with a factor $(1 - \xi)$, where ξ is a small positive value. Weight decay leads to much more stable learning results and thus is highly important for sensitivity analysis (see below), because otherwise no meaningful results can be obtained.

9.2.3.3 Sensitivity Analysis

Even though neural networks are essentially black boxes, which compute their output from the input through a series of numerical transformations that are usually difficult to interpret, there is a simple method to get at least some idea of how important different inputs are for the computation of the output. This method is called **sensitivity analysis** and consists in forming, for the training examples, the partial derivatives of the function computed by a trained neural network w.r.t. the different inputs. Given that the inputs are properly normalized, or these derivatives are put in proper relation with the range of input values, they provide an indication how strongly an input affects the output, because they describe how much the output changes if an input is changed. This may be used for feature selection if the relevant features are filtered out with an appropriately chosen threshold for the derivatives.

9.3 Support Vector Machines

Learning linear discriminant is rather simple as we have seen in Sect. 8.2.3.2. However, essentially all real-world problems exhibit more complex decision boundaries, which renders those nice types of models useless for real problems. Nevertheless, in recent years, a trick which allows one to use those simple classifiers for more complex problem settings has emerged, the so-called **support vector machines**, a special case of **kernel methods**. The models make use of the well-understood linear classifiers in combination with a projection into a higher-dimensional spaces where the original problem can, in fact, be solved—or at least reasonably well approximated—in a linear manner. Figure 9.9 illustrates this.

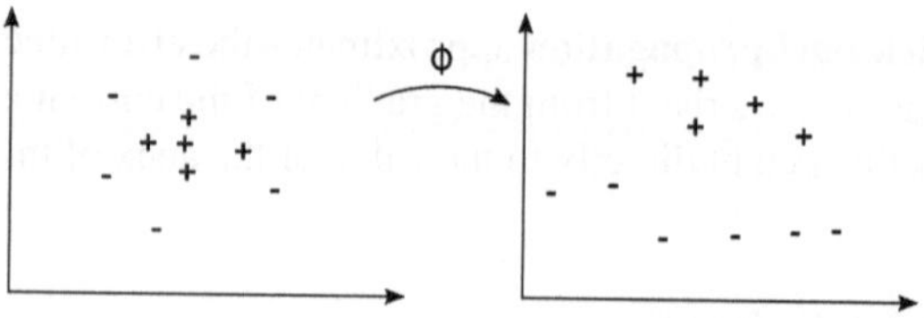

Fig. 9.9 An originally not linearly separable classification problem (*left*) is transformed into a new space (*right*) where the data points are linearly separable

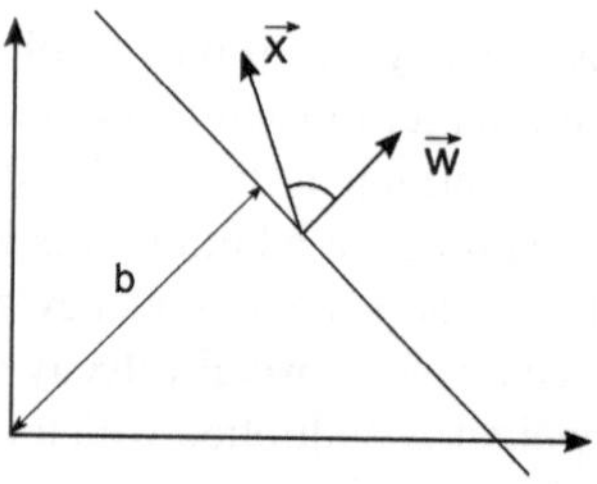

Fig. 9.10 A linear discriminant function defined through a weight vector **w** and an offset b. The inner product between **w** and an input vector **x** then computes a (scaled) cosine between those two vectors

Finding the embedding function Φ, so that the data is in the new space linearly separable is, of course, still a problem as we will discuss in more detail later. Nevertheless, kernel methods are a well-understood mechanism to build powerful classifiers—and also regression functions, as we will see later as well.

9.3.1 Overview

Before we look into linear separability again, let us first reformulate the problem slightly, to allow us to easier extend this later into other spaces. We are, as usual, considering a set of training examples $\mathcal{D} = \{(\mathbf{x}_j, y_j) | j = 1, \ldots, n\}$. The binary class information is encoded as ± 1 in contrast to the often used 0/1, which will allow us later to simplify equations considerably. Our goal is to find a linear discriminant function $f(\cdot)$ together with a decision function $h(\cdot)$. The latter reduces the continuous output of f to a binary class label, i.e., ± 1:

$$f(\mathbf{x}) = \langle \mathbf{w}, \mathbf{x} \rangle + b \quad \text{and} \quad h(x) = \text{sign}(f(\mathbf{x})).$$

The discriminant function $f(\cdot)$ returns the cosine of the angle between the weight vector **w** and the input vector **x**, under an offset b since

$$\cos \angle (\mathbf{w}, \mathbf{x}) = \frac{\langle \mathbf{w}, \mathbf{x} \rangle}{|\mathbf{w}||\mathbf{x}|}.$$

Figure 9.10 illustrates this.

Finding such linear discriminant functions has also attracted interest in the field of artificial neural networks. Before multilayer perceptrons, as discussed in Sect. 9.2, grew in popularity because the error backpropagation was developed, the majority of the interest lied on single perceptrons. A single perceptron can be seen as computing just this: the linear discriminant function shown above. In [26, 27] a learning rule for the single perceptron was introduced, which updates the parameters when sequentially presenting the training patterns one after the other:

$$\text{IF} \quad y_j \cdot (\langle \mathbf{w}, \mathbf{x}_j \rangle + b) < 0$$

$$\text{THEN} \quad \mathbf{w}_{t+1} = \mathbf{w}_t + y_j \cdot \mathbf{x}_j \quad \text{and} \quad \mathbf{b}_{t+1} = \mathbf{b}_t + y_j \cdot R^2$$

with $R = \max_j \|\mathbf{x}_j\|$. Whenever the product of the actual output value and the desired output is negative, we know that the signs of those two values were different and the classification was incorrect. We then update the weight vector using the learning rule. One nice property is that this process is guaranteed to converge to a solution if the training patterns are indeed perfectly classifiable with such a simple, linear discriminant function.

9.3.1.1 Dual Representation

One of the key observation is now that we can represent $f(\cdot)$ based on a weighted sum of the training examples instead of some arbitrary weight vector because

$$\mathbf{w} = \sum_{j=1}^{n} \alpha_j \cdot y_j \cdot \mathbf{x}_j.$$

The α_j essentially count how often the jth training pattern triggered an invocation of the learning rule above. Including the y_j, that is, the sign of the correct output, enables us to keep the α_j's to remain positive. A small trick, which will make our formalism a lot simpler later on.

Note that this representation does require that during initialization we do not assign random values to $\mathbf{w}$ but set it to 0 or at least a (random) linear combination of the training vectors. This, however, is not a substantial limitation.

From this it is straightforward to also represent the discriminant function based on the weighted training examples:

$$f(\mathbf{x}) = \langle \mathbf{w}, \mathbf{x} \rangle + b = \left(\sum_{j=1} \alpha_j \cdot y_j \cdot \langle \mathbf{x}_j, \mathbf{x} \rangle \right) + b.$$

Hence we can perform the classification of new patterns solely by computing the inner product between the new pattern $\mathbf{x}$ and the training pattern $(\mathbf{x}_j, y_j)$.

Finally, the update rule can also be represented based on inner products between training examples:

$$\text{IF} \quad y_j \cdot \left(\sum_{j'} \alpha_{j'} y_{j'} \langle \mathbf{x}_{j'}, \mathbf{x}_j \rangle + b \right) < 0$$

$$\text{THEN} \quad \alpha_j^{(t+1)} = \alpha_j^{(t)} + y_j \quad \text{and} \quad b^{(t+1)} = b^{(t)} + y_j \cdot R^2.$$

This representation uses only inner products with training examples. Note that suddenly we do not need to know anything about the input space anymore—as long we have some way to compute the inner product between the training instances, we can derive the corresponding α's. This representation is called **dual representation** in contrast to the **primal representation**, which represents the solution through a weight vector. A nice property of the dual representation is that the vector of α's expresses how much each training instance contributes to the solution, that is, how difficult to classify they were. In fact, in case of a nonlinearly separable problem, the α of the misclassified patterns will grow infinitely. However, on the other extreme, some α's will remain zero, since the corresponding patterns are never misclassified. Those are patterns that are easy to classify and we do not need to record their influence on the solution. We will return to this effect later.

9.3.1.2 Kernel Functions

The observation above indicates that we could ignore our input space if we had access to a function which returned the inner product for arbitrary vectors. So-called **kernel functions** offer, among others, precisely that property:

$$K(\mathbf{x}_1, \mathbf{x}_2) = \langle \Phi(\mathbf{x}_1), \Phi(\mathbf{x}_2) \rangle.$$

If we find a kernel for which $\Phi = I$, we can replace all our inner products with $K(\cdot, \cdot)$. This, however, is obviously boring: we can simply define K to compute the inner product. However, these kernels offer a very interesting perspective: we can suddenly compute inner products in spaces that we never really have to deal with. As long as, for the kernel K, there exists a function Φ which projects our original space into some other space, we can use the corresponding kernel K to compute the inner product directly. We can, of course, always define a kernel the way we see it above, e.g., as the inner product on the results of applying Φ to our original vectors. But what if we had much simpler kernel functions K that in effect computed an inner product in some other, Φ-induced space? Polynomial kernels allow us to demonstrate this nicely. Consider the function

$$\Phi\left(\begin{pmatrix} x_1 \\ x_2 \end{pmatrix}\right) = \left(x_1^2, x_2^2, \sqrt{2}x_1x_2\right)^T,$$

for which we can easily derive the corresponding kernel

$$K\left(\begin{pmatrix} x_1 \\ x_2 \end{pmatrix}, \begin{pmatrix} y_1 \\ y_2 \end{pmatrix}\right) = (x_1y_1)^2 + (x_2y_2)^2 + 2(x_1y_1x_2y_2) = \left(\left\langle \begin{pmatrix} x_1 \\ x_2 \end{pmatrix}, \begin{pmatrix} y_1 \\ y_2 \end{pmatrix}\right\rangle\right)^2.$$

This particular kernel has the additional twist that we can represent the inner product in our Φ-induced space through (among others) the inner product in the original space. This is, of course, not a requirement. More interestingly, the general kernel $K(\mathbf{x}, \mathbf{y}) = \langle \mathbf{x}, \mathbf{y} \rangle^d$ gives us an implicit induced space of dimension $\binom{n+d+1}{d}$. Calculating the resulting $\phi(\cdot)$ directly would very quickly become computationally very expensive. This kernel is a nice example how we can find a model in a very high-dimensional space without ever explicitly even dealing with the vectors in that space directly.

A few examples for other kernels are

$$K(x, y) = \langle x, y \rangle^d$$

and

$$K(x, y) = e^{-\frac{\|x-y\|^2}{2\sigma}}.$$

Since the set of kernels is closed under certain arithmetic operations, we can construct much more complex kernels based on simpler ones. Additionally, we can test for an arbitrary function K if it does indeed represent a kernel in some other space or not. There are, in fact, kernels which represent inner products in infinite-dimensional spaces, so we can find linear discriminant functions in spaces that we could not possibly deal with directly.

Even more interesting is the ability to define kernels for objects without numerical representations such as texts, images, graphs (such as molecular structures), or (biological) sequences. We can then create a classifier for such objects without ever entering an original or the derived space—all we need is a kernel function which returns a measure for the two objects which can then be used to derive the weighting factors α determining the weight vector $\mathbf{w}$.

Note that for the training, we do not even need access to the kernel function itself as long as we have the inner products of all training examples to each other. The resulting kernel (or Gram) matrix looks as follows:

$$\mathbf{K} = \begin{pmatrix} K(\mathbf{x}_1, \mathbf{x}_1) & K(\mathbf{x}_1, \mathbf{x}_2) & \cdots & K(\mathbf{x}_1, \mathbf{x}_m) \\ K(\mathbf{x}_2, \mathbf{x}_1) & K(\mathbf{x}_2, \mathbf{x}_2) & \cdots & K(\mathbf{x}_2, \mathbf{x}_m) \\ \vdots & \vdots & \ddots & \vdots \\ K(\mathbf{x}_m, \mathbf{x}_1) & K(\mathbf{x}_m, \mathbf{x}_2) & \cdots & K(\mathbf{x}_m, \mathbf{x}_m) \end{pmatrix}.$$

This matrix is a the center piece of kernel machines and contains all the information required during training. The matrix combines information of both, the training data and the chosen kernel. For kernel matrices, a couple of interesting observations hold:

- a kernel matrix is symmetric and positive definite;
- every positive definite, symmetric matrix is a kernel matrix, that is, it represents an inner product in some space; and
- the eigenvectors of the matrix correspond to the input vectors.

Note that it is still crucial to choose an appropriate kernel. If the Gram matrix is close to being a diagonal matrix, all points end up being essentially orthogonal to each other in the induced space, and finding a linear separation plane is very simple but also does not offer any generalization power.

9.3.1.3 Support Vectors and Margin of Error

If we can find a separating hyper plane, we can already visually motivate that in order to represent $\mathbf{w}$, we do not need to use all our original training data points. It is sufficient to use points which lie closest to this hyperplane. Those points are the ones ending up with an $\alpha \neq 0$ and are called support vectors. Figure 9.11 illustrates this, only the two boxed x's and the one boxed o are really needed to define the weight vector $\mathbf{w}$.

Fig. 9.11 The optimal hyperplane maximizing the margin of error for all training instances

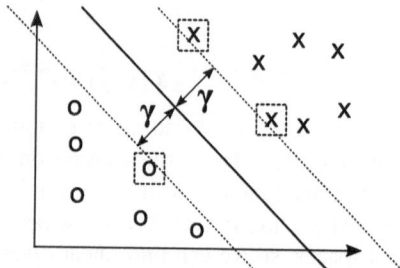

From this picture it also becomes evident that there are actually many different solutions for our classification problem—any line which correctly separates the points of different classes works just fine. However, there are lines that lie closer to training instances than others. If we were to maximize the minimum distance of any of the training instances to the separation line, we would create a solution with maximum distance to possibly making an error: the maximum margin classifier. Figure 9.11 show this distance for the optimal, that is, largest such margin of error:

$$\gamma = \max_{\mathbf{w}} \min_{j} \langle \mathbf{w}, \mathbf{x}_j \rangle.$$

There are solid theoretical explanations why this is indeed the best choice for the separating hyperplane. From statistical learning theory we can derive that the complexity of the class of all hyperplanes with constant margin is smaller than the class of hyperplanes with smaller margins. From this we can then derive upper bounds on the generalization error of the resulting SVM. We refer to [15] for a detailed treatment of these issues.

9.3.2 Construction

We will not describe the many training methods and their variants for SVMs in great detail here but instead refer to [15]. In a nutshell, the main idea reduces to solving a quadratic programming problem. In order to do this, we reformulate our constraint a bit. We now require our decision function to hold,

$$y_j \cdot (\langle \mathbf{w}, \mathbf{x}_j \rangle + h) \geq 1,$$

instead of merely being greater than zero. The decision line is still given by

$$\langle \mathbf{w}, \mathbf{x} \rangle + b = 0,$$

but we now can also describe the upper and lower margins by

$$\langle \mathbf{w}, \mathbf{x} \rangle + b = 1$$

and

$$\langle \mathbf{w}, \mathbf{x} \rangle + b = -1,$$

and the distance between those two hyperplanes is $2/\|\mathbf{w}\|$. Our goal of finding the maximum margin can now be formulated as the minimization problem minimize (in $\mathbf{w}, b$)

$\quad \|\mathbf{w}\|$

subject to (for any $j = 1, \ldots, n$)

$\quad y_j(\langle \mathbf{w}, \mathbf{x} \rangle - b) \geq 1$

This is fairly complex to solve because it depends on the norm of $\mathbf{w}$ which involves a square root. However, we can convert this into a quadratic form by substituting $\|\mathbf{w}\|$ with $\frac{1}{2}\|\mathbf{w}\|^2$ without changing the solution. After expressing this by means of Lagrange multipliers, this turns into a standard quadratic programming problem. In [15] more details and references to other, more memory- or time-efficient solutions are given.

9.3.3 Variations and Issues

Much work has been done on support vector machines. In the following we describe a few extensions which are essential for practical applications.

9.3.3.1 Slack Variables

We cannot always assume that we can find a linear hyperplane which cleanly separates our training examples. Especially for noisy training examples, this can also not be desirable as we would end up overfitting our data. So-called soft margin classifiers allow one to introduce slack variables which allow some of the training examples to be within the margin or even on the wrong side of the separation line. These slack variables end up expressing a degree of misclassification of the individual training examples. Our optimization problem in equation 9.3.2 is modified to

$$\forall j = 1, \ldots, n : y_j \cdot (\langle \mathbf{w}, \mathbf{x}_j \rangle + b) \geq 1 - \varepsilon_j,$$

and we need to introduce an additional penalty term to punish nonzero ε_j:

$$\arg\min \frac{1}{2}\|\mathbf{w}\|^2 + C \sum_j \varepsilon_j$$

$$\text{subject to} \quad y_j \cdot (\langle \mathbf{w}, \mathbf{x}_j \rangle + b) \geq 1 - \varepsilon_j \quad \text{for } 1 \leq j \leq n.$$

This can again be solved using Lagrange multipliers.

9.3.3.2 Multiclass Support Vector Machines

Not all real-world problems are binary classification tasks. In order to classify examples into more than two classes, one usually transforms the problem into a set of binary classification problems. Those can be classifiers which either separate one class from all others or separate pairs of classes from each other. In the former case, the class with the highest distance from the hyperplane wins; in the other case, the winners are counted, and the class which wins the most class-pair classifications determines the final classification.

9.3.3.3 Support Vector Regression

One interesting variation of support vector machines allows one to address regression problems instead of binary classifications. The key idea is to change the optimization to the following expression:

$$\arg\min \frac{1}{2}\|\mathbf{w}\|^2$$
$$\text{subject to} \quad y_j - (\langle \mathbf{w}, \mathbf{x}_j \rangle + b) \leq \varepsilon \quad \text{for } 1 \leq j \leq n.$$

So, instead of requiring the signs of the target variable and the prediction to match, we are requesting the prediction error to stay within a certain range (or margin) ε. We can, of course, also introduce slack variables to tolerate larger errors. Moreover, we can use the kernel trick to allow for not only linear regression functions. A very good tutorial of Support Vector Regression can be found in [32].

9.4 Ensemble Methods

Ensemble methods combine several predictors (classifiers or numeric predictors) in order to improve the prediction quality over the performance of the individual predictors. The core ingredients of ensemble methods are a procedure to construct different predictors and a rule how to combine their results. Depending on the choices that are made for these two ingredients, a large variety of different ensemble methods has been suggested. While usually yielding higher accuracy than individual models, the fact that sometimes very large ensembles are employed makes the ensemble prediction mechanism difficult to interpret (even if the elements are simple).

9.4.1 Overview

It is well known from psychological studies of problem solving activities (but also highly plausible without such scientific backing) that a committee of (human) experts with different, but complementary skills, usually produces better solutions than any individual. As a consequence, the idea suggests itself to combine several predictors (classifiers or regression models) in order to achieve a prediction accuracy exceeding the quality of the individual predictors. That is, instead of using a single model to predict the target value, we employ an ensemble of predictors and combine their predictions (for example, by majority voting for classification or by simple averaging for numeric targets) in order to obtain a joint prediction.

A necessary and sufficient condition for an ensemble of predictors to outperform the individuals it is made of is that the predictors are reasonably accurate and diverse. Technically, a predictor is already called **accurate** if it predicts the correct target value for a new input object better than random guessing. Hence this is a pretty weak requirement that is easy to meet in practice. Two predictors are called **diverse predictors** if they do not make the same mistakes on new input objects. It is obvious that this requirement is essential: if the predictors always made the same mistakes, no improvement could possibly result from combining them. As an extreme case, consider that the predictors in the ensemble are all identical: the combined prediction is necessarily the same as that of any individual predictor— regardless of how the individual predictions are combined. However, if the errors made by the individual predictors are uncorrelated, their combination will reduce these errors. For example, if we combine classifiers by majority voting and if we assume that the mistakes made by these classifiers are independent, the resulting ensemble yields a wrong result only if more than half of the classifiers misclassify the new input object, which is a lot less likely than any individual classifier assigning it to the wrong class. For instance, for five independent classifiers for a two-class problem, each of which has an error probability of 0.3, the probability that three or more yield a wrong result is

$$\sum_{i=3}^{5} \binom{5}{i} 0.3^i \cdot 0.7^{5-i} = 0.08748.$$

Note, however, that this holds only for the ideal case that the classifiers are fully independent, which is usually not the case in practice. Fortunately, though, improvements are also achieved if the dependence is sufficiently weak, although the gains are naturally smaller. Note also that even in the ideal case no gains result (but rather a degradation) if the error probability of an individual classifier exceeds 0.5, which substantiates the requirement that the individual predictors should be accurate.

According to [17], there are basically three reasons why ensemble methods work: statistical, computational, and representational. The statistical reason is that in practice any learning method has to work on a finite data set and thus may not be able to identify the correct predictor, even if this predictor lies within the set of models that the learning method can, in principle, return as a result (see also Sect. 5.4). Rather, it is to be expected that there are several predictors that yield similar accuracy. Since there is thus no sufficiently clear evidence which model is the correct or best one, there is a certain risk that the learning method selects a suboptimal model. By removing the requirement to produce a single model, it becomes possible to "average" over many or even all of the good models. This reduces the risk of excluding the best predictor and the influence of actually bad models.

The computational reason refers to the fact that learning algorithms usually cannot traverse the complete model space but must use certain heuristics (greedy, hill climbing, gradient descent, etc.) in order to find a model. Since these heuristics may yield suboptimal models (for example, local minima of the error function), a suboptimal model may be chosen (see also Sect. 5.4). However, if several models constructed with heuristics are combined in an ensemble, the result may be a better approximation of the true dependence between the inputs and the target variable.

The representational reason is that for basically all learning methods, even the most flexible ones, the class of models that can be learned is limited, and thus it may be that the true model cannot be represented accurately. By combining several models in a predictor ensemble, the model space can be enriched, that is, the ensemble may be able to represent a dependence between the inputs and the target variable that cannot be expressed by any of the individual models the learning method is able to produce. That is, from a representational point of view, ensemble methods make it possible to reduce the bias of a learning algorithm by extending its model space, while the statistical and computational reasons indicate that they can also reduce the variance. In this sense, ensemble methods are able to sever the usual link between bias and variance (see also Sect. 5.4.5).

9.4.2 Construction

As already mentioned, ensemble methods require basically two ingredients, a procedure to construct accurate and diverse classifiers and a rule how to combine their predictions. Depending on the choices made for these two ingredients, several specific methods can be distinguished.

Bayesian Voting In pure Bayesian voting the set of all possible models in a user-defined hypothesis space is enumerated to form the ensemble. The predictions of the individual models are combined weighted with the posterior probability of the model given the training data [17]. That is, models that are unlikely to be correct given the data have a low influence on the ensemble prediction, models that are likely to have a high influence. The likelihood of the model given the data can often be computed conveniently by exploiting $P(M \mid D) \propto P(D \mid M)P(M)$, where M is the model, D the data, $P(M)$ the prior probability of the model (often assumed to be the same for all models), and $P(D \mid M)$ the data likelihood given the model.

Theoretically, Bayesian voting is the optimal combination method, because all possible models are considered and their relative influence reflects their likelihood given the data. In practice, however, it suffers from several drawbacks. In the first place, it is rarely possible to actually enumerate all models in the hypothesis space defined by a learning method. For example, even if we restrict the tree size, it is usually infeasible to enumerate all decision trees that could be constructed for a given classification problem. In order to overcome this problem, model sampling methods are employed, which ideally should select a model with a probability that corresponds to its likelihood given the data. However, most such methods are biased and thus usually do not yield a representative sample of the total set of models, sometimes seriously degrading the ensemble performance.

Bagging   The method of bagging (bootstrap aggregating) predictors can be applied with basically any learning algorithm [7]. The basic idea is to select a single learning algorithm (most studied in this respect are decision tree inducers) and to learn several models with it by providing it each time with a different random sample of the training data. The sampling is carried out with replacement (bootstrapping) with the sample size commonly chosen as $n(1 - 1/e) \approx 0.632n$, where n is the number of training samples. Especially if the learning algorithm is unstable (like decision tree inducers, where a small change of the data can lead to a considerably different decision tree, see Sect. 8.1), the resulting models will usually be fairly diverse, thus satisfying one of the conditions needed for ensemble methods to work. The predictions of the individual models are then combined by simple majority voting or by averaging them (with the same weight for each model).

Bagging effectively yields predictions from an "average model," even though this model does not exist in simple form—it may not even lie in the hypothesis space of the learning algorithm (see Sect. 9.4.1). It has been shown that bagging reduces the risk of overfitting the training data (because each subsample has different special properties) and thus produces very robust predictions. Experiments show that bagging yields very good results especially for noisy data sets, where the sampling seems to be highly effective to avoid any adaptation to the noise data points.

A closely related alternative to bagging are **cross-validated committees**. Instead of resampling the training data with replacement (bootstrapping) to generate the predictors of an ensemble, the predictors learned during a cross validation run are combined with equal weights in a majority vote or by averaging.

Random Subspace Selection   While bagging obtains a set of diverse predictors by randomly varying the training data, random subspace selection employs a random selection of the features [22] for this purpose. That is, all data points are used in each training run, but the features the model construction algorithm can use are randomly selected. With a learning algorithm like a decision tree inducer (for which random subspace selection was first proposed), the available features may even be varied each time a split has to be chosen, so that the whole decision tree can potentially use all features. However, since the split selections are (randomly) constrained, several runs of the inducer yield diverse decision trees.

Combining random subspace selection with bagging—that is, varying the training data by bootstrapping and varying the available features by random selection— is a highly effective and strongly recommended method if accuracy is the main goal. Applied to decision trees, this combined method has been named **random forests** [8], which is known to be one of the most accurate classification methods to date. However, the fact that a huge number of trees may be employed, destroys the advantage decision trees usually have, namely that they are easy to interpret and can be checked for plausibility.

Injecting Randomness   Both bagging and random subspace selection employ random processes in order to obtain diverse predictors. This approach can of course be generalized to the principle of injecting randomness into the learning process. For example, such an approach is very natural and straightforward for artificial neural networks (see Sect. 9.2): different initialization of the connection weights often yield different learning results, which may then be used as the members of an ensemble. Alternatively, the network structure can be modified, for example, by randomly deleting a certain fraction of the connections between two consecutive layers.

Boosting While in all ensemble methods described so far the predictors can, in principle, be learned in parallel, **boosting** constructs them progressively, with the prediction results of the model learned last influencing the construction of the next model [18, 19, 28]. Like bagging, boosting varies the training data. However, instead of drawing random samples, boosting always works on the complete training data set. It rather maintains and manipulates a data point weight for each training example in order to generate diverse models. Boosting is usually described for classification problems with two classes, which are assumed to be coded by 1 and -1.

The best-known boosting approach is AdaBoost [18, 19, 28] and works as follows: Initially, all data point weights are equal and therefore set to $w_i = 1/n$, $i = 1, \ldots, n$, where n is the size of the data set. After a predictor M_t has been constructed in step t using the current weights $w_{i,t}$, $i = 1, \ldots, n$, it is applied to the training data, and

$$e_t = \frac{\sum_{i=1}^{n} w_{i,t} y_i M_t(\mathbf{x}_i)}{\sum_{i=1}^{n} w_{i,t}} \quad \text{and} \quad \alpha_t = \frac{1}{2} \ln\left(\frac{1 - e_t}{1 + e_t}\right)$$

are computed [29], where $\mathbf{x}_i$ is the input vector, y_i the class of the ith training example, and $M_t(\mathbf{x}_i)$ is the prediction of the model for the input $\mathbf{x}_i$. The data point weights are then updated according to

$$w_{i,t+1} = c \cdot w_{i,t} \exp(-\alpha_t y_i M_t(\mathbf{x}_i)),$$

where c is a normalization constant chosen so that $\sum_{i=1}^{n} w_{i,t+1} = 1$. The procedure of learning a predictor and updating the data point weights is repeated a user-specified number of times $t_{\max}$. The constructed ensemble classifies new data points by majority voting, with each model M_t weighted with α_t. That is, the joint prediction is

$$M_{\text{joint}}(\mathbf{x}_i) = \text{sign}\left(\sum_{t=1}^{t_{\max}} \alpha_t M_t(\mathbf{x}_i)\right).$$

Since there is no convergence guarantee and the performance of the ensemble classifier can even degrade after a certain number of steps, the inflection point of the error curve over t is often chosen as the ensemble size [17].

For low-noise data, boosting clearly outperforms bagging and random subspace selection in experimental studies [17]. However, if the training data contains noise, the performance of boosting can degrade quickly, because it tends to focus on the noise data points (which are necessarily difficult to classify and thus receive high weights after fairly few steps). As a consequence, boosting overfits the data. For noisy data, bagging and random subspace selection yield much better results [17].

Mixture of Experts   In the approach that is often referred to as **mixture of experts** the individual predictors to combine are assumed as already given, for example, selected by a user. They may be, for instance, results of different learning algorithms, like a decision tree, neural networks with different network structure, a support vector machine, etc.—whatever the user sees as promising to solve the application task. Alternatively, they may be the set of models obtained from any of the ensemble methods described so far. The focus is then placed on finding an optimal rule to combine the predictions of the individual models.

For classification tasks, for example, the input to this combination rule are the probability distributions over the classes that the individual classifiers yield. Note that this requires more than the simple (weighted) majority voting we employed up to now, which only asks each classifier to yield its best guess of the class of a new input object: each classifier must assign a probability to each class.

The most common rules to combine such class probabilities are the so-called **sum rule**, which simply averages, for each class, the probabilities provided by the individual classifiers, or the so-called **product rule**, which assumes conditional independence of the classifiers given the class and therefore multiplies, for each class, the probabilities provided by the different classifiers. In both cases the class with the largest sum or product is chosen as the prediction of the ensemble. Experiments show that the sum rule is usually preferable, likely because due to the product a class that is seen as (very) unlikely by a single classifier has little chance of being predicted, even if several other classifiers assign a high probability to it [35].

Both the sum rule and the product rule can be seen as special cases of a general family of combination rules that are known as f-means [35]. Other combinations rules include Dempster–Shafer combination and rank-based rules.

Stacking   Like a mixture of experts, **stacking** takes the set of predictors as already given and focuses on combining their individual predictions. The core idea is to view the outputs of the predictors as new features and to use a learning algorithm to find a model that combines them optimally [36]. Technically, a new data table is set up with one row for each training example, the columns of which contain the predictions of the different (level-1) models for training example. In addition, a final column states the true classes. With this new training data set a (level-2) model is learned, the output of which is the prediction of the ensemble.

Note that the level-2 model may be of the same or of a different type than the level-1 models. For example, the output of several regression trees (e.g., a random forest) may be combined with a linear regression [6] or with a neural network.

9.4.3 Further Reading

An extensive coverage of nearest-neighbor techniques can be found in [31]. For a deeper insight into artificial neural networks, the excellent and very extensive book [21] is one of the best on the market. The most accessible book on support vector machines is certainly [15], while [30] provides more details and a coverage of many special aspects. Good overviews of ensemble methods can be found in [17, 25].

9.5 Finding Predictors in Practice

Using predictors in practice is the classic application scenario for many data analysis, data mining, and machine learning toolkits. As discussed earlier in this chapter, there are two types of predictors: lazy learners, which require both training data and the records to assign labels to during prediction, and the model-based predictors, which first learn a model and then use this model later during prediction to determine the target value. This difference is also reflected in KNIME and R, as we will see in the next two sections.

9.5.1 Finding Predictors with KNIME

Nearest-neighbor algorithms are present in essentially all tools and are usually very straightforward to use. In KNIME, the kNN node allows one to set the number of neighbors to be considered and, if the distance should be used, to weight those neighbors. Figure 9.12 shows a small workflow reading in training data and data to determine the class for. Note that kNN is very sensitive to the chosen distance function, so we should make sure to normalize the data and use the exact same normalization procedure for both the training and test data: this can be achieved by using the normalizer node and "Normalizer (Apply)" node, which copies the settings from the first node (see also Sect. 6.6.1). We then feed those two data tables into the "K-Nearest-Neighbor" node which adds a column with the predicted class to the test data.

Both, artificial neural networks and support vector machines follow a different setup. Figure 9.13 shows the flow when training and applying a multilayer perceptron. KNIME uses two nodes, one creating the model based on training data and the second node applying the model to new data. Note that the connection between the model creation and model consumption (prediction) node has different port icons indicating that, along this connection, a model is transported instead of a data table.

Note that the workflow is also writing (node "PMML Writer") the network to file in a standardized XML dialect (Predictive Model Markup Language, PMML) which allows one to use this model in other learning environments but also databases and other prediction/scoring toolkits. KNIME offers other types of neural networks as

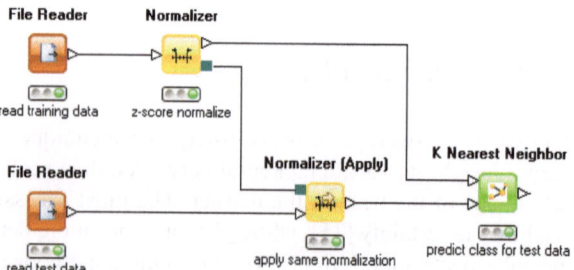

Fig. 9.12 The kNN node in KNIME allows one to apply the well-known k-Nearest-Neighbor method to the test data, given some training data with existing class labels

Fig. 9.13 A KNIME
workflow training and
applying a multi-layer
perceptron. The workflow
also exports the trained
network in the PMML format

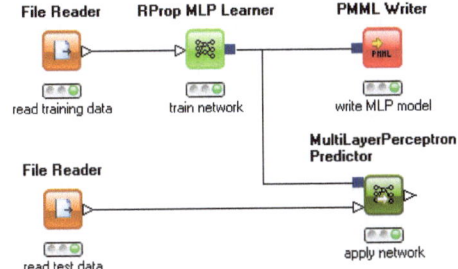

Fig. 9.14 The base SVM nodes in KNIME offer well-known kernels along with simply dialogs to
adjust the corresponding parameters

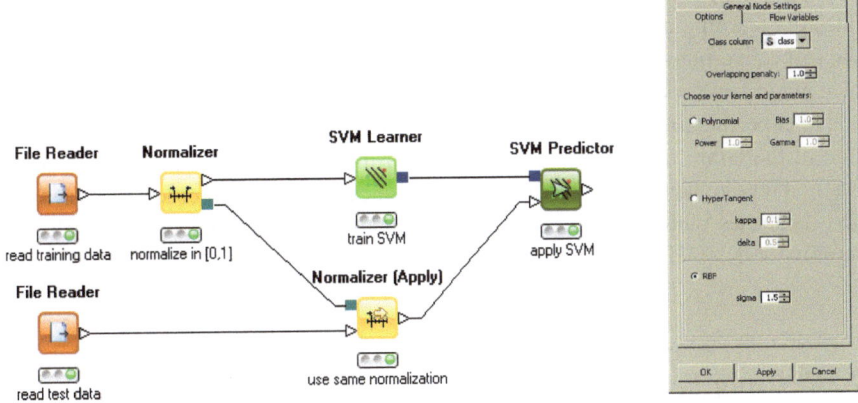

well; in particular a probabilistic version of radial basis function networks, called
Probabilistic Neural Network (PNN) along with an efficient, constructive training
algorithm is available. PNNs allow one to predict class probabilities and also assign
a "don't know" probability which is an interesting variation to many standard classifiers that classify new patterns to one of the existing classes, no matter how different
that pattern is to anything encountered during training.

Support vector machine algorithms have not been around as long as neural networks, but in the meantime, many implementations of support vector machines
(SVM) have shown up in commercial and open software packages. KNIME, in
fact, offers several own implementations but also integrates an open-source SVM
package in addition to the SVM classifier and regression models contained in Weka,
which is also accessible from within KNIME using the optional Weka integration.
The base KNIME implementation consists of two nodes, one learning the SVM
model offering the choice of a few well-known kernels and the second one allowing
one to apply the model to a second data set. Figure 9.14 shows part of a workflow
using these two nodes along with the parameter dialog for the learning node.

The native KNIME SVM implementation currently offers three different kernels:
a polynomial, hyper tangent, and RBF kernel. In contrast to, e.g., decision trees,

kernel functions (or at least the settings of their respective parameters) are rather sensitive to the normalization of the input data. For instance, the sigma parameter of the RBF kernels controls the width of the basis functions in relation to the Euclidean distance of the input vectors to the support vectors. Hence it is critical to adjust this parameter accordingly. In the above example workflow we again use the normalization module to avoid problems with different normalizations.

Using a different implementation, for instance, the Weka SVM classifier or regression nodes works analogously: one simply replaces the learner and predictor nodes in the above workflow with their respective alternatives. Especially using the LibSVM [10], integration is of interest since it is computationally substantially more efficient than both the native KNIME and Weka nodes. LibSVM offers numerous different learning methods in addition to a number of kernel functions. We refer to the respective literature for more information on the LibSVM library, where also a nice guide for classification using SVMs can be found.

An experimental extension to KNIME also allows one to separate the computation of the Gram matrix from the linear discriminant learning. This allows one to easily integrate new kernel functions, for instance, to compute kernels on graphs or text databases without having to recode the entire learning procedure as well. We refer to the KNIME Labs webpage[9] for more details.

9.5.2 Using Predictors in R

9.5.2.1 Nearest Neighbor Classifiers

A nearest-neighbor classifier based on the Euclidean distance is implemented in the package class in R. To show how to use the nearest-neighbor classifier in R, we use splitting of the Iris data set into a training set iris.training and test set iris.test as it was demonstrated in Sect. 5.6.2. The function knn requires a training and a set with only numerical attributes and a vector containing the classifications for the training set. The parameter k determines how many nearest neighbors are considered for the classification decision.

```
> library(class)
> iris.knn <- knn(iris.training[,1:4],iris.test[,1:4],
                  iris.training[,5],k=3)
> table(iris.knn,iris.test[,5])
```

The last line prints the confusion matrix.

---

[9]KNIME Labs: http://labs.knime.org.

9.5.2.2 Neural Networks

For the example of multilayer perceptrons in R, we use the same training and test data as for the nearest-neighbor classifier above. The multilayer perceptron can only process numerical values. Therefore, we first have to transform the categorical attribute *Species* into a numerical attribute:

```
> x <- iris.training
> x$Species <- as.numeric(x$Species)
```

The multilayer perceptron is constructed and trained in the following way, where the library neuralnet needs to be installed first:

```
> library(neuralnet)
> iris.nn <- neuralnet(x$Species + x$Sepal.Length ~
                       x$Sepal.Width + x$Petal.Length
                       + x$Petal.Width, x,
                       hidden=c(3))
```

The first argument of neuralnet defines that the attributes *species* and sepal length correspond to the output neurons. The other three attributes correspond to the input neurons. x specifies the training data set. The parameter hidden defines how many hidden layers the multilayer perceptron should have and how many neurons in each hidden layer should be. In the above example, there is only one hidden layer with three neurons. When we replace c(3) by c(4,2), there would be two hidden layers, one with four and one with two neurons.

The training of the multilayer perceptron can take some time, especially for larger data sets.

When the training is finished, the multilayer perceptron can be visualized:

```
> plot(iris.nn)
```

The visualization includes also dummy neurons as shown in Fig. 9.4.

The output of the multilayer perceptron for the test set can be calculated in the following way. Note that we first have to remove the output attributes from the test set:

```
> y <- iris.test
> y <- y[-5]
> y <- y[-1]
> y.out <- compute(iris.nn,y)
```

We can then compare the target outputs for the training set with the outputs from the multilayer perceptron. If we want to compute the squared errors for the second output neuron—the *sepal length*—we can do this in the following way:

```
> y.sqerr <- (y[1] - y.out$net.result[,2])^2
```

9.5.2.3 Support Vector Machines

For support vector machine, we use the same training and test data as already for the nearest-neighbor classifier and for the neural networks. A support vector machine to predict the *species* in the Iris data set based on the other attributes can be constructed in the following way. The package e1071 is needed and should be installed first if it has not been installed before:

```
> library(e1071)
> iris.svm <- svm(Species ~ ., data = iris.training)
> table(predict(iris.svm,iris.test[1:4]),iris.test[,5])
```

The last line prints the confusion matrix for the test data set.

The function svm works also for support vector regression. We could, for instance, use

```
> iris.svm <- svm(Petal.Width ~ ., data = iris.training)
> sqerr <- (predict(iris.svm,iris.test[-4])
                                    -iris.test[4])^2
```

to predict the numerical attribute *petal width* based on the other attributes and to compute the squared errors for the test set.

9.5.2.4 Ensemble Methods

As an example for ensemble methods, we consider random forest with the training and test data of the Iris data set as before. The package randomForest needs to be installed first:

```
> library(randomForest)
> iris.rf <- randomForest(Species ~., iris.training)
> table(predict(iris.rf,iris.test[1:4]),iris.test[,5])
```

In this way, a random forest is constructed to predict the *species* in the Iris data set based on the other attributes. The last line of the code prints the confusion matrix for the test data set.

References

1. Aha, D.W., Kibler, D., Albert, M.K.: Instance-based learning algorithms. Mach. Learn. **6**(1), 37–66 (1991)
2. Aurenhammer, F.: Voronoi diagrams—a survey of a fundamental geometric data structure. ACM Comput. Surv. **23**(3), 345–405 (1991)

3. Beckmann, N., Beckmann, H.-N., Kriegel, H.-P., Schneider, R., Seeger, B.: The R*-tree: an efficient and robust access method for points and rectangles. In: Proc. ACM SIGMOD Conference on Management of Data (Atlantic City, NJ), pp. 322–331. ACM Press, New York (1990)

4. Bentley, J.L.: Multidimensional divide and conquer. Commun. ACM **23**(4), 214–229 (1980)

5. Blum, M., Floyd, R.W., Pratt, V., Rivest, R., Tarjan, R.: Time bounds for selection. J. Comput. Syst. Sci. **7**, 448–461 (1973)

6. Breiman, L.: Stacked regressions. Mach. Learn. **24**(1), 49–64 (1996)

7. Breiman, L.: Bagging predictors. Mach. Learn. **24**(2), 123–140 (1996)

8. Breiman, L.: Random forests. Mach. Learn. **45**, 5–32 (2001)

9. Chang, C.-L.: Finding prototypes for nearest neighbor classifiers. IEEE Trans. Comput. **23**(11), 1179–1184 (1974)

10. Chang, C.-C., Lin, C.-L.: LIBSVM: a library for support vector machines. Manual (2001). http://www.csie.ntu.edu.tw/~cjlin/libsvm

11. Cleveland, W.S.: Robust locally weighted regression and smoothing scatterplots. J. Am. Stat. Assoc. **74**(368), 829–836 (1979)

12. Cleveland, W.S., Devlin, S.J.: Locally-weighted regression: an approach to regression analysis by local fitting. J. Am. Stat. Assoc. **83**(403), 596–610 (1988)

13. Cover, T., Hart, P.: Nearest neighbor pattern classification. IEEE Trans. Inf. Theory **13**(1), 21–27 (1967)

14. Cormen, T.H., Stein, C., Leiserson, C.E., Rivest, R.L.: Introduction to Algorithms, 2nd edn. MIT Press/McGraw-Hill, Cambridge/New York (2001)

15. Cristianini, N., Shawe-Taylor, J.: Kernel Methods for Pattern Analysis. Cambridge University Press, Cambridge (2004)

16. Cybenko, G.V.: Approximation by superpositions of a sigmoidal function. Math. Control Signals Syst. **2**, 303–314 (1989)

17. Dietterich, T.G.: Ensemble methods in machine learning. In: Proc. 1st Int. Workshop on Multiple Classifier Systems (MCS 2000, Cagliari, Italy). Lecture Notes in Computer Science, vol. 1857, pp. 1–15. Springer, Heidelberg (2000)

18. Freund, Y.: Boosting a weak learning algorithm by majority. In: Proc. 3rd Ann. Workshop on Computational Learning Theory (COLT'90, Rochester, NY), pp. 202–216. Morgan Kaufmann, San Mateo (1990)

19. Freund, Y., Schapire, R.E.: A decision-theoretic generalization of on-line learning and an application to boosting. J. Comput. Syst. Sci. **55**(1), 119–139 (1997)

20. Friedman, J.H., Bentley, J.L., Finkel, R.A.: An algorithm for finding best matches in logarithmic expected time. ACM Trans. Math. Softw. **3**(3), 209–226 (1977)

21. Haykin, S.: Neural Networks and Learning Machines. Prentice Hall, Englewood Cliffs (2008)

22. Ho, T.K.: The random subspace method for constructing decision forests. IEEE Trans. Pattern Anal. Mach. Intell. **20**, 832–644 (1998)

23. Mitchell, T.: Machine Learning. McGraw-Hill, New York (1997)

24. Manolopoulos, Y., Nanopoulos, A., Papadopoulos, A.N., Theodoridis, Y.: R-Trees: Theory and Applications. Springer, Heidelberg (2005)

25. Polikar, R.: Ensemble based systems in decision making. IEEE Circuits Syst. Mag. **6**, 21–45 (2006)

26. Rosenblatt, F.: The Perceptron: a probabilistic model for information storage and organization in the brain. Psychol. Rev. **65**, 386–408 (1958)

27. Rosenblatt, F.: Principles of Neurodynamics. Spartan Books, New York (1962)

28. Schapire, R.E.: Strength of weak learnability. Mach. Learn. **5**, 197–227 (1990)

29. Schapire, R.E., Singer, Y.: Improved boosting algorithms using confidence-rated predictors. Mach. Learn. **37**(3), 297–336 (1999)

30. Schölkopf, B., Smola, A.J.: Learning with Kernels: Support Vector Machines, Regularization, Optimization, and Beyond. MIT Press, Cambridge (2001)

31. Shakhnarovich, G., Darrel, T., Indyk, P. (eds.): Nearest Neighbor Methods in Learning and Vision: Theory and Practice. MIT Press, Cambridge (2006)

32. Smola, A.J., Schölkopf, B.: A tutorial on support vector regression. Technical Report (2003). http://eprints.pascal-network.org/archive/00002057/01/SmoSch03b.pdf
33. Tropf, H., Herzog, H.: Multidimensional range search in dynamically balanced trees. Angew. Inform. **1981**(2), 71–77 (1981)
34. Voronoi, G.: Nouvelles applications des paramètres continus à la théorie des formes quadratiques. J. Reine Angew. Math. **133**, 97–178 (1907)
35. Xu, L., Amari, S.-I.: Combining classifiers and learning mixture of experts. In: Encyclopedia of Artificial Intelligence, pp. 318–326. IGI Global, Hershey (2008)
36. Wolpert, D.: Stacked generalization. Neural Netw. **5**(2), 241–259 (1992)

Chapter 10
Evaluation and Deployment

The models generated by techniques from Chaps. 7–9 have already been evaluated during modeling (as discussed in Chap. 5). Performance on technical measures such as classification accuracy has been checked routinely whenever changes to the model were made to judge the advantageousness of the modifications. The models were also interpreted to gain new insights for feature construction (or even data acquisition). Once we are satisfied with the technical performance, what remains is to judge the potential impact the model will have in the projects domain should we implement and deploy it. We will tackle these two steps only briefly in the following two sections. A deployment in the form of, say, a software system for decision support involves several planning and coordination tasks, which are out of the scope of this book.

10.1 Evaluation

The analyst should assure herself that the interpretations of the data and models and the conclusions drawn are conform to the knowledge of the problem owner or, if available, domain expert. In particular the evidences, findings, and conclusions (not only a final model) need to be documented throughout the process and presented at this evaluation phase, where the results are discussed to decide if and how they may be deployed later. There are at least three more reasons for documentation: First, it is important that all steps are revised from the perspective of the project's owner to guarantee flawless interpretations and conclusions. Secondly, the findings represent important resources for future projects, where data understanding and data preparation phases may exhibit a large overlap with the current project. For instance, the cognitive map may be revised or extended and possibly linked to evidence found in the data. Finally, the drawbacks and problems faced may initiate improved data entering procedures (to improve data quality where it is crucial) or establish new data collections (data acquisition) such that future projects can benefit from an improved situation.

M.R. Berthold et al., *Guide to Intelligent Data Analysis,*
Texts in Computer Science 42,
DOI 10.1007/978-1-84882-260-3_10, © Springer-Verlag London Limited 2010

Documentation The necessity of documenting the evidences, findings, and generally all the steps that were carried out during the analysis are often underestimated. Complaints about the difficulty of reverse engineering what has been done and why can even be found in the literature [1, 5] and are the reason why CRISP-DM includes a detailed list of outputs (Sect. IV of [3]). Why is the documentation so important? The modeling tools greatly depend on the input, which is generated in the data preparation phase. As already mentioned, the phases of the data analysis process are by no means linear, and in order to improve the model further, the analyst may want to test one or another sudden inspiration to see if it impacts the resulting model. Often this requires a different preprocessing and thus the construction of a new input data set for the modeling tool of her choice. Some ideas appear promising, and others do not. After some time, many different versions exist, and it becomes increasingly difficult to keep track of what has been done already and why. Usually the (important) results are well remembered, but when it comes to reproducibility of results, the unorganized analyst starts a cumbersome reconstruction. Documentation is important, and modern tools support the analyst by modeling the full preprocessing rather than just supporting her with the individual steps. Thus documentation is indispensable to achieve a reproducible process—and writing down a few lines what we are about to do next and why let us pause for a moment and helps to avoid cheap shots.

Testbed Evaluation If possible, a further evaluation that comes close to the final operation area may be carried out. This is particularly important if several models, which may perform well individually, but not much is known about their orchestration and possible interdependencies, are needed to achieve the projects objective. In a narrow sense, we have seen this already in the third example of the overview chapter (Sect. 2.4): the effects of Stan's strategy are observed in the database by Laura to judge its successfulness. Depending on the project goal, such an evaluation may be costly but offers a realistic assessment of the performance in practice. We have, however, to plan this evaluation in the same way as we have done earlier with the technical model assessment: we need a baseline to compare with. For instance, if a new product is introduced and at the same time we start the new marketing campaign, there is no way to separate the effect of marketing. Either we use data from different points in time (which is not present in this example, because the product was not available earlier) or we have to generate a control group for comparison (people excluded from the marketing campaign).

To avoid self-fulfilling prophecies, the decision about the deployment should not be based on observation of extreme cases. Suppose that we investigate the sales figures of individual customers and identify the, say, 10% best- and worst-performing customers. Now, we just developed and started a marketing campaign and want to convince others that we were really successful by showing the sales figures of the same customers during the weeks following the campaign. Which group (best- or worst-performer) would you choose? Which would somebody choose who does not believe in the success of your campaign? Retail is not deterministic, and the needs of people may vary from week to week substantially. By cherry-picking the worst performing customers first and observing how they perform next week, it is very likely that *not all of them* will perform as poor as before (eventually some other will do), simply because of the inherent arbitrariness of many shopping decisions. So we are *guaranteed* to perform better (regardless of the marketing campaign) when picking the lowest 10%. Likewise for the 10% top-performers: Customers who had an extremely full cart last time probably still have some stockpile at home and need less next time. So these customers will underperform. This effect will be the more prominent the smaller the groups are chosen and the more both measurements are correlated. If we identify the worst-performing customers before and after the campaign *independently* from the first selection (rather than sticking to the same customers), the observed differences level off.

10.2 Deployment and Monitoring

Arriving at a fine-tuned, approved, ready-to-deploy model takes considerable time. During evaluation, we have double-checked that the model will deliver the expected results on newly incoming data. It is time to benefit from the undertaken effort. Predictive models are often deployed in the form of a software system that accomplishes some kind of decision support, like the signal light in the customer relationship management indicating the likeliness of churn to the operator, the prediction of the most promising customers for direct mailing, the detection of fraudulent transactions, etc. This process of putting a system to use in a productive environment is called deployment. Deployment can have many forms. We can use the server version of our analytics tool and allow others to call the model using a web server or other Software-as-a-Service (SaaS) setups. Alternatively, we can use an exchange format to export our model into a different system which is optimized for efficient model application—a prominent format for the exchange for formats is PMML, the Predictive Model Markup Language. PMML is an XML dialect which allows one to represent a wide range of predictive models along with some aspects of the required preprocessing. Quite a number of database systems actually allow one to load PMML models and apply them to the tables within the DB. Increasingly, we also see offerings to use cloud or other types of cluster computing to score your data given a specific model in PMML.

If we have identified (nominal) variables with a dynamic domain, we have to make sure that newly occurring values are handled appropriately and not just passed over to the model, which is unaware of such entries (see Sect. 6.3).

Deployment almost always also means that the problem or question is not singular and hence to be applied continuously. This can, for instance, be the case where the developed model is used for autonomous decision making repeatedly. An important question is then how we can make sure that the model is still valid. The point is, of course, that we need to recognize an invalid model as soon as possible, because otherwise the model will propose costly incorrect decisions. Monitoring the system during operation aims at recognizing a drop in performance. In predictive tasks the continuous tracing of the, say, accuracy measure suggests itself: A (sudden or slowly drifting) decline in the model's accuracy indicates divergence of the model and the world, and we must consider a recalibration. The dataset that has been used for modeling has, however, been constructed in a laborious preprocessing phase, which sometimes also includes the target attribute. (See, for example, the introductory example in Sect. 2.4, where the target attribute has been generated from a preceding experiment.) Providing the information whether the prediction was correct or not is thus highly important but often easier said than done.

Additionally, we may look directly for changes in the world that could possibly invalidate the model. Some changes may be detected (semi-)automatically, and others require external input. Possible reasons include:

- **Objectives change**. First of all, obviously the objectives can change because of, say, a different business strategy. This may lead to revised goals and modified problem statements. Such a management decision is typically propagated through the whole company, but we must not forget to check the impact on the deployed model and problem statement.
- **Invalidated assumptions**. Starting in the project understanding phase, we have carefully collected all assumptions we made. We have checked, during data understanding and validation, that these assumptions hold in principle. So our model may implicity or explicitly rely on these assumptions, and if one of them does no longer hold, it may perform poor because it was not designed for this kind of setting. At least for those assumptions we were able to verify in the data, we could periodically repeat this check to give at least a warning that the model performance might be affected. Sometimes this is easily accomplished by statistical tests that look for deviations in data distributions (such as the chi-square of goodness of fit test, Sect. A.4.3.4).
- **The world changes**. In the super market domain new products are offered, customer habits adapt, other products become affordable, competitors may open (or close) new stores, etc. In manufacturing newly installed machines (and measurement devices that provide the data) may have different characteristics (e.g., precision, operation point), machines deteriorate over time, operators may be replaced, etc. The data that has been used to construct the model may be affected by any of these factors, distributions of arbitrary variables may change. We have seen that many models estimate distribution parameters, derive optimal thresholds, etc. from the data. So if these distributions change, the parameters and thresholds may no longer be optimal. Again, statistical test can be applied to detect changes in the variables distributions.
- **Shift from inter- to extrapolation**. There is one particular situation (typically in modeling technical systems) which deserve special care, namely when the model is applied to data that is atypical for the data that has been used for learning. Unlike humans, who would recognize (and possibly complain) about situations that never occurred before, most of the models deliver *always* some result—some do not even carry information about the range of data for which they were designed. A polynomial of a higher degree, fitted to the data, usually yields poor predictions outside the range of the training data (see, for example, Fig. 8.8 on page 230).

Thus the detection of such cases is extremely important, because otherwise the model's behavior becomes objectionable.

Compared to the training data, such cases of extrapolation represent outliers. Therefore a feasible approach to overcome this problem is to employ clustering algorithms on the training data. As in outlier detection, the clusters may be used as a coarse representation of the data distribution (say, k prototypes of k-means, where k is usually chosen relatively large as the prototypes shall not find some *true* clusters in the data but distribute themselves over the whole dataset[1]). Then, for every new incoming data, its *typicality* is determined by finding the closest cluster (e.g., prototype). If it is too far away from the *known* data, a case of extrapolation is on hand.

The objective of monitoring is in the first place to avoid situations where the model is still applied even if the model may be inappropriate. However, detecting changes is often interesting by itself, as it may indicate the need to derive, say, new marketing strategies. We have already seen such a case in the example of Sect. 2.3, where Laura planned to compare models learned from data of different periods of time. Such questions are investigated in a relatively new field called *change mining* [2]. Prominent methods integrate the detection of change and an automatic adaption. For instance, in [4] each node of a decision tree monitors the distribution of newly arriving data and compares it with the training data distribution. If the deviation is significant, the node is replaced with a tree that has been learned in the meantime with more recent examples.

References

1. Becker, K., Ghedini, C.: A documentation infrastructure for the management of data mining projects. Inf. Softw. Technol. **47**, 95–111 (2005)
2. Böttcher, M., Höppner, F., Spiliopoulou, M.: On exploiting the power of time in data mining. SIGKDD Explorations **10**(2), 3–11 (2008)
3. Chapman, P., Clinton, J., Kerber, R., Khabaza, T., Reinartz, T., Shearer, C., Wirth, R.: Cross Industry Standard Process for Data Mining 1.0, Step-by-step Data Mining Guide. CRSIP-DM consortium (2000)
4. Hulten, G., Spencer, L., Domingos, P.: Mining time changing data streams. In: Proc. Int. ACM SIGKDD Conf. on Knowledge Discovery and Data Mining (KDD'01, San Francisco, CA), pp. 97–106. ACM Press, New York (2001)
5. Wirth, R., Hipp, J.: CRISP-DM: towards a standard process model for data mining. In: Proc. 4th Int. Conf. on the Practical Application of Knowledge Discovery and Data Mining, pp. 29–39. London, United Kingdom (2000)

---

[1] Sometimes a k-means cluster consists of a single outlier—such clusters should be, of course, rejected.

Appendix A
Statistics

Since classical statistics provides many data analysis methods and supports and justifies a lot of others, we provide in this appendix a brief review of some basics of
statistics. We discuss descriptive statistics, inferential statistics, and needed fundamentals from probability theory. Since we strove to make this appendix as self-
contained as possible, some overlap with the chapters of this book is unavoidable.
However, material is not simply repeated here but presented from a slightly different
point of view, emphasizing different aspects and using different examples.

In [14] (classical) **statistics** is characterized as follows:

Statistics is the art to acquire and collect data, to depict them, and to analyze
and interpret them in order to gain new knowledge.

This characterization already indicates that statistics is very important for data analysis. Indeed: there is a vast collection of statistical procedures with which the tasks
described in Sect. 1.3 (see page 11) can be tackled or which are needed to support or
justify other methods. Some of these methods are discussed in this appendix. However, we do not claim to have provided a complete overview. For a more detailed
review of (classical) statistics and its procedures, an interested reader is referred to
standard textbooks and references like [3, 4, 10].

The statistical concepts and procedures we are going to discuss can roughly be
divided into two categories corresponding to the two main areas of statistics:

- **descriptive statistics** (Sect. A.2) and
- **inferential statistics** (Sect. A.4).

In descriptive statistics it is tried to make data more comprehensible and interpretable by representing them in tables, charts, and diagrams, and to summarize
them by computing certain characteristic measures. In inferential statistics, on the
other hand, it is tried to draw inferences about the data generating process, like estimating the parameters of the process or selecting a model that fits it. The basis of
many procedures of inferential statistics is **probability theory** (see Sect. A.3); its
goal is usually to prepare for and to support decision making.

M.R. Berthold et al., *Guide to Intelligent Data Analysis,*
Texts in Computer Science 42,
DOI 10.1007/978-1-84882-260-3, © Springer-Verlag London Limited 2010

A.1 Terms and Notation

Before we can turn to statistical procedures, we have to introduce some terms and notions, together with some basic notation, with which we can refer to data.

- **object, case**
  Data describe objects, cases, people etc. For example, medical data usually describes patients, stockkeeping data usually describes components, devices or generally products, etc. The objects or cases are sometimes called the *statistical units*.
- **(random) sample**
  The set of objects or cases that are described by a data set is called a *sample*, its size (number of elements) is called the *sample size*. If the objects or cases are the outcomes of a random experiment (for example, drawing the numbers in a lottery), we call the sample a *random sample*.
- **attribute, feature**
  The objects or cases of the sample are characterized by attributes or features that refer to different properties of these objects or cases. Patients, for example, may be described by the attributes sex, age, weight, blood group, etc., component parts may have features like their physical dimensions or electrical parameters.
- **attribute value**
  The attributes, by which the objects/cases are described, can take different *values*. For example, the sex of a patient can be *male* or *female*, its age can be a positive integer number, etc. The set of values an attribute can take is called its *domain*.

Depending on the kind of the attribute values, one distinguishes different **scale types** (also called *attribute types*). This distinction is important, because certain *characteristic measures* (which we study in Sect. A.2.3) can be computed only for certain scale types. Furthermore, certain statistical procedures presuppose attributes of specific scale types. Table A.1 shows the most important scale types **nominal**, **ordinal**, and **metric** (or **numerical**), together with the core operations that are possible on them and a few examples of attributes having the scale type.

Table A.1 The most important scale types

Scale type	Possible operations	Examples
nominal (categorical, qualitative)	test for equality	sex blood type
ordinal (rank scale, comparative)	test for equality greater/less than	school grade wind strength
metric (numerical) (interval scale, quantitative)	test for equality greater/less than difference maybe ratio	length weight time temperature

Within nominal scales, one sometimes distinguishes according to the number of possible values. Attributes with only two values are called *dichotomous, alternatives* or *binary*, while attributes with more than two values are called *polytomous*. Within metric scales, one distinguishes whether only differences (temperature, calender time) are meaningful or whether it makes sense to compute ratios (length, weight, duration). One calls the former case *interval scale* and the latter *ratio scale*. In the following, however, we will not make much use of these additional distinctions.

From the above explanations of notions and expressions it already follows that a data set is the joint statement of attribute values for the objects or cases of a sample. The number of attributes that is used to describe the sample is called its *dimension*. One-dimensional data sets will be denoted by lowercase letters from the end of the alphabet, that is, for example x, y, z. These letters denote the attribute that is used to describe the objects or cases. The elements of the data set (the sample values are denoted by the same lowercase letter, with an index that states their position in the data set. For instance, we write $x = (x_1, x_2, \ldots, x_n)$ for a sample of size n. (A data set is written as a vector and not as a set, since several objects or cases may have the same sample value.) Multidimensional data sets are written as vectors of lowercase letters from the end of the alphabet. The elements of such data sets are vectors themselves. For example, a two-dimensional data set is written as $(x, y) = ((x_1, y_1), (x_2, y_2), \ldots, (x_n, y_n))$, where x and y are the two attributes by which the sample is described.

A.2 Descriptive Statistics

The task of descriptive statistics is to describe states and processes on the basis of observed data. The main tools to tackle this task are tabular and graphical representations and the computation of characteristic measures.

A.2.1 Tabular Representations

Tables are used to display observed data in a clearly arranged form, and also to collect and display characteristic measures. The simplest tabular representation of a (one-dimensional) data set is the **frequency table**, which is the basis for many graphical representations. A frequency table records for every attribute value its (absolute and/or relative) frequency in a sample, where the **absolute frequency** f_k is simply the occurrence frequency of an attribute value a_k in the sample, and the **relative frequency** r_k is defined as $r_k = \frac{f_k}{n}$ with the sample size n. In addition, columns for the cumulated (absolute and/or relative) frequencies (also simply referred to as *frequency sums*) may be present. As an example, we consider the data set

$$x = (3, 4, 3, 2, 5, 3, 1, 2, 4, 3, 3, 4, 4, 1, 5, 2, 2, 3, 5, 3, 2, 4, 3, 2, 3),$$

Table A.2 A simple frequency table showing the absolute frequencies f_k, the relative frequencies r_k, and the cumulated absolute and relative frequencies $\sum_{i=1}^{k} h_i$ and $\sum_{i=1}^{k} r_i$, respectively

a_k	f_k	r_k	$\sum_{i=1}^{k} f_i$	$\sum_{i=1}^{k} r_i$
1	2	$\frac{2}{25} = 0.08$	2	$\frac{2}{25} = 0.08$
2	6	$\frac{6}{25} = 0.24$	8	$\frac{8}{25} = 0.32$
3	9	$\frac{9}{25} = 0.36$	17	$\frac{17}{25} = 0.68$
4	5	$\frac{5}{25} = 0.20$	22	$\frac{22}{25} = 0.88$
5	3	$\frac{3}{25} = 0.12$	25	$\frac{25}{25} = 1.00$

Table A.3 A contingency table for two attributes A and B

	a_1	a_2	a_3	a_4	$\sum$
b_1	8	3	5	2	18
b_2	2	6	1	3	12
b_3	4	1	2	7	14
$\sum$	14	10	8	12	44

which may be, for instance, the grades of a written exam at school.[1] A frequency table for this data set is shown in Table A.2. Obviously, this table provides a much better view of the data than the raw data set as it is shown above, which only lists the sample values (an not even in a sorted fashion).

A two- or generally multidimensional frequency table, into which the (relative and/or absolute) frequency of every attribute value *combinations* is entered, is also called a **contingency table**. An example of a contingency table for two attribute A and B (with absolute frequencies), which also records the row and column sums, that is, the frequencies of the values of the individual attributes, is shown in Table A.3.

A.2.2 Graphical Representations

Graphical representations serve the purpose to make tabular data more easily comprehensible. The main tool to achieve this is to use geometric quantities—like lengths, areas, and angles—to represent numbers, since such geometric properties are more quickly interpretable for humans than abstract numbers. The most important types of graphical representations are:

---

[1]In most of Europe it is more common to use numbers for grades, with 1 being the best and 6 being the worst possible, while in the United States it is more common to use letters, with A being the best and F being the worst possible. However, there is an obvious mapping between the two scales. We chose numbers here to emphasize that nominal scales may use numbers and thus may look deceptively metric.

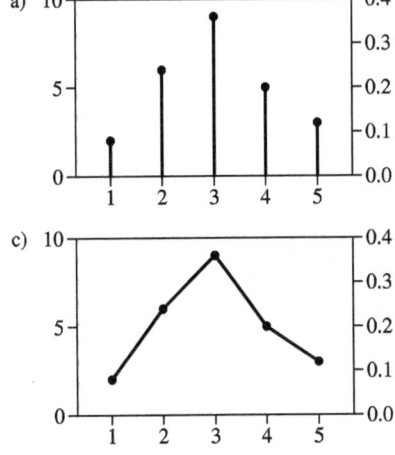

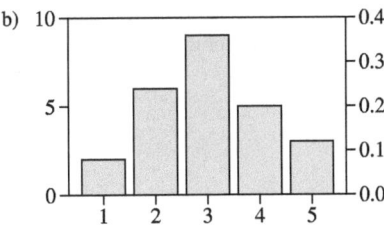

Fig. A.1 Pole (**a**) and bar chart (**b**) and frequency polygons (**c**) for the data shown in Table A.2

Fig. A.2 Area chart for the
data shown in Table A.2

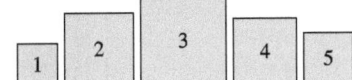

- **pole/stick/bar chart**
  Numbers, which may be, for instance, the frequencies of different attribute values
  in a sample, are represented by the lengths of poles, sticks, or bars. In this way
  a good impression especially of ratios can be achieved (see Figs. A.1a and b, in
  which the frequencies of Table A.2 are displayed).
- **area and volume charts**
  Area and volume charts are closely related to pole and bar charts: the difference
  is merely that they use areas and volumes instead of lengths to represent numbers
  and their ratios (see Fig. A.2, which again shows the frequencies of Table A.2).
  However, area and volume charts are usually less comprehensive (maybe except
  if the represented quantities are actually areas and volumes), since human beings usually have trouble comparing areas and volumes and often misjudge their
  numerical ratios. This can already be seen in Fig. A.2: only very few people correctly estimate that the area of the square for the value 3 (frequency 9) is three
  times as large as that of the square for the value 5 (frequency 3).
- **frequency polygons** and **line chart**
  A *frequency polygon* results if the ends of the poles of pole diagram are connected
  by lines, so that a polygonal course results. This can be advantageous if the attribute values have an inherent order and one wants to show the development of
  the frequency along this order (see Fig. A.1c). In particular, it can be used if
  numbers are to be represented that depend on time. This particular case is usually
  referred to as a *line chart*, even though the name is not exclusively reserved for
  this case.

Fig. A.3 A pie chart (**a**) and a stripe chart (**b**) for the data shown in Table A.2

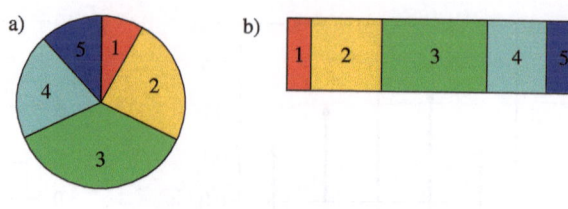

Fig. A.4 A mosaic chart for the contingency table of Table A.3

Fig. A.5 A bar chart for the contingency table of Table A.3

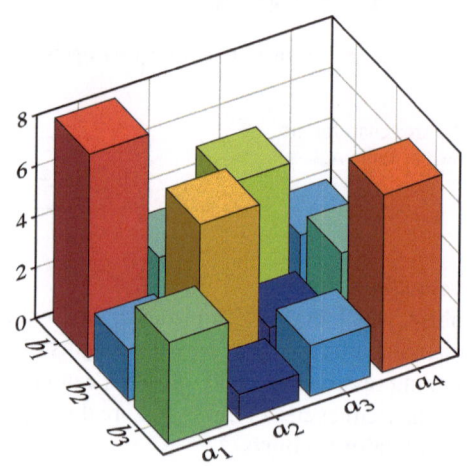

- **pie and stripe chart**

  Pie and stripe charts are particularly well suited if proportions or fractions of a total, for instance, relative frequencies, are to be displayed. In a pie chart proportions are represented by angles, and in a stripe chart by lengths (see Fig. A.3).

- **mosaic chart**

  Contingency tables (that is, two- or generally multidimensional frequency tables) can nicely be represented as mosaic charts. For the first attribute, the horizontal direction is divided like in a stripe diagram. Each section is then divided according to the second attribute along the vertical direction—again like in a stripe diagram (see Fig. A.4). Mosaic charts can have advantages over two-dimensional bar charts, because bars at the front can hide bars at the back, making it difficult to see their height, as shown in Fig. A.5. In principle, arbitrarily many attributes can be displayed by subdividing the resulting mosaic pieces alternatingly along the horizontal and vertical axis. However, even if one uses the widths of the gaps

Fig. A.6 A simple scatter plot

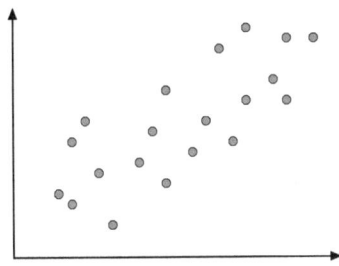

and colors in order to help a viewer to identify attribute values, mosaic charts can easily become confusing if it is tried to use to many attributes.

- **histogram**
  In principle, a histogram looks like a bar chart, with the only difference that the domain of the underlying attribute is metric (numerical). As a consequence, it is usually impossible to simply enumerate the frequencies of the individual attribute values (because there are usually too many different values), but one has to form counting intervals, which are usually called *bins* or *buckets*. The width (or, if the domain is fixed, equivalently the number) of these bins has to be chosen by a user. All bins should have the same width, since histograms with varying bin widths are usually more difficult to read—for the same reasons why area charts are more difficult to interpret than bar charts (see above). In addition, a histogram may only provide a good impression of the data if an appropriate bin width has been chosen and onto which values the borders of the bins fall (see Sect. 4.3.1).

- **scatter plot**
  A scatter plot displays a two-dimensional data set of metric attributes by interpreting the sample values as coordinates of a point in a metric space (see Fig. A.6). A scatter plot is very well suited if one wants to see whether the two represented quantities depend on each other or vary independently (see also Sects. A.2.4 and 8.3).

Examples how graphical representations can be misleading—a property that is sometimes (all too often actually) exploited to convey a deceptively favorable or unfavorable impression, in particular in the press and in advertisements—can be found in the highly recommended books [6, 8].

A.2.3 Characteristic Measures for One-Dimensional Data

The goal of computing characteristic measures is to summarize the data set, that is, to capture characteristic and relevant properties in as few quantities as possible. There are basically three types of characteristic measures:

- **location measures**
  As their name clearly indicates, location measures specify the location of the (majority of) the data in the domain of an attribute by a single number or attribute values. Thus location measures summarize the data heavily.

- **dispersion measures**
  Given the value of a location measure, dispersion measures specify how much the data scatter around this value (how much they deviate from it) and thus characterize how well the location measure captures the location of the data.
- **shape measures**
  Given the values of a location and a dispersion measure, shape measures characterize the distribution of the data by comparing its shape to a reference shape. The most common reference shape is the normal distribution (see Sect. A.3.5.7).

In the following we study these characteristic measures, which will turn out to be very useful in inferential statistics (see Sect. A.4), in more detail.

A.2.3.1 Location Measures

As already mentioned, a location measure characterizes the location of the data in the domain of the underlying attribute (recall that we are currently concerned only with one-dimensional data sets) by a single attribute value. This value should be as representative for the data as possible. We may require, for example, that the sum of the deviations of the individual sample values from the value of the location measure should be as small as possible. The most important location measures are the *mode*, the *median* (also called the *central value*), and its generalization, the so-called *quantiles* and the *mean*, which is the most common location measure.

Mode An attribute value is called the (empirical) *mode* x^* of a data set if it is the value that occurs most frequently in the sample. As a consequence, it need not be uniquely determined, since there may be several values that have the frequency. Modes can be determined for any scale type, because the only operation needed to compute them is a test for equality. Therefore the mode is the most general location measure. However, for metric data, it is most of the time less well suited than other measures, because the usually large number of possible values of metric attributes obviously poses some problems. However, in many cases one can amend this situation by choosing the middle of the highest bar in an appropriate histogram as the mode of the distribution of a numerical attribute.

Median (Central Value) The (empirical) *median* or *central value* $\tilde{x}$ can be introduced as a value that minimizes the sum of the absolute deviations. That is, a median $\tilde{x}$ is any value that satisfies

$$\sum_{i=1}^{n} |x_i - \tilde{x}| = \min.$$

In order to find a value for $\tilde{x}$, we take the derivative of the left-hand side and equate the result to zero (since the derivative must vanish at the minimum). In this way we obtain

$$\sum_{i=1}^{n} \operatorname{sgn}(x_i - \tilde{x}) = 0,$$

where sgn is the sign function (which is -1 if its argument is negative, $+1$ if its argument is positive, and 0 if its argument is 0).[2] Therefore a median is a value that lies "in the middle of the data." That is, in the data set there are as many values greater than $\tilde{x}$ as smaller than $\tilde{x}$ (this justifies the expression *central value* as an alternative to *median*).

With the above characterization, the median is not always uniquely determined. For example, if all sample values are distinct, there is only a unique middle element if the sample size is odd. If it is even, there may be several values that satisfy the above defining equations. As an example, consider the data set $(1, 2, 3, 4)$. Any value in the interval $[2, 3]$ minimizes the sum of absolute deviations. In order to obtain a unique value, one usually defines the median as the arithmetic mean of the two sample values in the middle of the (sorted) data set in such a case. In the above example, this would result in $\tilde{x} = \frac{2+3}{2} = \frac{5}{2}$. Note that the median is always uniquely determined for even sample size if the two middle values are equal.

Formally the median is defined as follows: let $x = (x_{(1)}, \ldots, x_{(n)})$ be a sorted data set, that is, we have $\forall i, j : (j > i) \rightarrow (x_{(j)} \geq x_{(i)})$. Then

$$\tilde{x} = \begin{cases} x_{(\frac{n+1}{2})} & \text{if } n \text{ is odd,} \\ \frac{1}{2}\left(x_{(\frac{n}{2})} + x_{(\frac{n}{2}+1)}\right) & \text{if } n \text{ is even,} \end{cases}$$

is called the *(empirical) median* of the data set x.

The median can be computed for ordinal and metric attributes, because all it requires is a test for greater or less than, namely for sorting the sample values. For ordinal values, computing the arithmetic mean of the two middle values for even sample size is replaced by simply choosing one of them, thus eliminating the need for the computation. Note, however, that the above characterization of the median as the minimizer of the absolute deviations can, of course, not be used for ordinal attributes as they do not allow for computing differences. We used it here nevertheless in order to show the analogy to the mean, which is considered below.

Quantiles We have seen in the preceding section that the median is an attribute value such that half of the sample values are less than it, and the other half is greater. This idea can easily be generalized by finding an attribute value such that a certain fraction p, $0 < p < 1$, of the sample values is less than this attribute value (and a fraction of $1 - p$ of the sample values are greater). These values are called (empirical) p-quantiles. The median in particular is the (empirical) $\frac{1}{2}$-quantile of a data set.

Other important quantiles are the first, second, and third quartiles, for which $p = \frac{1}{4}, \frac{2}{4}$, and $\frac{3}{4}$, respectively, of the data set are smaller (therefore the median is also identical to the second quartile), and the deciles (k tenths of the data set are smaller) and the percentiles (k hundredths of the data set are smaller).

Note that for metric attributes, it may be necessary, depending on the sample size and the exact sample values, to introduce adaptations that are analogous to the computation of the arithmetic mean of the middle values for the median.

---

[2]Note that, with the standard definition of the sign function, this equation cannot always be satisfied. In this case one confines oneself with the closest possible approximation to zero.

Mean While the median minimizes the *absolute deviations* of the sample values, the (empirical) mean $\bar{x}$ can be defined as the value that minimizes the sum of the squares of the deviations of the sample values. That is, the mean is the attribute value that satisfies

$$\sum_{i=1}^{n}(x_i - \bar{x})^2 = \min.$$

In order to find a value for $\bar{x}$, we take the derivative of the left-hand side and equate it to zero (since the derivative must vanish at a minimum). In this way we obtain

$$\sum_{i=1}^{n}(x_i - \bar{x}) = \sum_{i=1}^{n}x_i - n\bar{x} = 0,$$

and thus

$$\bar{x} = \frac{1}{n}\sum_{i=1}^{n}x_i.$$

Therefore the mean of a sample is the arithmetic mean of the sample values.

Even though the mean is the most commonly used location measure for metric attributes (note that it cannot be applied for ordinal attributes as it requires summation and thus an interval scale), the median should be preferred for

- few measurement values,
- asymmetric (skewed) data distributions, and
- likely presence of outliers,

since the median is more robust in these cases and conveys a better impression of the data. In order to make the mean more robust against outliers, it is often computed by eliminating the largest and the smallest sample values (a typical procedure for averaging the ratings of the judges in sports events) or even multiple extreme values, like all values before the 1st and beyond the 99th percentile.

A.2.3.2 Dispersion Measures

As already mentioned in the general overview of characteristic measures, dispersion measures specify how broadly the sample values scatter around a location parameter. Hence they characterize how well the data are captured by the location parameter. The reason for introducing dispersion measures is that a location measure alone does not tell us anything about the size of the deviations and thus one may be deceived about the true situation. This possibility is captured well in the old statistics joke:

A man with his head in the freezer and feet in the oven is *on the average* quite comfortable.

Range The range of a data set is simply the difference between that largest and the smallest sample value:

$$R = x_{\max} - x_{\min} = \max_{i=1}^{n} x_i - \min_{i=1}^{n} x_i.$$

The range is a very intuitive dispersion measure. However, it is very sensitive against outliers, which tend to corrupt one or even both of the values it is computed from.

Interquantile Range The difference between the (empirical) $(1-p)$- and the (empirical) p-quantiles of a data set is called the p-interquantile range, $0 < p < \frac{1}{2}$. Commonly used interquantile ranges are the interquartile range ($p = \frac{1}{4}$, difference between the third and first quartiles), the interdecile range ($p = \frac{1}{10}$, difference between the 9th and the 1st deciles), and the interpercentile range ($p = \frac{1}{100}$, difference between the 99th and the 1st percentiles). For small p, the p-interquantile range transfers the idea to make the mean more robust by eliminating extreme values to the range.

Mean Absolute Deviation The mean absolute deviation is the arithmetic mean of the absolute deviations of the sample values from the (empirical) median or mean:

$$d_{\tilde{x}} = \frac{1}{n} \sum_{i=1}^{n} |x_i - \tilde{x}|$$

is the mean absolute deviation from the median, and

$$d_{\bar{x}} = \frac{1}{n} \sum_{i=1}^{n} |x_i - \bar{x}|$$

is the mean absolute deviation from the mean. It is always $d_{\tilde{x}} \leq d_{\bar{x}}$, because the median minimizes the sum and thus also the mean of the absolute deviations.

Variance and Standard Deviation In analogy to the absolute deviation, one may also compute the mean squared deviation. (Recall that the mean minimizes the sum of the squared deviations.) However, instead of

$$m^2 = \frac{1}{n} \sum_{i=1}^{n} (x_i - \bar{x})^2,$$

it is more common to employ

$$s^2 = \frac{1}{n-1} \sum_{i=1}^{n} (x_i - \bar{x})^2$$

as a dispersion measure, which is called the *(empirical) variance* of the sample. The reason for the value $n-1$ in the denominator is provided by inferential statistics, in which the characteristic measures of descriptive statistics are related to certain parameters of probability distributions and density functions (see Sect. A.4).

A detailed explanation will be provided in Sect. A.4.2, which deals with parameter estimation (unbiasedness of an estimator for the variance of a normal distribution).

The positive square root of the variance, that is,

$$s = \sqrt{s^2} = \sqrt{\frac{1}{n-1} \sum_{i=1}^{n} (x_i - \bar{x})^2},$$

is called the *(empirical) standard deviation* of the sample.

Not that the (empirical) variance can often be computed more conveniently with formula that is obtained with the following transformation:

$$s^2 = \frac{1}{n-1} \sum_{i=1}^{n} (x_i - \bar{x})^2 = \frac{1}{n-1} \sum_{i=1}^{n} (x_i^2 - 2x_i\bar{x} + \bar{x}^2)$$

$$= \frac{1}{n-1} \left(\sum_{i=1}^{n} x_i^2 - 2\bar{x} \sum_{i=1}^{n} x_i + \sum_{i=1}^{n} \bar{x}^2 \right) = \frac{1}{n-1} \left(\sum_{i=1}^{n} x_i^2 - 2n\bar{x}^2 + n\bar{x}^2 \right)$$

$$= \frac{1}{n-1} \left(\sum_{i=1}^{n} x_i^2 - n\bar{x}^2 \right) = \frac{1}{n-1} \left(\sum_{i=1}^{n} x_i^2 - \frac{1}{n} \left(\sum_{i=1}^{n} x_i \right)^2 \right).$$

The advantage of this formula is that it allows us to compute both the mean and the variance of a sample with one pass through the data, by computing the sum of sample values and the sum of their squares. A computation via the original formula, on the other hand, needs two passes: in the first pass the mean is computed, and in the second pass the variance is computed from the sum of the squared deviations.

A.2.3.3 Shape Measures

If one plots a histogram of observed metric data, one often obtains a bell shape. In practice, this bell shape usually differs more or less from the reference of an ideal Gaussian bell curve (normal distribution, see Sect. A.3.5.7). For example, the empirical distribution, as shown by the histogram, is asymmetric or differently curved. With shape measures one tries to capture these deviations.

Skewness The skewness or simply skew α_3 states whether, and if yes, by how much a distribution differs from a symmetric distribution.[3] The skewness is computed as

$$\alpha_3 = \frac{1}{n \cdot s^3} \sum_{i=1}^{n} (x_i - \bar{x})^3 = \frac{1}{n} \sum_{i=1}^{n} z_i^3 \quad \text{with} \quad z_i = \frac{x_i - \bar{x}}{s},$$

that is, z is the z-score normalized variable. For a symmetric distribution, $\alpha_3 = 0$. If the skew is positive, the distribution is steeper on the left, and if it is negative, the distribution is steeper on the right (see Fig. A.7).

---

[3]The index 3 indicates that the skew is the 3rd moment of the sample around the mean—see the defining formula, which employs a third power.

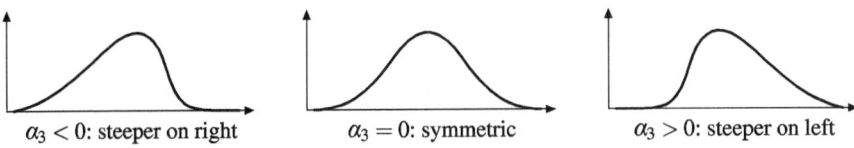

Fig. A.7   Illustration of the shape measure *skewness*

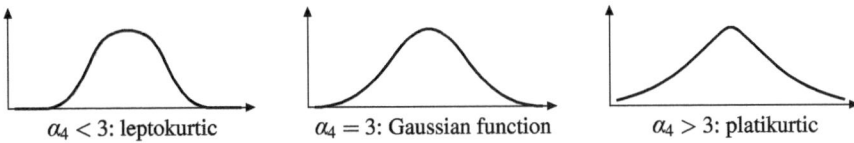

Fig. A.8   Illustration of the shape measure *kurtosis*

Kurtosis   The kurtosis α_4 describes how strongly a bell-shaped distribution is curved and thus how steep the peak is[4] (compared to the ideal Gaussian bell curve). The kurtosis is computed as

$$\alpha_4 = \frac{1}{n \cdot s^4} \sum_{i=1}^{n} (x_i - \bar{x})^4 = \frac{1}{n} \sum_{i=1}^{n} z_i^4 \quad \text{with} \quad z_i = \frac{x_i - \bar{x}}{s}.$$

An ideal Gaussian function as a kurtosis of 3. If the kurtosis is smaller than 3, the distribution is more peaked (leptokurtic) than a Gaussian bell curve; if it is greater than 3, the distribution is less peaked (platikurtic) than a Gaussian bell curve (see Fig. A.8). Sometimes the kurtosis is defined as $\alpha_4 - 3$, so that the Gaussian functions has a kurtosis of 0 and the sign indicates whether the distribution under consideration is more (negative, leptokurtic) or less peaked (positive, platikurtic) than a Gaussian function.

A.2.3.4 Box Plots

Some characteristic measures, namely the median, the mean, the range, and the interquartile range are often displayed jointly in a so-called box plot (see Fig. A.9): the outer lines show the range and the box in the middle, which gives this diagram form its name, indicates the interquartile range. Into the box the median is drawn as a solid, and the mean as a dashed line (alternatively mean and median can be drawn in different colors). The range may be replaced by the interpercentile range. In this case the extreme values outside this range are depicted as individual dots. Sometimes the box that represents the interquartile range is drawn constricted at the location of the mean in order to emphasize the location of the mean. Obviously this simple diagram provides a good compact impression of the rough shape of the data distribution.

---

[4]The index 4 indicates that the kurtosis is the 4th moment around the mean—see the defining formula, which employs a fourth power.

Fig. A.9 A box plot is a simple diagram that captures the most important characteristic measures

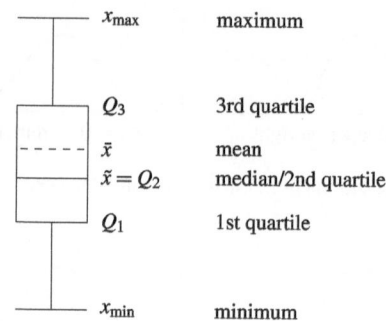

x_{max}	maximum
Q_3	3rd quartile
$\bar{x}$	mean
$\tilde{x} = Q_2$	median/2nd quartile
Q_1	1st quartile
x_{min}	minimum

A.2.4 Characteristic Measures for Multidimensional Data

Several of the characteristic measures that we introduced in the preceding section for one-dimensional data sets can easily be transferred to multidimensional data by simply executing the computations with vectors instead of scalars (simple numbers). Here we consider as examples the transfer of the mean and the variance, which will lead us to the covariance (matrix). By normalizing the covariance, we obtain the important measure of the correlation coefficient.

Mean For multidimensional data, the mean turns into the vector mean of the data points. For example, for two-dimensional data, we have

$$\overline{(x, y)} = \frac{1}{n} \sum_{i=1}^{n} (x_i, y_i) = (\bar{x}, \bar{y}).$$

It should be noted that one obtains the same result if one forms the vector that consists of the means of the individual attributes. Hence, for computing the mean, the attributes can be treated independently.

Covariance and Correlation It is equally simple to transfer dispersion measure *variance* to multidimensional data by simply executing the computations with vectors instead of scalars. The only problem consists in squaring the differences between the sample data points and the mean vector, since these differences are now vectors. In order to compute this square, the so-called outer product or matrix product of the difference vector with itself is computed. This outer product is defined as $\mathbf{v}\mathbf{v}^\top$ (where $\top$ denotes a transposed vector) and yields a square matrix. These matrices (one for each sample data point) are summed and (like the standard scalar variance) divided by $n - 1$, where n is the sample size. The result is a square, symmetric,[5] and positive definite[6] matrix, the so-called covariance matrix. For two-dimensional data, the covariance matrix is defined as

---

[5]A square matrix $\mathbf{M} = (m_{ij})_{1 \leq i \leq m, 1 \leq j \leq m}$ is called symmetric if $\forall i, j : m_{ij} = m_{ji}$.
[6]A matrix $\mathbf{M}$ is called *positive definite* if for all vectors $\mathbf{v} \neq \mathbf{0}$, $\mathbf{v}^\top \mathbf{M}\mathbf{v} > 0$.

$$\Sigma = \frac{1}{n-1} \sum_{i=1}^{n} \left(\begin{pmatrix} x_i \\ y_i \end{pmatrix} - \begin{pmatrix} \bar{x} \\ \bar{y} \end{pmatrix} \right) \left(\begin{pmatrix} x_i \\ y_i \end{pmatrix} - \begin{pmatrix} \bar{x} \\ \bar{y} \end{pmatrix} \right)^{\top} = \begin{pmatrix} s_x^2 & s_{xy} \\ s_{xy} & s_y^2 \end{pmatrix},$$

where

$$s_x^2 = \frac{1}{n-1} \left(\sum_{i=1}^{n} x_i^2 - n\bar{x}^2 \right) \quad \text{(variance of } x\text{)},$$

$$s_y^2 = \frac{1}{n-1} \left(\sum_{i=1}^{n} y_i^2 - n\bar{y}^2 \right) \quad \text{(variance of } y\text{)},$$

$$s_{xy} = \frac{1}{n-1} \left(\sum_{i=1}^{n} x_i y_i - n\bar{x}\bar{y} \right) \quad \text{(covariance of } x \text{ and } y\text{)}.$$

In addition to the variances of the individual attributes, a covariance matrix contains an additional quantity, the so-called *covariance*. It yields information about the strength of the (linear) dependence of the two attributes. However, since its value also depends on the variance of the individual attributes, it is normalized by dividing it by the standard deviations of the individual attributes, which yields the so-called correlation coefficient (more precisely, Pearson's product moment correlation coefficient; see Sect. 4.4 for alternatives, especially for ordinal attributes),

$$r = \frac{s_{xy}}{s_x s_y}.$$

It should be noted that the correlation coefficient is identical to the covariance of the two attributes if their values are first normalized to mean 0 and standard deviation 1. The correlation coefficient has a value between -1 and $+1$ and characterizes the strength of the *linear dependence* of two metric quantities: if all data points lie exactly on an ascending straight line, its value is $+1$. If they lie exactly on a descending straight line, its value is -1. In order to convey an intuition of intermediate values, Fig. A.10 shows some examples.

Note that it does not mean that two measures are (stochastically) independent if their correlation coefficient vanishes. For example, if the data points lie symmetrically on a parabola, the correlation coefficient is $r = 0$. Nevertheless there is, of course, a clear and exact functional dependence of the two measures. If the correlation coefficient is zero, this only means that this dependence is not linear.

Since the covariance and correlation describe the linear dependence of two measures, it is not surprising that they can be used to fit a straight line to the data, that is, to determine a so-called regression line. This line is defined as

$$(y - \bar{y}) = \frac{s_{xy}}{s_x^2}(x - \bar{x}) \quad \text{or} \quad y = \frac{s_{xy}}{s_x^2}(x - \bar{x}) + \bar{y}.$$

The regression line can be seen as kind of mean function, which assigns a mean of the y-values to each of the x-values (conditional mean). This interpretation is supported by the fact that the regression line minimizes the sum of the squares of the deviations of the data points (in y-direction), just like the mean. More details about regression and the method of least squares, together with generalizations to larger function classes, can be found in Sect. 8.3.

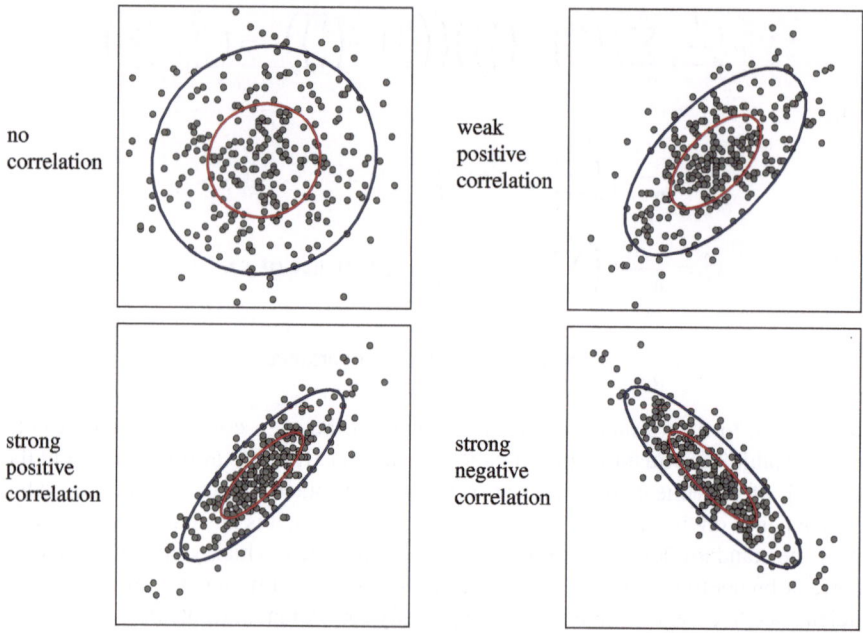

no
correlation

weak
positive
correlation

strong
positive
correlation

strong
negative
correlation

Fig. A.10 Illustration of the meaning of the correlations coefficient

A.2.5 Principal Component Analysis

Correlations between the attributes of a data set can be used to reduce its dimension: if an attribute is (strongly) correlated with another, then this attribute is essentially a linear function of the other (plus some noise). In such a case it is often sufficient to consider only one of the two attributes, since the other can be reconstructed (approximately) via the regression line. However, this approach has the disadvantage that it is not trivial to decide which of several correlated attributes should be kept and which can be discarded.

A better approach to reduce the dimension of a data set is the so-called principal component analysis (PCA; see also Sect. 4.3.2.1). The basic idea of this procedure is not to select a subset of the features of the data set, but to construct a small number of new features as linear combinations of the original ones. These new quantities are supposed to capture the greater part of the information in the data set, where the information content is measured by the (properly normalized) variance: the larger the variance, the more important the (constructed) feature.

In order to find the linear combinations that define the new features, the data is first normalized to mean 0 and standard deviation 1 in all original attributes, so that the scale of the attributes (that is, for example, the units in which they were measured) does not influence the result. In the next step one tries to find a new basis for the data space, that is, perpendicular directions. This is done in such a way that the first direction is the one in which the (normalized) data exhibits the largest variance. The second direction is the one which is perpendicular to the first and in which the data exhibits the largest variance among all directions perpendicular to the first, and so on. Finally, the data is transformed to the new basis of the data space, and some of the constructed features are discarded, namely those, for which the transformed data exhibits the lowest variances. How many features are discarded is decided based on the sum of the variances of the kept features relative to the total sum of the variance of all features.

Formally, the perpendicular directions referred to above can be found with a mathematical method that is known as principal axes transformation. This transformation is applied to the correlation matrix, that is, the covariance matrix of the data set that has been normalized to mean 0 and standard deviation 1 in all features. That is, one finds a rotation of the coordinate system such that the correlation matrix becomes a diagonal matrix. The elements of this diagonal matrix are the variances of the data set w.r.t. the new basis of the data space, while all covariances vanish. This is also a fundamental goal of principal component analysis: one wants to obtain features that are linearly independent. As is well known from linear algebra (see, for example, [5, 11]), a principal axes transformation consists basically in computing the eigenvalues and eigenvectors of a matrix. The eigenvalues show up on the diagonal of the transformed correlation matrix, the eigenvectors (which can be obtained in parallel with appropriate methods) indicate the desired directions in the data space. The directions w.r.t. which the data is now described are selected based on the eigenvalues, and finally the data is projected to the subspace chosen in this way.

The following physical analog may make the idea of principal axes transformation clearer: how a solid body reacts to a rotation around a given axis, can be described by the so-called tensor of inertia [9]. Formally, this tensor is a symmetric 3×3 matrix,

$$
\Theta = \begin{pmatrix} \Theta_{xx} & \Theta_{xy} & \Theta_{xz} \\ \Theta_{xy} & \Theta_{yy} & \Theta_{yz} \\ \Theta_{xz} & \Theta_{yz} & \Theta_{zz} \end{pmatrix}.
$$

The diagonal elements of this matrix are the *moments of inertia* of the body w.r.t. the axes that pass through its center of gravity[7] and are parallel to the axes of the coordinate system that we use to describe the rotation. The remaining (off-diagonal) elements of the matrix are called deviation moments and describe the forces that act perpendicular to the axes during the rotation.[8] However, for any solid body, regardless of its shape, there are three axes w.r.t. which the deviation moments vanish, the

---

[7]The inertia behavior of axes that do not pass through the center of gravity can easily be described with Steiner's law. However, this goes beyond the scope of this discussion, see standard textbooks on theoretical mechanics like [9] for details.

[8]These forces result from the fact that generally the vector of angular momentum is not parallel to the vector of angular velocity. However, this again leads beyond the scope of this discussion.

Fig. A.11 The principal axes
of inertia of a box

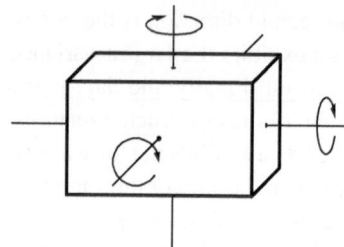

so-called *principal axes of inertia.*[9] As an example, Fig. A.11 shows the principal axes of inertia of a box. The principal axes of inertia are always perpendicular to each other. In the coordinate system that is spanned by the principal axes of inertia, the tensor of inertia is a diagonal matrix.

Formally, the principal axes of inertia are found by carrying out a principal axes transformation of the tensor of inertia (given w.r.t. an arbitrary coordinate system): its eigenvectors are the directions of the principal axes of inertia.

In the real world, the deviation moments cause shear forces, which lead to vibrations and jogs in the bearings of the axis. Since such vibrations and jogs naturally lead to quick abrasion of the bearings, it is tried to minimize the deviation moments. As a consequence, a car mechanic who balances a wheel can be seen as carrying out a principal axes transformation (though not in mathematical form), because he/she tries to equate the rotation axis with a principal axis of inertia. However, he/she does not do so by changing the direction of the rotation axis, as this is fixed in the wheel. Rather, he/she changes the mass distribution by adding, removing, and shifting small weights so that the deviation moments vanish.

Based on this analog, we may say that a statistician looks, in the first step of a principal component analysis, for axes around which a mass distribution with unit weights at the locations of the data points can be rotated without vibrations or jogs in the bearings. Afterwards, he selects a subset of the axes by removing those axes around which the rotation needs the most energy, that is, those axes for which the moments of inertia are largest (in the direction of these axes, the variance is smallest, and perpendicularly to them, the variance is largest).

Formally, the axes are selected via the percentage of explained variance. It can be shown that the sum of the eigenvalues of a correlation matrix equals the dimension m of the data set, that is, it is equal to the number of features (see, for example, [5, 11]). In this case it is plausible to define that the proportion of the total variance that is captured by the jth principal axis as

$$p_j = \frac{\lambda_j}{m} \cdot 100\%,$$

where λ_j is the eigenvalue corresponding to the jth principal axis.

---

[9]Note that a body may possess more than three axes w.r.t. which the deviation moments vanish. For a sphere with homogeneous mass distribution, for example, any axis that passes through the center is such an axis. However, any body, regardless of its shape, has at least three such axes.

Let $p_{(1)}, \ldots, p_{(m)}$ a sequence of these percentages that is sorted descendingly. For this sequence, one determines the smallest value k such that

$$\sum_{j=1}^{k} p_{(j)} \geq \alpha \cdot 100\%$$

with a proportion α that has to be chosen by a user (for example, $\alpha = 0.9$). The corresponding k principal axes are chosen as the new features, and the data points are projected to them. Alternatively, one may specify to how many features one desires to reduce the data set to and then chooses the axes following a descending proportion $\frac{\lambda_j}{m}$. In this case the above sum provides information about how much information contained in the original data is lost.

A.2.5.1 Example of a Principal Component Analysis

As a very simple example of a principal component analysis, we consider the reduction of a data set of two correlated quantities to a one-dimensional data set. Let the following data be given:

x	5	15	21	29	31	43	49	51	61	65
y	33	35	24	21	27	16	18	10	4	12

Even a quick look at this table already reveals that the two features are strongly correlated, an even clearer impression is provided by Fig. A.12. As a consequence, it can be expected that this data set can be reduced to a one-dimensional data set without much loss of information.

In the first step we compute the correlation matrix. To this end we normalize the data to mean 0 and standard deviation 1 and compute the covariance matrix of the normalized data. Thus we obtain the normalized data set

x'	−1.6	−1.1	−0.8	−0.4	−0.3	0.3	0.6	0.7	1.2	1.4
y'	1.3	1.5	0.4	0.1	0.7	−0.4	−0.2	−1.0	−1.6	−0.8

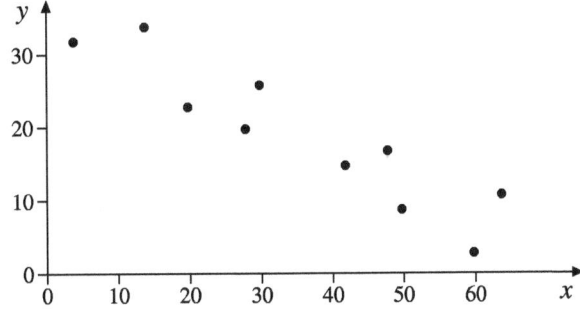

Fig. A.12 Data of an example for the principal component analysis as a scatter plot

The covariance of these normalized features is the correlation coefficient

$$r = s_{x'y'} = \frac{1}{9}\sum_{i=1}^{10} x_i y_i = \frac{-8.28}{9} = -\frac{23}{25} = -0.92.$$

Therefore the correlation matrix is

$$\Sigma = \frac{1}{9}\begin{pmatrix} 9 & -8.28 \\ -8.28 & 9 \end{pmatrix} = \begin{pmatrix} 1 & -\frac{23}{25} \\ -\frac{23}{25} & 1 \end{pmatrix}.$$

(Note that the diagonal elements of this matrix are the correlation coefficients of the features with themselves. Hence they are necessarily 1.)

For this correlation matrix, we have to carry out a principal axes transformation. In order to do so, we compute the eigenvalues and eigenvectors of this matrix, that is, those values λ_i and vectors v_i, $i = 1, 2$, for which

$$\Sigma v_i = \lambda_i v_i \quad \text{or} \quad (\Sigma - \lambda_i I)v_i = 0,$$

where I is the unit matrix. In order to compute the eigenvalues λ_i, we rely here on the simple method of finding the roots of the characteristic polynomial[10]

$$c(\lambda) = |\Sigma - \lambda E| = (1 - \lambda)^2 - \frac{529}{625}.$$

The roots of this polynomial are the eigenvalues

$$\lambda_{1/2} = 1 \pm \sqrt{\frac{529}{625}} = 1 \pm \frac{23}{25}; \quad \text{hence} \quad \lambda_1 = \frac{48}{25} \text{ and } \lambda_2 = \frac{2}{25}.$$

The corresponding eigenvectors are

$$v_1 = \left(\frac{\sqrt{2}}{2}, -\frac{\sqrt{2}}{2}\right) \quad \text{and} \quad v_2 = \left(\frac{\sqrt{2}}{2}, \frac{\sqrt{2}}{2}\right)$$

that can be found by simply inserting the eigenvalues into

$$(\Sigma - \lambda_i E)v_i = 0$$

and solving the resulting underdetermined linear equation system.[11] Therefore the principal axes transformation is given by the orthogonal matrix that has the eigenvectors as its columns, that is,

$$T = \begin{pmatrix} \frac{\sqrt{2}}{2} & \frac{\sqrt{2}}{2} \\ -\frac{\sqrt{2}}{2} & \frac{\sqrt{2}}{2} \end{pmatrix}.$$

---

[10]Note, however, that for larger matrices, this method is numerically unstable and should be replaced by some other approach; see, for example, [13].

[11]Note that for two variables, due to the special form of the characteristic polynomial, the eigenvectors are always exactly these two vectors (for the normalized data), independent of original data.

This matrix satisfies

$$\mathbf{T}^\top \Sigma \mathbf{T} = \begin{pmatrix} \lambda_1 & 0 \\ 0 & \lambda_2 \end{pmatrix}.$$

However, in order to obtain simple numbers for the transformed data, we multiply the data points with $\sqrt{2}\mathbf{T}^\top$ instead of $\mathbf{T}^\top$, that is, we compute

$$\begin{pmatrix} x'' \\ y'' \end{pmatrix} = \sqrt{2} \cdot \mathbf{T}^\top \cdot \begin{pmatrix} x' \\ y' \end{pmatrix}.$$

With the help of this transformation, the data points are projected to the principal axes. Intuitively, the multiplication with the matrix is equivalent to dropping perpendiculars from each data point to the axes that are given by the eigenvectors and using the distances of the feet of the perpendiculars from the origin as new coordinates. The transformed data set is

x''	−2.9	−2.6	−1.2	−0.5	−1.0	0.7	0.8	1.7	2.8	2.2
y''	−0.3	0.4	−0.4	−0.3	0.4	−0.1	0.4	−0.3	−0.4	0.6

This data set describes the data points in the space that is spanned by the principal axes. Since the y'' values vary only fairly little compared to the x'' values (without the factor $\sqrt{2}$ that we added to the transformation, the eigenvalues $\lambda_1 = \frac{23}{25}$ and $\lambda_2 = \frac{2}{25}$ would be the variances in these dimensions, with the factor they are twice as large), we can confine ourselves to considering the x'' values. Thus we have reduced the data set to one dimension, without losing much information.

Note that the final data set shown above can be obtained directly from the original data set by the transformation

$$\begin{pmatrix} x'' \\ y'' \end{pmatrix} = \sqrt{2} \cdot \mathbf{T}^\top \cdot \begin{pmatrix} s_x^{-1} & 0 \\ 0 & s_y^{-1} \end{pmatrix} \cdot \left(\begin{pmatrix} x \\ y \end{pmatrix} - \begin{pmatrix} \bar{x} \\ \bar{y} \end{pmatrix} \right),$$

which combines the normalization and the projection of the data to the principal axes. This clearly demonstrates that the new features are merely linear combinations of the original features—as stated at the beginning.

A.3 Probability Theory

Inferential statistics, as it is considered in Sect. A.4, is firmly based on probability theory. It uses notions and methods from probability theory in order to obtain conclusions and judgments about the data generating process. Therefore this section recalls some essential notions and theorems of probability theory.

A.3.1 Probability

Probability theory is concerned with **random events**. That is, it is known what specific events can occur in principle, but it is uncertain which of the possible events will actually occur in any given instance. Examples are the throw of a coin or the cast of die. In probability theory a numerical quantity, called *probability*, is assigned to random events, which is intended to capture the chance or the tendency of the event to actually occur, at least in relation to the other possible outcomes.

A.3.1.1 Intuitive Notions of Probability

In order to achieve a first intuition, we start by considering the classical definition of probability. This definition is actually a procedure to compute probabilities and was developed out of analyses of the possible outcomes of gambles:

The probability of an event A is the ratio of the number N_A of favorable events (where an event is called favorable for A if A occurs with it) to the total number N of mutually exclusive, equally possible events:

$$P(A) = \frac{N_A}{N}.$$

That the events are equally possible is usually derived from symmetry considerations. The mutually exclusive, equally possible events are called **elementary events**. Together they form the **sample space**. Other random events can be represented as combinations of elementary events, generally as sets of elementary events. An essential set of tools for determining probabilities on the basis of the above definition are the methods of **combinatorics**, which can be used to count the number of all possible and the number of favorable events (see Sect. A.3.2.1).

As an example, we consider a symmetric die consisting of homogeneous material. The sample space consists of the elementary events "The number x was cast," $x \in \{1, 2, 3, 4, 5, 6\}$. The event "An even number was cast" consists of the elementary events "The number x was cast," $x \in \{2, 4, 6\}$, and thus occurs in three out of six equally possible cases. Therefore its probability is $\frac{3}{6} = \frac{1}{2}$.

The classical notion of probability can be generalized by dropping the requirement that the elementary events are equally possible. Instead, a sample space is defined, and each elementary event is associated with an **elementary probability**. Obviously, the classical notion then results as a special case of equal elementary events, demonstrating the term "equally possible" is actually a trick to avoid having to speak of "equally probable," which would result in a circular definition.

Another intuitive concept related to the probability of an event is the relative frequency of the event. Suppose that some experiment with a random outcome is executed n times under equal conditions. If A is a random event, then A either occurs or does not occur in each execution of the experiment. The number $f_n(A)$ of executions of the experiment in which A occurs is called the **absolute frequency**, and the ratio $r_n(A) = \frac{f_n(A)}{n}$ is called the **relative frequency** of A (cf. page 305).

In [12] von Mises tried to define the probability $P(A)$ of an event as the limit that is approached by the relative frequencies as the sample size n goes to infinity, that is,

$$P(A) = \lim_{n \to \infty} r_n(A).$$

However, this does not work. Even though it is necessary for the above definition to be valid, it is not possible to ensure that sequences of experiments occur in which

$$\forall n \geq n_0(\varepsilon): \quad P(A) - \varepsilon \leq r_n(A) \leq P(A) + \varepsilon$$

does not hold, regardless of how large n_0 is chosen. (Even though it is highly unlikely, it is not impossible that repeated throws of a die result only in, say, ones.) Nevertheless the intuition of a probability as a relative frequency is helpful, especially if it is interpreted as our estimate of the relative frequency of the event (in future executions of the same experiment).

As an example, consider the sex of a newborn child. Usually the number of girls roughly equals the number of boys. Hence the relative frequency of a girl being born is equal to the relative frequency of a boy being born. Therefore we say that the probability that a girl is born (and also the probability that a boy is born) equals $\frac{1}{2}$. Note that this probability cannot be derived from considerations of symmetry (as we can derive the probabilities of heads and tails when throwing a coin).[12]

A.3.1.2 The Formal Definition of Probability

The classical notion of probability and the interpretation as a relative frequency are deeply rooted in our intuition. However, modern mathematics is based on the axiomatic method, because one has realized that relying too much on intuition can introduce hidden assumptions, which can reduce the generality of obtained results or even invalidate them. The axiomatic method abstracts from the meaning of the objects that one considers. It takes the objects as given and having initially no other properties than their identity (that is, we can somehow distinguish one object from another). Then it studies merely the structure of the relations between these objects as they follow from the axioms (basic statements) that are laid down.

As a consequence, modern probability theory is also based on an axiomatic approach, which relies on **Kolmogorov's axioms** [7]. In this approach an event is simply a set of elementary events, which are distinguishable, that is, which have an identity. A probability is a number that is assigned to such events, so that the resulting system of numbers satisfies certain conditions, which are specified in the axioms. However, before we take a look at these axioms, we define the more basic notions of event algebra and σ-algebra, which are referred to in the axioms.

Definition A.1 Let Ω be a base set of elementary events (the sample space). Any subset $E \subseteq \Omega$ is called an **event**. A system $\mathcal{S} \subseteq 2^\Omega$ of events is called an **event algebra** iff

1. $\mathcal{S}$ contains the **certain event** Ω and the **impossible event**.
2. If an event A belongs to $\mathcal{S}$, then the event $\overline{A} = \Omega - A$ also belongs to S.

---

[12]It should be noted, though, that explanations from evolutionary biology for the usually equal probabilities of the two sexes indeed make references to symmetry.

3. If events A and B belong to $\mathcal{S}$, then the events $A \cap B$ and $A \cup B$ also belong to $\mathcal{S}$.

In addition the following condition may be satisfied:

$3'$. If for all $i \in \mathbb{N}$, the event A_i belongs to $\mathcal{S}$,
   then the events $\bigcup_{i=1}^{\infty} A_i$ and $\bigcap_{i=1}^{\infty} A_i$ also belong to $\mathcal{S}$.

In this case $\mathcal{S}$ is called a σ-**algebra**.

The notion of a *probability* is now defined by the following axioms:

Definition A.2 (Kolmogorov's axioms) Let $\mathcal{S}$ be an event algebra that is defined on a finite sample space Ω.

1. The **probability** $P(A)$ of an event $A \in \mathcal{S}$ is a uniquely determined, nonnegative real number, which can be at most one, that is, $0 \le P(A) \le 1$.
2. The certain event Ω has probability 1, that is, $P(\Omega) = 1$.
3. **Additivity**: If A and B are two mutually exclusive events (that is, $A \cap B = \emptyset$), then $P(A \cup B) = P(A) + P(B)$.

If the sample space Ω has infinitely many elements, $\mathcal{S}$ must be a σ-algebra, and Axiom 3 must be replaced by:

$3'$. **Extended additivity**: If $A_1, A_2, \ldots$ are countably infinitely many, pairwise mutually exclusive events, then

$$P\left(\bigcup_{i=1}^{\infty} A_i\right) = \sum_{i=1}^{\infty} A_i.$$

Hence the probability $P(A)$ can be seen as a function that is defined on an event algebra or on a σ-algebra and that has certain properties. From the above definition several immediate consequences follow:

1. For every event A, we have $P(\overline{A}) = 1 - P(A)$.
2. The impossible event has probability 0, that is, $P(\emptyset) = 0$.
3. From $A \subseteq B$ it follows that $P(A) \le P(B)$.
4. For arbitrary events A and B, we have
   $P(A - B) = P(A \cap \overline{B}) = P(A) - P(A \cap B)$.
5. For arbitrary events A and B, we have
   $P(A \cup B) = P(A) + P(B) - P(A \cap B)$.
6. Since $P(A \cap B)$ cannot be negative, it follows that
   $P(A \cup B) \le P(A) + P(B)$.
7. By simple induction we infer from the additivity axiom:
   if $A_1, \ldots, A_m$ are (finitely many) pairwise mutually exclusive events, we have
   $P(A_1 \cap \cdots \cap A_m) = \sum_{i=1}^{m} P(A_i)$.

Kolmogorov's system of axioms is consistent, because there are actual systems that satisfy these axioms. This system of axioms allows us to construct probability theory as part of measure theory and to interpret a probability as a nonnegative, normalized, additive set function and thus as a measure (in the sense of measure theory).

Definition A.3 Let Ω be a set of elementary events, $\mathcal{S}$ a σ-algebra defined on a sample space Ω, and P a probability that is defined on $\mathcal{S}$. Then the triple $(\Omega, \mathcal{S}, P)$ is called a **probability space**.

A.3.2 Basic Methods and Theorems

After we have defined probability, we now turn to basic methods and theorems of probability theory, which we explain with simple examples. Among these are the computation of probabilities with combinatorial and geometrical approaches, the notions of conditional probability and (conditionally) independent events, the product law, the theorem of total probability, and finally the very important Bayes' rule.

A.3.2.1 Combinatorial Methods

As already mentioned, on the basis of the classical definition of probability, combinatorics provides many tools to compute probabilities. We confine ourselves to a simple example, namely the famous birthday problem:

Let m people be randomly chosen. What is the probability of the event A_m that at least two of these people have their birthday on the same day (day and month in a year, but not the same year)? In order not to over-complicate things, we neglect leap years. In addition, we assume that every day of a year is equally possible as a birthday.[13] It is immediately clear that for $m \geq 366$, at least two people must have their birthday on the same day. Hence we have

$$\forall m \geq 366: \quad P(A_m) = 1.$$

For $m \leq 365$, we consider the complementary event $\overline{A_m}$, which occurs if all m people have their birthday on different days. This switch to the complementary event is a very common technique when computing with probabilities, which often leads to considerable simplifications. If we number the people, the first person may have its birthday on 365, the second on 364 (since the birthday of the first is excluded), for the mth, $365 - m + 1$ days (since the birthdays of the $m - 1$ preceding people are excluded. This yields the number of favorable cases for $\overline{A_m}$. In all, there are 365^m possible cases (because any person could have its birthday on any day of the year). Therefore we have

$$P(\overline{A_m}) = \frac{365 \cdot 364 \cdots (365 - m + 1)}{365^m},$$

and consequently

$$P(A_m) = 1 - \frac{365 \cdot 364 \cdots (365 - m + 1)}{365^m}.$$

---

[13]This is not quite correct, since surveys show that the frequency of births is not quite uniformly distributed over the year, but a reasonable approximation.

The surprising property of this formula is that for m as low as 23, we already have $P(A_{23}) \approx 0.507$. Hence for 23 or more people, it is more likely that two people have their birthday on the same day than that all have their birthdays on different days.

A.3.2.2 Geometric Probabilities

Geometric probabilities are a generalization of the classical definition of probability. One no longer counts the favorable and all possible cases and then forms their ratio. Rather, the counts are replaced by geometric quantities, like lengths or areas.

As an example, we consider the game franc-carreau as it was studied in [2]: In this game a coin is thrown onto an area that is divided into rectangles of equal shape and size. Let the coin have the radius r, and the rectangles the side lengths a and b, with $2r \le a$ and $2r \le b$, so that the coin fits completely into a rectangle. We desire to find the probability that the coin lies on at least one of the two sides of a rectangle. If we inscribe into each rectangle a smaller one with side lengths $a - 2r$ and $b - 2r$ in such a way that the centers of the rectangles coincide and the sides are parallel, it is clear that the coin does *not* lie on any side of a rectangle if and only if its center lies inside the inner rectangle. Since the area of the inner rectangle is $(a - 2r)(b - 2r)$ and the area of the outer ab, the desired probability is

$$P(A) = 1 - \frac{(a - 2r)(b - 2r)}{ab}.$$

A.3.2.3 Conditional Probability and Independent Events

In many cases one has to determine the probability of an event A when it is already known that some other event B has occurred. Such probabilities are called conditional probabilities and denoted $P(A \mid B)$. In a strict sense the "unconditional" probabilities we considered up to now are also conditional, because they always refer to specific frame conditions and circumstances. For example, we assumed that the die we throw is symmetric and made of homogeneous material. Only under these, and possibly other, silently adopted frame conditions (like no electromagnetic influence on the die, etc.) we stated that the probability of each number is $\frac{1}{6}$.

Definition A.4 Let A and B be two arbitrary events with $P(B) > 0$. Then

$$P(A \mid B) = \frac{P(A \cap B)}{P(B)}$$

is called the **conditional probability** of A given B.

A simple example: two dice are cast. What is the probability that one of the dice displays a five if it is known that the sum of the pips is eight. If two dice are cast, we have 36 elementary events, five of which satisfy that the sum of the pips is eight $(4 + 4, 5 + 3,$ and $6 + 2$, where the last two have to be counted twice due to the two possible distributions of the numbers to the two dice). That is, we have $P(B) = \frac{5}{36}$. The event "The sum of the pips is eight, and one of the dice shows a five" can be obtained from two elementary events: either the first die shows a five, and the second a three, or vice versa. Therefore $P(A \cap B) = \frac{2}{36}$, and thus the desired conditional probability is $P(A \mid B) = \frac{2}{5}$.

Theorem A.1 (product law)   *For arbitrary events A and B, we have*

$$P(A \cap B) = P(A \mid B) \cdot P(B).$$

This theorem follows immediately from the definition of conditional probability together with the obvious relation that $P(A \cap B) = 0$ if $P(B) = 0$.[14] By simple induction over the number of events, we obtain the generalization for m events:

$$P\left(\bigcap_{i=1}^{m} A_i\right) = \prod_{i=1}^{m} P\left(A_i \,\middle|\, \bigcap_{k=1}^{i-1} A_k\right).$$

A conditional probability has all properties of a normal probability, that is, it satisfies Kolmogorv's Axioms. Therefore we have:

Theorem A.2 *For an arbitrary, but fixed, event B with $P(B) > 0$, the function P_B that is defined as $P_B(A) = P(A \mid B)$ is a probability which satisfies $P_B(\overline{B}) = 0$.*

With the help of the notion of a conditional probability, we can now define the notion of (stochastic) independence of events. This notion can be motivated as follows: if, for example, smoking had no influence on the development of lung cancer, then the proportion people with lung cancer among smokers should be (roughly) equal to the proportion of people with lung cancer among nonsmokers.

Definition A.5 Let B be an event with $0 < P(B) < 1$. Then an event A is called **(stochastically) independent** of B iff

$$P(A \mid B) = P(A \mid \overline{B}).$$

The following two relations, which are usually easier to handle, are equivalent:

Theorem A.3 *An event A is (stochastically) independent of an event B with $0 < P(B) < 1$ iff*

$$P(A \mid B) = P(A)$$

or equivalently iff

$$P(A \cap B) = P(A) \cdot P(B).$$

---

[14]Formally this argument is not quite valid, though, since for $P(B) = 0$, the conditional probability $P(A \mid B)$ is undefined (see Definition A.4). However, since the equation holds for any value that may be fixed for $P(A \mid B)$ in case that $P(B) = 0$, we allow ourselves to be slightly sloppy here.

Note that the relation of (stochastic) independence is symmetric, that is, if A is (stochastically) independent of B, then B is also (stochastically) independent of A (provided that $0 < P(A) < 1$). In addition, the notion of (stochastic) independence can easily be extended to more than two events:

Definition A.6 m events $A_1, \ldots, A_m$ are called completely (stochastically) independent if for any selection $A_{i_1}, \ldots, A_{i_t}$ of t events with $\forall r, s; 1 \le r, s \le t : i_r \ne i_s$, we have

$$P\left(\bigcap_{k=1}^{t} A_{i_k}\right) = \prod_{k=1}^{t} P(A_{i_k}).$$

Note that for the complete (stochastic) independence of more than two events, their pairwise independence is necessary but not sufficient.

Let us consider a simple example: A white and a red die are cast. Let A be the event "The number of pips shown by the white die is even," B the event "The number of pips shown by the red die is odd," and C the event "The sum of the pips is even." It is easy to check that A and B are pairwise (stochastically) independent, as well as B and C, and also A and C. However, due to $P(A \cap B \cap C) = 0$ (since the sum of an even number and an odd number must be odd), they are not completely independent.

Another generalization of (stochastic) independence can be achieved by introducing another condition for all involved probabilities:

Definition A.7 (conditionally (stochastically) independent) Two events A and B are called conditionally (stochastically) independent given a third event C with $0 < P(C) < 1$ iff

$$P(A \cap B \mid C) = P(A \mid C) \cdot P(B \mid C).$$

Note that two events A and B may be conditionally independent but not unconditionally independent and vice versa. To see this, consider again the example of the red and white die discussed above. A and B are independent but not conditionally independent given C, because if C holds, only one of A and B can be true, even though either of them can be true (provided that the other is false). Hence the joint probability $P(A \cap B \mid C) = 0$, while $P(A \mid C) \cdot P(B \mid C) > 0$. Examples for the reverse case (conditional independence, but unconditional dependence) are also easy to find.

A.3.2.4 Total Probability and Bayes' Rule

Often we face situations in which the probabilities of disjoint events A_i are known, which together cover the whole sample space. In addition, we know the conditional probabilities of an event B given these A_i. Desired is the (unconditional) probability of the event B. As an example, consider a plant that has a certain number of machines to produce the same product. Suppose that the capacities of the individual machines and their probabilities (rates) of producing faulty products are known. The total probability (rate) of faulty products is to be computed. This rate can easily be found by using the law of total probability. However, before we turn to it, we formally define the notion of an event partition.

Definition A.8 m events $A_1, \ldots, A_m$ form an **event partition** iff all pairs A_i, A_k, $i \neq k$, are mutually exclusive (that is, $A_i \cap A_k = \emptyset$ for $i \neq k$) and if $A_1 \cup \cdots \cup A_m = \Omega$, that is, all events together cover the whole sample space.

Theorem A.4 (law of total probability) *Let* $A_1, \ldots, A_m$ *be an event partition with* $\forall i; 1 \leq i \leq m : P(A_i) > 0$ *(and, as follows from the additivity axiom,* $\sum_{i=1}^{m} P(A_i) = 1$*). Then the probability of an arbitrary event* B *is*

$$P(B) = \sum_{i=1}^{m} P(B \mid A_i) P(A_i).$$

The law of total probability can be derived by applying the product rule (see Theorem A.1) to the relation

$$P(B) = P(B \cap \Omega) = P\left(B \cap \bigcup_{i=1}^{m} A_i\right) = P\left(\bigcup_{i=1}^{m} (B \cap A_i)\right) = \sum_{i=1}^{m} P(B \cap A_i),$$

the last step of which follows from the additivity axiom.

With the help of this theorem we can easily derive the important Bayes' rule. To do so, it is merely necessary to realize that the product rule can be applied in two ways to the simultaneous occurrence of two events A and B:

$$P(A \cap B) = P(A \mid B) \cdot P(B) = P(B \mid A) \cdot P(A).$$

Dividing the right-hand side by $P(B)$ (which, of course, must be positive to be able to do so), yields the simple form of Bayes' rule. By applying the law of total probability to the denominator, we obtain the extended form.

Theorem A.5 (Bayes' rule) *Let* $A_1, \ldots, A_m$ *be an event partition with* $\forall i; 1 \leq i \leq m : P(A_i) > 0$, *and* B *an arbitrary event with* $P(B) > 0$. *Then*

$$P(A_i \mid B) = \frac{P(B \mid A_i) P(A_i)}{P(B)} = \frac{P(B \mid A_i) P(A_i)}{\sum_{k=1}^{m} P(B \mid A_k) P(A_k)}.$$

This rule[15] is also called the formula for the probability of hypotheses, since it can be used to compute the probability of hypotheses (for example, the probability that a patient suffers from a certain disease) if the probabilities are known with

---

[15]Note that Thomas Bayes (1702–1761) did not derive this formula, despite the fact that it bears his name. In the form given here it was stated only later by Pierre-Simon de Laplace (1749–1827). This supports a basic law of the history of science: a law or an effect that bears the name of a person was found by somebody else.

which the hypotheses lead to the considered events A_i (for example, medical symptoms).

As an example, we consider five urns with the following contents:

- two urns with the contents A_1 with two white and three black balls each,
- two urns with the contents A_2 with one white and four black balls each,
- one urn with the contents A_3 with four white and one black ball.

Suppose that an urn is chosen at random and a ball is drawn from it, also at random. Let this ball be white: this is the event B. What is the (posterior) probability that the ball stems from an urn with the contents A_3?

According to our presuppositions, we have:

$$P(A_1) = \frac{2}{5}, \qquad P(A_2) = \frac{2}{5}, \qquad P(A_3) = \frac{1}{5},$$

$$P(B \mid A_1) = \frac{2}{5}, \quad P(B \mid A_2) = \frac{1}{5}, \quad P(B \mid A_3) = \frac{4}{5}.$$

We start by applying the law of total probability in order to find the probability $P(B)$:

$$P(B) = P(B \mid A_1)P(A_1) + P(B \mid A_2)P(A_2) + P(B \mid A_2)P(A_3)$$
$$= \frac{2}{5} \cdot \frac{2}{5} + \frac{1}{5} \cdot \frac{2}{5} + \frac{4}{5} \cdot \frac{1}{5} = \frac{10}{25}.$$

Using Bayes' rule, we then obtain

$$P(A_3 \mid B) = \frac{P(B \mid A_3)P(A_3)}{P(B)} = \frac{\frac{4}{5} \cdot \frac{1}{5}}{\frac{10}{25}} = \frac{2}{5}.$$

Likewise, we can obtain

$$P(A_1 \mid B) = \frac{2}{5} \quad \text{and} \quad P(A_2 \mid B) = \frac{1}{5}.$$

A.3.2.5 Bernoulli's Law of Large Numbers

In Sect. A.3.1 we already considered the relation between the probability $P(A)$ and the relative frequency $r_n(A) = \frac{h_n(A)}{n}$ of an event A, where $h_n(A)$ is the absolute frequency of this event in n trials. We saw that it is not possible to define the probability of A as the limit of the relative frequency as $n \to \infty$. However, a slightly weaker statement holds, namely the famous law of large numbers.

Definition A.9 A random experiment in which the event A can occur is repeated n times. Let A_i be the event that A occurs in the ith trial. Then the sequence of experiments of length n is called a **Bernoulli experiment**[16] for the event A iff the following conditions are satisfied:

---

[16]The notion "Bernoulli experiment" was introduced in recognition of the Swiss mathematician Jakob Bernoulli (1654–1705).

1. $\forall 1 \leq i \leq n : P(A_i) = p$.
2. The events $A_1, \ldots, A_n$ are fully independent.

Theorem A.6 (Bernoulli's law of large numbers) *Let $h_n(A)$ be the number of oc-currences of an event A in n independent trials of a Bernoulli experiment, where in each of the trials the probability $P(A)$ of the occurrence of A equals an arbitrary, but fixed, value p, $0 \leq p \leq 1$. Then for any $\varepsilon > 0$,*

$$\lim_{n \to \infty} P\left(\left| \frac{h_n(A)}{n} - p \right| < \varepsilon \right) = \frac{1}{\sqrt{2\pi}} \int e^{-\frac{z^2}{2}} \, dz = 1.$$

This property of the relative frequency $r_n(A) = \frac{h_n(A)}{n}$ can be interpreted as fol-lows: even though $p = P(A)$ is not the limit of the relative frequency for infinite sample size, as von Mises [12] tried to define, but it can be seen as very probable (practically certain) that in a Bernoulli experiment of sufficiently large size n, the relative frequency $r_n(A)$ differs only very little from a fixed value, the probabil-ity p. This is often referred to by saying that the relative frequency $r_n(A)$ **converges in probability** to the probability $P(A) = p$. With the above law we have the funda-mental relationship between the relative frequency and the probability of an event A.

A.3.3 Random Variables

In many situations we are not interested in the complete set of elementary events and their individual probabilities, but in the probabilities of events that result from a par-tition of the sample space. That is, these events are mutually exclusive, but together cover the whole sample space (an *event partition*, see above). The probabilities of such events are commonly described by so-called *random variables*, which can be seen as transformations from one sample space into another [10].

Definition A.10 A function X that is defined on a sample space Ω and has the do-main $\text{dom}(X)$ is called **random variable** if the preimage of any subset of its domain possesses a probability. Here the preimage of a subset $U \subseteq \text{dom}(X)$ is defined as

$$X^{-1}(U) = \{\omega \in \Omega \mid X(\omega) \in U\}.$$

A.3.3.1 Real-Valued Random Variables

The simplest random variable is obviously one that has the elementary events as its possible values. However, in principle, any set can be the domain of a random variable. However, most often the domain is the set of real numbers.

Definition A.11 A function X that maps a sample space Ω to the real numbers is called a **real-valued random variable** if it possesses the following properties: for any $x \in \mathbb{R}$ and any interval $(a, b]$, $a < b$, (where $a = -\infty$ is possible), the events $A_x = \{\omega \in \Omega \mid X(\omega) = x\}$ and $A_{(a,b]} = \{\omega \in \Omega \mid a < X(\omega) \leq b\}$ possess probabilities.

Sometimes the required properties are stated with an interval $[a, b)$ that is open on the right. This does not lead to any significant differences.

Definition A.12 Let X be a real-valued random variable. The real-valued function

$$F(x) = P(X \leq x)$$

is called the **distribution function** of X.

A.3.3.2 Discrete Random Variables

Definition A.13 A random variable X with finite or only countable infinite domain $\mathrm{dom}(X)$ is called a **discrete random variable**. The total of all pairs $(x, P(X = x))$, $x \in \mathrm{dom}(X)$, is called the **(probability) distribution** of a discrete random variable X.

If the probabilities $P(X = x)$ can be stated as a function, the distribution of a discrete random variable is often stated as a function $v_X(x) = P(X = x)$. Possible parameters (in order to select a function from a parameterized family of functions) are separated by a semicolon from the function argument. For example, the binomial distribution

$$b_X(x; p, n) = \binom{n}{x} p^x (1 - p)^{n-x}$$

has the probability p of the occurrence of the considered event in a single trial and the size n of the sample as parameters. (The binomial distribution is considered in more detail in Sect. A.3.5.1; later sections consider other important distributions.)

The values of the distribution function of a discrete, real-valued random variable can be computed as follows from the values of the probability distribution:

$$F(x) = P(X \leq x) = \sum_{x' \leq x} P(X = x').$$

Every discrete real-valued random variable X has a step function F as its distribution function, which has jumps of height $P(X = x')$ only at those values x' that are in the domain $\mathrm{dom}(X)$. From $x < y$ it follows that $F(x) \leq F(y)$, that is, F is monotone nondecreasing. The function values $F(x)$ become arbitrarily small if only x is chosen small enough, while the values $F(x)$ get arbitrarily close to 1 for growing x. Therefore we have

$$\lim_{x \to -\infty} F(x) = 0 \quad \text{and} \quad \lim_{x \to \infty} F(x) = 1.$$

Vice versa, from any step function F, which satisfies the above conditions, the distribution $(x, P(X = x))$, $x \in \text{dom}(X)$, of a real-valued random variable can be derived.

With the help of a distribution function F it is very simple to compute the probability that X assumes a value from a given interval:

Theorem A.7 *Let F be the distribution function of a discrete real-valued random variable X. Then the following relations hold:*

$$P(a < X \le b) = F(b) - F(a),$$
$$P(a \le X \le b) = F(b) - F_L(a),$$
$$P(a < X) = 1 - F(a),$$

where $F_L(a)$ is the left limit of $F(x)$ at the location a. This limit equals $F(a)$, if a is not the location of a discontinuity, and otherwise equal to the value of the step directly to the left of a, or formally, $F_L(a) = \sup_{x<a} F(x)$.

A.3.3.3 Continuous Random Variables

In contrast to discrete random variables, continuous (real-valued) random variables are defined as random variables with a super-countably infinite domain. Obviously, this requires to replace the sum in the distribution function by an integral.

Definition A.14 A real-valued random variable X is called **continuous** if there exists a nonnegative, integrable function f such that for its distribution function $F(x) = P(X \le x)$, the integral representation

$$F(x) = P(X \le x) = \int_{-\infty}^{x} f(u)\, du$$

holds. The function f is called the **(probability) density function** (or simply density) of the random variable X.

Since $P(-\infty < X < \infty) = P(\{\omega \in \Omega \mid -\infty < X(\omega) < \infty\}) = P(\Omega)$, the density function f satisfies

$$\int_{-\infty}^{\infty} f(u)\, du = 1.$$

For continuous random variables, similar relations hold as for discrete real-valued random variables.

Theorem A.8 *If X is a continuous random variable with density function f, then for arbitrary numbers $a, b, c \in \mathbb{R}$ with $a < b$, we have*

$$P(a < X \le b) = F(b) - F(a) = \int_{a}^{b} f(u)\, du,$$
$$P(X > c) = 1 - F(c) = \int_{c}^{\infty} f(u)\, du.$$

A.3.3.4 Random Vectors

Up to now we have only considered single random variables. In the following we expand our study to several random variables and their interaction, that is, their joint distribution and their dependence or independence. In order to do so, we define the notion of a random vector or, equivalently, of a multidimensional random variable.

Definition A.15 Let $X_1, \ldots, X_m$ be m random variables that are defined on the same probability space $(\Omega, \mathcal{S}, P)$, that is, on the same sample space Ω with the same σ-algebra $\mathcal{S}$ and probability P. In this case the vector $(X_1, \ldots, X_m)$ is called a **random vector** or an m**-dimensional random variable**.

In order to keep things simple, we consider here only two-dimensional random variables. However, all definitions and theorems can easily be transferred to multi-dimensional random variables (random vectors with finite length m). In addition we confine ourselves, as in the preceding sections, to real-valued random variables.

Definition A.16 Let X and Y be two real-valued random variables. The function F which is defined for all pairs $(x, y) \in \mathbb{R}^2$ as

$$F(x, y) = P(X \leq x, Y \leq y)$$

is called the **distribution function** of the two-dimensional random variable (X, Y). The one-dimensional distribution functions

$$F_1(x) = P(X \leq x) \quad \text{and} \quad F_2(y) = P(Y \leq y)$$

are called **marginal distribution functions**.

For discrete random variables, the notion of their joint distribution is defined in an analogous way.

Definition A.17 Let X and Y be two discrete random variables. Then the total of pairs $\forall x \in \text{dom}(X) : \forall y \in \text{dom}(Y) : ((x, y), P(X = x, Y = y))$ is called the **joint distribution** of X and Y. The one-dimensional distributions $\forall x \in \text{dom}(X) :$ $(x, \sum_y P(X = x, Y = y))$ and $\forall y \in \text{dom}(Y) : (y, \sum_x P(X = x, Y = y))$ are **marginal distributions**.

Continuous random variables are treated in a similar way, by simply replacing the joint distribution with the joint density function.

Definition A.18 The two-dimensional random variable (X, Y) is called continuous if there exists a nonnegative function $f(x, y)$ such that for every $(x, y) \in \mathbb{R}^2$, we have

$$F(x, y) = P(X \leq x, Y \leq y) = \int_{-\infty}^{x} \int_{-\infty}^{y} f(u, v) \, du \, dv.$$

The function $f(x, y)$ is called the **joint density function** or simply **joint density** of the random variables X and Y. The one-dimensional density functions

$$f_1(x) = \int_{-\infty}^{+\infty} f(x, y)\, dy \quad \text{and} \quad f_2(y) = \int_{-\infty}^{+\infty} f(x, y)\, dx$$

are called **marginal density functions** or simply **marginal densities**.

By extending the notion of the independence of events one can define the notion of the independence of random variables.

Definition A.19 Two continuous real-valued random variables X and Y with two-dimensional distribution function $F(x, y)$ and marginal distribution functions $F_1(x)$ and $F_2(y)$ are called **(stochastically) independent** if for all pairs of values $(x, y) \in \mathbb{R}^2$, we have

$$F(x, y) = P(X \le x, Y \le y) = P(X \le x) \cdot P(Y \le y) = F_1(x) \cdot F_2(y).$$

Definition A.20 Two discrete random variables X and Y with joint distribution $\forall x \in \text{dom}(X) : \forall y \in \text{dom}(Y) : ((x, y), P(X = x, Y = y))$ and marginal distributions $\forall x \in \text{dom}(X) : (x, P(X = x))$ and $\forall y \in \text{dom}(Y) : (y, P(Y = y))$ are called **(stochastically) independent**, if

$$\forall x \in \text{dom}(X) : \forall y \in \text{dom}(Y): \quad P(X = x, Y = y) = P(X = x) \cdot P(Y = y).$$

As for events, the notion of (stochastic) independence can easily be generalized to conditional (stochastic) independence. In this case the distributions and distribution functions are replaced by conditional distributions and conditional distribution functions. For example, for discrete random variables, we obtain:

Definition A.21 Let X, Y, and Z be three discrete random variables. Let the X and Y have the conditional joint distribution $\forall x \in \text{dom}(X) : \forall y \in \text{dom}(Y) : \forall z \in \text{dom}(Z) : ((x, y, z), P(X = x, Y = y \mid Z = z))$ given Z and the conditional marginal distributions $\forall x \in \text{dom}(X) : \forall z \in \text{dom}(Z) : ((x, z), P(X = x \mid Z))$ and $\forall y \in \text{dom}(Y) : \forall z \in \text{dom}(Z) : ((y, z), P(Y = y \mid Z))$. X and Y are **conditionally (stochastically) independent** if

$$\forall x \in \text{dom}(X) : \forall y\, in\, \text{dom}(Y) : \forall z \in \text{dom}(Z):$$
$$P(X = x, Y = y \mid Z = z) = P(X = x \mid Z = z) \cdot P(Y = y \mid Z = z).$$

For continuous random variables, the definition is generalized in an analogous way but is formally a bit tricky, and that is why we omit it here. In order to distinguish normal (stochastic) independence from conditional (stochastic) independence, the former is often also called **marginal independence**.

Since the notion of conditional independence is a very important concept, we try to convey a better intuition of it with a simple example. Consider the scatter

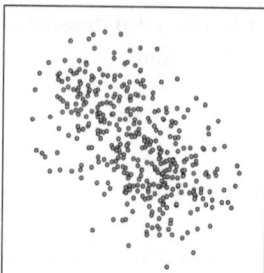 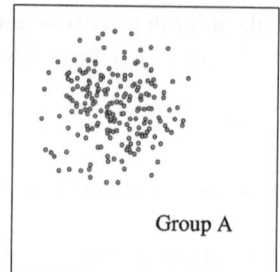 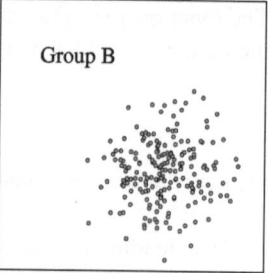

Fig. A.13 Illustration of marginal dependence and conditional independence: the two measures describing the points are marginally dependent (*left*), but if a third, binary variable is fixed and thus the data points are split into two groups, the dependence vanishes (*middle* and *right*)

plot of a sample from two continuous random variables that is shown in Fig. A.13 on the left. Obviously, the two quantities are not independent, because there is a clear tendency for the Y variable (vertical axis) to take lower values if X (horizontal axis) has higher values. Hence X and Y are not marginally independent. To make the example more vivid, we can interpret the horizontal axis as the average number of cigarettes a person smoked per day and the vertical axis as the age of death of that person. Of course, this is a fictitious example, with artificially generated data points, but medical surveys usually show such a dependence.[17] From such an observed dependence it is usually concluded that smoking is a health hazard.

However, this need not be conclusive. There may be a third variable that couples the two, which, if we fix its value, renders the two quantities independent. A passionate smoker may claim, for example, that this third variable is whether a person is exposed to severe stress at work. Such stress is certainly a health hazard and it causes, as our passionate smoker may argue, both: a shorter life span (due to the strain on the person's health by the stress it is exposed to) and a higher cigarette consumption (due to the fact that smoking has a calming effect and thus can help to cope with stress). If this argument were correct,[18] the dependence should vanish if we consider people that are exposed to stress at work and those who are not separately. That is, if the argument were correct, we should see the separate data as depicted in Fig. A.13 in the middle (people that are not exposed to stress at work and thus smoke less and live longer) and on the right (people that are exposed to stress at work and thus smoke more and live less long). In both cases the dependence between the two quantities has vanished, and thus they are conditionally independent given the third variable (stress at work or not).

---

[17]However, we do not claim that the actual dependence looks like our data.

[18]We do not believe it is. The claim that smoking harms your health is much better supported than just by an observation of a correlation like the one depicted in Fig. A.13, even though such correlations are part of the argument.

A.3.4 Characteristic Measures of Random Variables

In analogy to Sect. A.2.3, where we defined characteristic measures for data sets, random variables can be described by analogous measures. While for data sets, these measures are derived from the sample values, measures for random variables are derived from their distributions and distribution functions. The analogy is actually very close: the unit weight of each sample data point is replaced by the probability mass that is assigned to the different values in the domain of a random variable.

A.3.4.1 Expected Value

If the notion of a random variable is applied to gambling, where the value of the random variable could be, for example, the gains or losses connected to different outcomes, the idea suggests itself to consider a kind of average or expected win (or loss) if a sufficiently large number of gambles is played. This idea leads to the notion of an expected value.

Definition A.22 Let X be a discrete real-valued random variable with distribution $(x, P(X = x))$, $x \in \mathrm{dom}(X)$. If $\sum_x |x| P(X = x)$ is finite, the limit (which must exist in this case)

$$\mu = E(X) = \sum_{i=1}^{\infty} x_i P(X = x_i)$$

is called the **expected value** of X.

As an example, consider the expected value of the winnings in the classical game of roulette. We do not bet on a so-called *simple chance* (*Rouge* vs. *Noir*, *Pair* vs. *Impair*, *Manque* vs. *Passe*), in order to avoid the difficulties that result from the special rules applying to them, but bet on a column (one of the sets of numbers 1–12, 13–24, and 14–36). In case of a win we receive three times our wager. Since in roulette 37 numbers can occur (0–36),[19] all of which are equally likely if we assume perfect conditions, winning with a bet on a column has the probability $\frac{12}{37}$, and losing has the probability $\frac{25}{37}$. Let us assume that the wager consists of m chips. In case of a win we have twice the wager as a net gain (the paid out win has to be reduced by the initially waged m chips), that is, $2m$ chips, whereas in case of a failure we lose m chips. As a consequence, the expected value is

$$E(X) = 2m \cdot \frac{12}{37} - m \cdot \frac{25}{37} = -\frac{1}{37}m \approx -0.027m.$$

On average we thus lose 2.7% of our wager in every gamble.

In order to define the expected value of continuous random variables, we only have to replace the sum by an integral, and the probabilities of the distribution by the density function.

---

[19]In certain types of American roulette even 38, as these have 0 and 00 (double zero).

Definition A.23 If X be a continuous random variable with density function f and if the integral

$$\int_{-\infty}^{\infty} |x|f(x)\,dx = \lim_{\substack{a\to-\infty \\ b\to\infty}} \int_{a}^{b} |x|f(x)\,dx$$

exists, then the integral (which must exist in this case)

$$\mu = E(X) = \int_{-\infty}^{\infty} xf(x)\,dx$$

is called the expected value of X.

A.3.4.2 Properties of the Expected Value

In this section some properties of the expected value are collected, which can often be exploited when one has to compute the expected value.

Theorem A.9 *Let X be a discrete random variable that takes no other values than some constant c. Then its expected value is equal to c: $\mu = E(X) = c$.*

Theorem A.10 (linearity of the expected value) *Let X be a (discrete or continuous) real-valued random variable with expected value $E(X)$. Then the expected value of the random variable $Y = aX + b$, $a, b \in \mathbb{R}$, is*

$$E(Y) = E(aX + b) = aE(X) + b.$$

The statement can easily be checked by simply inserting the expression $aX + b$ into the definition of the expected value, once for a discrete and once for a continuous random variable.

As an example, we consider the distribution of a random variable X, which describes the sum of the pips of two dice. This sum is clearly symmetric w.r.t. the value 7, that is, $\forall k \in \{0, 1, \ldots, 5\}: P(X = 7 + k) = P(X = 7 - k)$. It follows that the expected values of the random variables $Y_1 = X - 7$ and $Y_2 = -(X - 7) = 7 - X$ must be identical, since they have, due to the symmetry, the same distribution. Therefore,

$$E(Y_1) = E(X - 7) = E(7 - X) = E(Y_2).$$

Applying Theorem A.10 yields $E(X) - 7 = 7 - E(X)$, and therefore $E(X) = 7$. From this example we can conclude generally that the point of symmetry of a distribution, provided that it has one, must be its expected value. This also holds for continuous random variables.

Next, we turn to the expected value of functions of two random variables, namely their sum and their product.

Theorem A.11 (expected value of a sum of random variables) *The expected value of a sum $Z = X + Y$ of two arbitrary real-valued random variables X and Y, whose* *expected values* $E(X)$ *and* $E(Y)$ *both exist, is equal to the sum of their expected values,*

$$E(Z) = E(X + Y) = E(X) + E(Y).$$

Theorem A.12 (expected value of a product of random variables) *The expected value of a product* $Z = X \cdot Y$ *of two independent real-valued random variables X and Y, whose expected values* $E(X)$ *and* $E(Y)$ *exist, is equal to the product of their expected values,*

$$E(Z) = E(X \cdot Y) = E(X) \cdot E(Y).$$

Again the validity of these theorems can easily be checked by inserting the sum/product into the definition of the expected value, in the case of a product of random variables by also exploiting the definition of independence. It should be clear that both theorems can easily be generalized to sums and products of finitely many (independent) random variables. Never forget about the presupposition of independence in the second theorem, since it does not hold for dependent random variables.

A.3.4.3 Variance and Standard Deviation

The expected value alone does not sufficiently characterize a random variable. We must consider also what deviation from the expected value can occur on average (see Sect. A.2.3.2). This dispersion is described by variance and standard deviation.

Definition A.24 If μ is the expected value of a discrete real-valued random variable X, then the value (provided that it exists)

$$\sigma^2 = D^2(X) = E\big([X - \mu]^2\big) = \sum_{i=1}^{\infty} (x_i - \mu)^2 P(X = x_i)$$

is called the **variance**, and the positive square root $\sigma = D(X) = +\sqrt{\sigma^2}$ is called the **standard deviation** of X

Let us again consider roulette as an example. If we bet m chips on a column, the variance is

$$D^2(X) = \left(2m + \frac{1}{37}m\right)^2 \cdot \frac{12}{37} + \left(-m + \frac{1}{37}m\right)^2 \cdot \frac{25}{37} = \frac{99900}{50653}m^2 \approx 1.97m^2.$$

Hence the standard deviation $D(X)$ is about $1.40m$. In comparison, the variance of a bet on a *plain chance*, that is, on a single number, has the same expected value, but the variance

$$D^2(X) = \left(35m + \frac{1}{37}m\right)^2 \cdot \frac{1}{37} + \left(-m + \frac{1}{37}m\right)^2 \cdot \frac{36}{37} = \frac{1726272}{50653}m^2 \approx 34.1m^2,$$

and thus a standard deviation $D(X)$ of about $5.84m$. Despite the same expected value, the average deviation from the expected value is about 4 times as large for a bet on a plain chance than for a bet on a column.

In order to define the variance of a continuous random variable, we only have to replace the sum by an integral—just as we did for the expected value.

Definition A.25 If μ is the expected value of a continuous random variable X, then the value (provided that it exists)

$$\sigma^2 = D^2(X) = \int_{-\infty}^{\infty} (x - \mu)^2 f(x) \, dx$$

is called the **variance** of X, and $\sigma = D(X) = +\sqrt{\sigma^2}$ is called the **standard deviation** of X.

A.3.4.4 Properties of the Variance

In this section we collect a few useful properties of the variance.

Theorem A.13 *Let X be a discrete random variable which takes no other values than a constant c. Then its variance is 0: $\sigma^2 = D^2(X) = 0$.*

Theorem A.14 *Let X be a (discrete or continuous) real-valued random variable with variance $D^2(X)$. Then the variance of the random variable $Y = aX + b, a, b \in \mathbb{R}$, is*

$$D^2(Y) = D^2(aX + b) = a^2 D^2(X),$$

and therefore, for the standard deviation, we have

$$D(Y) = D(aX + b) = |a| D(X).$$

The validity of this theorem (like the validity of the next theorem) can easily be checked by inserting the given expressions into the definition of the variance, once for discrete and once for continuous random variables.

Theorem A.15 *The variance σ^2 of a (discrete or continuous) real-valued random variable satisfies*

$$\sigma^2 = E(X^2) - \mu^2.$$

Theorem A.16 (variance of a sum of random variables, covariance) *If X and Y are two (discrete or continuous) real-valued random variables, whose variances $D^2(X)$ and $D^2(Y)$ exist, then*

$$D^2(X + Y) = D^2(X) + D^2(Y) + 2[E(X \cdot Y) - E(X) \cdot E(Y)].$$

The expression $E(X \cdot Y) - E(X) \cdot E(Y) = E[(X - E(X))(Y - E(Y))]$ is called the **covariance** *of X and Y. From the (stochastic) independence of X and Y it follows that*

$$D^2(Z) = D^2(X + Y) = D^2(X) + D^2(Y),$$

that is, the covariance of independent random variables vanishes.

Again the validity of this theorem can easily be checked by inserting the sum into the definition of the variance. By simple induction it can easily be generalized to finitely many random variables.

A.3.4.5 Quantiles

Quantiles are defined in direct analogy to the quantiles of a data set, with the fraction of the data set replaced by the fraction of the probability mass. For continuous random variables, quantiles are often also called **percentage points**.

Definition A.26 Let X be a real-valued random variable. Then any value x_α, $0 < \alpha < 1$, with

$$P(X \le x_\alpha) \ge \alpha \quad \text{and} \quad P(X \ge x_\alpha) \ge 1 - \alpha$$

is called an α-**quantile** of X (or of its distribution).

Note that for discrete random variables, several values may satisfy both inequalities, because their distribution function is piecewise constant. It should also be noted that the pair of inequalities is equivalent to the double inequality

$$\alpha - P(X = x) \le F_X(x) \le \alpha,$$

where $F_X(x)$ is the distribution function of a random variable X. For a continuous random variable X, it is usually more convenient to define that the α-quantile is the value x that satisfies $F_X(x) = \alpha$. In this case a quantile can be computed from the inverse of the distribution function F_X (provided that it exists and can be specified in closed form).

A.3.5 Some Special Distributions

In this section we study some special distributions, which are often needed in applications (see Sect. A.4 about inferential statistics).

A.3.5.1 The Binomial Distribution

Let X be a random variable that describes the number of trials of a Bernoulli experiment of size n in which an event A occurs with probability $p = P(A)$ in each trial.

Then X has the distribution $\forall x \in \mathbb{N} : (x; P(X = x))$ with

$$P(X = x) = b_X(x; p, n) = \binom{n}{x} p^x (1 - p)^{n-x}$$

and is said to be **binomially distributed** with parameters p and n. This formula is also known as **Bernoulli's formula**. The expression $\binom{n}{x} = \frac{n!}{x!(n-x)!}$ (pronounced "n choose x") is called a **binomial coefficient**.

The distribution satisfies the recursive relation

$$\forall x \in \mathbb{N}_0 : b_X(k + 1; p, n) = \frac{(n - x)p}{(x + 1)(1 - p)} b_X(x; p, n)$$

with $\quad b_X(0; p, n) = (1 - p)^n$.

For the expected value and variance, we have

$$\mu = E(X) = np; \qquad \sigma^2 = D^2(X) = np(1 - p).$$

A.3.5.2 The Polynomial Distribution

Bernoulli experiments can easily be generalized to more than two mutually exclusive events. In this way one obtains the polynomial distribution, which is a multidimensional distribution: a random experiment is executed independently n times. Let $A_1, \ldots, A_k$ be mutually exclusive events, of which in each trial exactly one must occur, that is, let $A_1, \ldots, A_k$ be an event partition. In every trial each event A_i occurs with constant probability $p_i = P(A_i)$, $1 \leq i \leq k$. Then the probability that in n trials the event A_i, $i = 1, \ldots, k$, occurs x_i times, $\sum_{i=1}^{k} x_i = n$, is equal to

$$P(X_1 = x_1, \ldots, X_k = x_k) = \binom{n}{x_1 \ldots x_k} p_1^{x_1} \cdots p_k^{x_k} = \frac{n!}{x_1! \cdots x_k!} p_1^{x_1} \cdots p_k^{x_k}.$$

The total of all probabilities of all vectors $(x_1, \ldots, x_k)$ with $\sum_{i=1}^{k} x_i = n$ is called the (k-dimensional) **polynomial distribution** with parameters $p_1, \ldots, p_k$ and n. The binomial distribution is obviously a special case of for $k = 2$. The expression $\binom{n}{x_1 \ldots x_k} = \frac{n!}{x_1! \cdots x_k!}$ is called a **polynomial coefficient**, in analogy to the binomial coefficient $\binom{n}{x} = \frac{n!}{x!(n-x)!}$.

A.3.5.3 The Geometric Distribution

Let X be a random variable that describes the number of trials in a Bernoulli experiment that are needed until an event A, which occurs with $p = P(A) > 0$ in each trial, occurs for the first time. Then X has the distribution $\forall x \in \mathbb{N} : (x; P(X = x))$ with

$$P(X = x) = g_X(x; p) = p(1 - p)^{x-1}$$

and is said to be **geometrically distributed** with parameter p. In order to compute the probabilities the recursive relation

$$\forall x \in \mathbb{N}: P(X = x + 1) = (1 - p)P(X = x) \quad \text{with} \quad P(X = 1) = p$$

can be useful. The expected value and variance are

$$\mu = E(X) = \frac{1}{p}; \qquad \sigma^2 = D^2(X) = \frac{1 - p}{p^2}.$$

A.3.5.4 The Hypergeometric Distribution

From an urn which contains M black and $N - M$ white, and thus in total N balls, n balls are drawn without replacement. Let X be the random variable that describes the number of black balls that have been drawn. Then X has the distribution $\forall x; \max(0, n - (N - M)) \le x \le \min(n, M): (x; P(X = x))$ with

$$P(X = x) = h_X(x; n, M, N) = \frac{\binom{M}{x}\binom{N-M}{n-x}}{\binom{N}{n}}$$

and is said to be **hypergeometrically distributed** with parameters n, M and N. This distribution satisfies the recursive relation

$$\forall x; \max(0, n - (N - M)) \le x \le \min(n, M):$$

$$h_X(x + 1; n, M, N) = \frac{(M - x)(n - x)}{(x + 1)(N - M - n + x + 1)} h_X(x; n, M, N)$$

$$\text{with} \quad h_X(1; n, M, N) = \frac{M}{N}.$$

With $p = \frac{M}{N}$ and $q = 1 - p$, the expected value and variance are

$$\mu = E(X) = np; \qquad \sigma^2 = D^2(X) = npq\frac{N - n}{N - 1}.$$

A.3.5.5 The Poisson Distribution

A random variable X with the distribution $\forall x \in \mathbb{N}: (x; P(X = x))$ where

$$P(X = x) = \Lambda_X(x; \lambda) = \frac{\lambda^x}{x!}e^{-\lambda}$$

is said to be **Poisson distributed**[20] with parameter λ. This distribution satisfies the recursive relation

$$\forall x \in \mathbb{N}_0: \Lambda_X(x + 1; \lambda) = \frac{\lambda}{x + 1}\Lambda_X(x; \lambda) \quad \text{with} \quad \Lambda_X(0; \lambda) = e^{-\lambda}.$$

---

[20]This distribution bears its name in recognition of the French mathematician Siméon-Denis Poisson (1781–1840).

The expected value and variance are

$$\mu = E(X) = \lambda; \qquad \sigma^2 = D^2(X) = \lambda.$$

A Poisson distribution describes the occurrence frequency of a certain type of events in a fixed duration of time, for example, the number of deadly traffic accidents.

For rare events A and a large number n of trials, the binomial distribution can be approximated by a Poisson distribution, because the following relation holds: if in a binomial distribution, n goes to infinity so that $np = \lambda$ stays constant, then

$$\forall x \in \mathbb{N}: \lim_{\substack{n \to \infty \\ np = \lambda}} b_X(x; p, n) = \frac{\lambda^x}{x!} e^{-\lambda},$$

and thus for large n and small p, we obtain the approximation

$$\forall x \in \mathbb{N}: b_X(x; p, n) \approx \frac{np^x}{x!} e^{-np}.$$

For Poisson distributed random variables, the following reproduction law holds:

Theorem A.17 *Let X and Y be two (stochastically) independent Poisson-distributed random variables with parameters λ_X and λ_Y, respectively. Then the sum $Z = X + Y$ is also Poisson distributed with parameter $\lambda_Z = \lambda_X + \lambda_Y$.*

A.3.5.6 The Uniform Distribution

A random variable X with the density function

$$f_X(x; a, b) = \begin{cases} \frac{1}{b-a} & \text{for } x \in [a, b], \\ 0 & \text{otherwise,} \end{cases}$$

with $a, b \in \mathbb{R}$, $a < b$, is said to be **uniformly distributed** in $[a, b]$. Its distribution function F_X is

$$F_X(x; a, b) = \begin{cases} 0 & \text{for } x \leq a, \\ \frac{x-a}{b-a} & \text{for } a \leq x \leq b, \\ 1 & \text{for } x \geq b. \end{cases}$$

The expected value and variance are

$$\mu = E(X) = \frac{a+b}{2}; \qquad \sigma^2 = D^2(X) = \frac{(b-a)^2}{12}.$$

A.3.5.7 The Normal Distribution

A random variable X with the density function

$$N_X(x; \mu, \sigma^2) = \frac{1}{\sqrt{2\pi\sigma^2}} e^{\frac{(x-\mu)^2}{2\sigma^2}}$$

is said to be **normally distributed** with parameters μ and σ^2.

The expected value and variance are

$$E(X) = \mu; \qquad D^2(X) = \sigma^2.$$

The normal distribution with expected value $\mu = 0$ and variance $\sigma^2 = 1$ is called standard normal distribution.

The density function $f_X(x; \mu, \sigma^2)$ has its maximum at μ and inflection points at $x = \mu \pm \sigma$. The distribution function of X does not possess a closed-form representation. As a consequence, it is usually tabulated, most commonly for the standard normal distribution, from which the values of arbitrary normal distributions can be easily obtained by simple linear transformations.

However, in practice one often faces the reversed problem, namely to find the argument of the distribution function of the standard normal distribution for which is has a given value (or, in other words, one desires to find a quantile of the normal distribution). In order to solve this problem one may just as well use tabulated values. However, the inverse function can be approximated fairly well by the ratio of two polynomials, which is usually employed in computer programs (e.g., [15]).

The normal distribution is certainly the most important continuous distribution, since many random processes, especially measurements of physical quantities, can be described well by this distribution. The theoretical justification for this observed fact is the important central limit theorem:

Theorem A.18 (central limit theorem) *Let $X_1, \ldots, X_m$ be m (stochastically) independent real-valued random variables. In addition, let them satisfy the so-called Lindeberg condition, that is, if $F_i(x)$ are the distribution functions of the random variables X_i, $i = 1, \ldots, m$, μ_i their expected values, and σ_i^2 their variances, then for every $\varepsilon > 0$, it is*

$$\lim_{m \to \infty} \frac{1}{V_m^2} \sum_{i=1}^{m} \int_{|x_i - \mu_i| > \varepsilon V_m^2} (x_i - \mu_i)^2 \, dF_i(x) = 0$$

with $V_m^2 = \sum_{i=1}^{m} \sigma_i^2$. Then the standardized sums

$$S_m = \frac{\sum_{i=1}^{m}(X_i - \mu_i)}{\sqrt{\sum_{i=1}^{m} \sigma_i^2}}$$

(that is, standardized to expected value 0 and variance 1) satisfy

$$\forall x \in \mathbb{R}: \quad \lim_{m \to \infty} P(S_m \le x) = \Phi(x) = \frac{1}{\sqrt{2\pi}} \int_{-\infty}^{x} e^{-\frac{t^2}{2}} \, dt,$$

where $\Phi(x)$ is the distribution function of the standard normal distribution.

Intuitively this theorem says that the sum of a large number of almost arbitrarily distributed random variables (the Lindeberg condition is a very weak restriction) is approximately normally distributed. Since physical measurements are usually affected by a large number of random influences from several independent sources,

which all add up to form the total measurement error, the result is often approximately normally distributed. The central limit theorem thus explains why normally
distributed quantities are so common in practice.

Like the Poisson distribution, the normal distribution can be used as an approximation of the binomial distribution, even if the probabilities p are not small.

Theorem A.19 (limit theorem of de Moivre–Laplace)[21] *If the probability of the
occurrence of an event A in n independent trials is constant and equal to p, $0 <
p < 1$, then the probability $P(X = x)$ that, in these trials, the event A occurs exactly
x times satisfies as $n \to \infty$ the relation*

$$\sqrt{np(1-p)}P(X=x)\frac{1}{\sqrt{2\pi}}e^{-\frac{1}{2}y^2} \to 1, \quad where \quad y = \frac{x-np}{\sqrt{np(1-p)}}.$$

*The convergence is uniform for all x for which y lies in an arbitrary finite inter-
val (a, b).*

This theorem allows us to approximate the probabilities of the binomial distribution for large n by

$\forall x; 0 \le x \le n:$

$$P(X=x) = \binom{n}{x}p^x(1-p)^{n-x}$$

$$\approx \frac{1}{\sqrt{2\pi np(1-p)}}\exp\left(-\frac{(x-np)^2}{2np(1-p)}\right) \quad \text{or}$$

$$P(X=x) \approx \Phi\left(\frac{x-np+\frac{1}{2}}{\sqrt{np(1-p)}}\right) - \Phi\left(\frac{x-np-\frac{1}{2}}{\sqrt{np(1-p)}}\right) \quad \text{and}$$

$\forall x_1, x_2; 0 \le x_1 \le x_2 \le n:$

$$P(x_1 \le X \le x_2) \approx \Phi\left(\frac{x_2-np+\frac{1}{2}}{\sqrt{np(1-p)}}\right) - \Phi\left(\frac{x_1-np-\frac{1}{2}}{\sqrt{np(1-p)}}\right),$$

where Φ is the distribution function of the standard normal distribution. The approximation is reasonably good for $np(1-p) > 9$.

A.3.5.8 The χ^2 Distribution

If one forms the sum of m independent, standard normally distributed random variables (expected value 0 and variance 1), one obtains a random variable X with the
density function

$$f_X(x; m) = \begin{cases} 0 & \text{for } x < 0, \\ \frac{1}{2^{\frac{m}{2}} \cdot \Gamma(\frac{m}{2})} \cdot x^{\frac{m}{2}-1} \cdot e^{-\frac{x}{2}} & \text{for } x \ge 0, \end{cases}$$

---

[21]This theorem bears its name in recognition of the French mathematicians Abraham de Moivre
(1667–1754) and Pierre-Simon de Laplace (1749–1827).

where Γ is the so-called **Gamma function** (generalization of the factorial)

$$\Gamma(x) = \int_0^\infty t^{x-1} e^t \, dt$$

for $x > 0$. This random variable is said to be χ^2-distributed with m degrees of freedom. The expected value and variance are

$$E(X) = m; \qquad D^2(X) = 2m.$$

The χ^2 distribution plays an important role in the statistical theory of hypothesis testing (see Sect. A.4.3), for example, for independence tests.

A.3.5.9 The Exponential Distribution

A random variable X with the density function

$$f_X(x; \alpha) = \begin{cases} 0 & \text{for } x \le 0, \\ \alpha e^{-\alpha x} & \text{for } x > 0, \end{cases}$$

with $\alpha > 0$ is said to be **exponentially distributed** with parameter α. Its distribution function F_X is

$$F_X(x; \alpha) = \begin{cases} 0 & \text{for } x \le 0, \\ 1 - e^{-\alpha x} & \text{for } x > 0. \end{cases}$$

The expected value and variance are

$$\mu = E(X) = \frac{1}{\alpha}; \qquad \sigma^2 = D^2(X) = \frac{1}{\alpha^2}.$$

The exponential distribution is commonly used to model the durations between the arrivals of people or jobs that enter a queue to wait for service or processing.

A.4 Inferential Statistics

With inferential statistics one tries to answer the question whether observed phenomena are typical or regular or whether they may also be caused by random influences. In addition, one strives to find probability distributions that are good models of the data-generating process and tries to estimate their parameters. Therefore inferential statistics is concerned with the following important tasks:

- **parameter estimation**
  Given a model of the data-generating process, especially an assumption about the family of distributions of the underlying random variables, the parameters of this model are estimated from the data.
- **hypothesis testing**
  One or more hypotheses about the data-generating process are checked on the basis of the given data. Special types of hypothesis tests are:

- parameter test: a test whether a parameter of a distribution can have a certain value or whether the parameter of the distributions underlying two different data sets are equal.
- goodness-of-fit test: a test whether a certain distribution assumption fits the data or whether observed deviations from the expected characteristics of the data (given the distribution assumption) can be explained by random influences.
- dependence test: a test whether two features are dependent, or whether observed deviations from an independent distribution can be explained by random influences.

• **model selection**

From several models, which could be used to explain the data, select the best fitting one, paying attention to the complexity of the models.

Note that parameter estimation can be seen as a special case of model selection, in which the class of models from which one can select is highly restricted, so that they differ only in the value(s) of a (set of) parameter(s). The goodness-of-fit test resides on the border between hypothesis testing and model selection, because it serves the purpose to check whether a model of a certain class is appropriate to explain the data. We do not discuss more complex model selection in this appendix.

A.4.1 Random Samples

In Sect. A.2 we considered so-called *samples* (vectors of observed or measured feature values) and represented them in a clearly arranged form in tables and charts. In inferential statistics we also consider samples: we employ the mathematical tools of probability calculus in order to gain (new) knowledge or to check hypotheses on the basis of data samples. In order for this to be possible, the sample values must be obtained as the outcomes of random experiments, so that probability theory is applicable. Such specific samples are called **random samples**.

The random variable which yields the sample value x_i when carrying out the corresponding random experiment is denoted X_i. The value x_i is called a **realization** of the random variable X_i, $i = 1, \ldots, n$. Therefore a random sample $x = (x_1, \ldots, x_n)$ can be seen as a *realization of the random vector* $X = (X_1, \ldots, X_n)$. A random sample is called **independent** if the random variables $X_1, \ldots, X_n$ are (stochastically) independent, that is, if

$$\forall c_1, \ldots, c_n \in \mathbb{R}: \quad P\left(\bigwedge_{i=1}^{n} X_i \leq c_i\right) = \prod_{i=1}^{n} P(X_i \leq c_i).$$

An independent random sample is called **simple** if all random variables $X_1, \ldots, X_n$ have the same distribution function. Likewise we will call the corresponding random vector *independent* and *simple*, respectively.

A.4.2 Parameter Estimation

As already pointed out, parameter estimation rests on an assumption of a model for the data-generating process, in particular an assumption about the family of distribution functions of the underlying random variables. Given this assumption, it estimates the parameters of this model from the given data.

Given: • A data set and
        • a family of equally shaped, parameterized distribution functions $f_X(x; \theta_1, \ldots, \theta_k)$.

        Such families may be, for example:
        The family of binomial distributions $b_X(x; p, n)$ with parameters p, $0 \leq p \leq 1$, and $n \in \mathbb{N}$, where n, however, is already implicitly given by the sample size.
        The family of Poisson distributions $\Lambda_X(x; \lambda, n)$ with parameters $\lambda > 0$ and $n \in \mathbb{N}$, where n is again given by the sample size.
        The family of normal distributions $N_X(x; \mu, \sigma^2)$ with parameters μ (expected value) and σ^2 (variance).

Assumption: The process that has generated the given data can appropriately be described by an element of the considered family of distribution functions: distribution assumption.

Desired: There is an element of the considered family of distribution functions (determined by the values of the parameters) that is the best model of the given data (w.r.t. certain quality criteria).

Estimators for the parameters are **statistics**, that is, functions of the sample values of a given data set. Hence they are functions of (realizations of) random variables and therefore (realizations of) random variables themselves. As a consequence, the whole set of tools that probability theory provides for examining the properties of random variables can be applied to estimators.

One distinguishes mainly two types of parameter estimation:

• **point estimators**
  determine the best individual value of a parameter w.r.t. the given data and certain quality criteria;
• **interval estimators**
  yield a so-called confidence interval, in which the true value of the parameter lies with high certainty, with the degree of certainty to be chosen by a user.

A.4.2.1 Point Estimation

It is immediately clear that not every statistic, that is, not every function computed from the sample values, is a usable point estimator for an examined parameter θ. Rather, a statistic should have certain properties, in order to be a reasonable estimator. Desirable properties are:

- **consistency**
  If the amount of available data grows, the estimated value should become closer and closer to the actual value of the estimated parameter, at least with higher and higher probability. This can be formalized by requiring that for growing sample size, the estimation function converges in probability to the true value of the parameter. For example, if T is an estimator for the parameter θ, it should be

  $$\forall \varepsilon > 0: \quad \lim_{n \to \infty} P(|T - \theta| < \varepsilon) = 1,$$

  where n is the sample size. This condition should be satisfied by every point estimator; otherwise we have no reason to assume that the estimated value is in any way related to the true value.

- **unbiasedness**
  An estimator should not tend to generally under- or over-estimate the parameter, but should, on average, yield the right value. Formally, this means that the expected value of the estimator should coincide with the true value of the parameter. For example, if T is an estimator for the parameter θ, it should be

  $$E(T) = \theta,$$

  independently of the sample size.

- **efficiency**
  The estimation should be as precise as possible, that is, the deviation from the true value of the parameter should be as small as possible. Formally, one requires that the variance of the estimator should be as small as possible, since the variance is a natural measure for the precision of the estimation. For example, let T and U be unbiased estimators for a parameter θ. Then T is called more efficient than U iff

  $$D^2(T) < D^2(U).$$

  However, it is rarely possible to show that an estimator achieves the highest possible efficiency for a given estimation problem.

- **sufficiency**
  An estimation function should exploit the information that is contained in the data in an optimal way. This can be made more precise by requiring that different samples, which yield the same estimated value, should be equally likely (given the estimated value of the parameter). The reason is that if they are not equally likely, it must be possible to derive additional information about the parameter value from the data. Formally, this means that an estimator T for a parameter θ is called *sufficient* if for all random samples $x = (x_1, \ldots, x_n)$ with $T(x) = t$, the expression

  $$\frac{f_{X_1}(x_1; \theta) \cdots f_{X_n}(x_n; \theta)}{f_T(t; \theta)}$$

  does not depend on θ [10].

Note that the estimators used in the definition of efficiency must be unbiased, since otherwise arbitrary constants (variance $D^2 = 0$) would be efficient estimators. Consistency, on the other hand, can often be neglected as an additional condition, since an unbiased estimator T for a parameter θ which also satisfies

$$\lim_{n\to\infty} D^2(T) = 0$$

is consistent (not surprisingly).

A.4.2.2 Point Estimation Examples

Given: A family of uniform distributions on the interval $[0, \theta]$, that is,

$$f_X(x; \theta) = \begin{cases} \frac{1}{\theta} & \text{if } 0 \leq x \leq \theta, \\ 0 & \text{otherwise.} \end{cases}$$

Desired: An estimator for the unknown parameter θ.

(a) Estimate the parameter θ as the maximum of the sample values, that is, choose $T = \max\{X_1, \ldots, X_n\}$ as the estimation function.
In order to check the properties of this estimator, we first determine its probability density,[22] from which we can then derive other properties:

$$f_T(t; \theta) = \frac{d}{dt} F_T(t; \theta) = \frac{d}{dt} P(T \leq t)$$

$$= \frac{d}{dt} P(\max\{X_1, \ldots, X_n\} \leq t)$$

$$= \frac{d}{dt} P\left(\bigwedge_{i=1}^n X_i \leq t\right) = \frac{d}{dt} \prod_{i=1}^n P(X_i \leq t)$$

$$= \frac{d}{dt} (F_X(t; \theta))^n = n \cdot (F_X(t; \theta))^{n-1} f_X(t, \theta),$$

where

$$F_X(x; \theta) = \int_{-\infty}^x f_X(x; \theta)\, dx = \begin{cases} 0 & \text{if } x \leq 0, \\ \frac{x}{\theta} & \text{if } 0 \leq x \leq \theta, \\ 1 & \text{if } x \geq \theta. \end{cases}$$

It follows that

$$f_T(t; \theta) = \frac{n \cdot t^{n-1}}{\theta^n} \quad \text{for } 0 \leq t \leq \theta.$$

T is a consistent estimator for θ:

$$\lim_{n\to\infty} P(|T - \theta| < \varepsilon) = \lim_{n\to\infty} P(T > \theta - \varepsilon)$$

$$= \lim_{n\to\infty} \int_{\theta-\varepsilon}^{\theta} \frac{n \cdot t^{n-1}}{\theta^n}\, dt = \lim_{n\to\infty} \left[\frac{t^n}{\theta^n}\right]_{\theta-\varepsilon}^{\theta}$$

---

[22]Recall that estimators are functions of random variables and thus random variables themselves. As a consequence, they have a probability density.

$$= \lim_{n\to\infty}\left(\frac{\theta^n}{\theta^n} - \frac{(\theta-\varepsilon)^n}{\theta^n}\right)$$

$$= \lim_{n\to\infty}\left(1 - \left(\frac{\theta-\varepsilon}{\theta}\right)^n\right) = 1.$$

However, T is not unbiased. This is already intuitively clear, since T can only underestimate the value of θ, but never overestimate it. Therefore the estimated value will "almost always" be too small. Formally, we have:

$$E(T) = \int_{-\infty}^{\infty} t \cdot f_T(t;\theta)\,dt = \int_0^\theta t \cdot \frac{n \cdot t^{n-1}}{\theta^n}\,dt$$

$$= \left[\frac{n \cdot t^{n+1}}{(n+1)\theta^n}\right]_0^\theta = \frac{n}{n+1}\theta < \theta \quad \text{for } n < \infty.$$

(b) Choose $U = \frac{n+1}{n}T = \frac{n+1}{n}\max\{X_1,\ldots,X_n\}$ as the estimation function. The statistic U is a consistent and unbiased estimator for the parameter θ. We omit a formal proof, which can be obtained along the same lines as in a). However, we will later need the probability density of this estimator, which is

$$f_U(u;\theta) = \frac{n^{n+1}}{(n+1)^n} \frac{u^{n-1}}{\theta^n}.$$

Given:　A family of normal distributions $N(x;\mu,\sigma)$, i.e.,

$$f_X(x;\mu,\sigma^2) = \frac{1}{\sqrt{2\pi\sigma^2}}\exp\left(-\frac{(x-\mu)^2}{2\sigma^2}\right).$$

Desired: Estimators for the unknown parameters μ and σ^2.

(a) The (empirical) median and the (empirical) mean are both consistent and unbiased estimators for the parameter μ. The median is less efficient than the mean. Since it only exploits order information from the data, it is naturally also not sufficient. Even though the median is preferable for small sample size due to the fact that it is less sensitive to outliers, its variance is larger than that of the mean (provided that the sample size is sufficiently large).

(b) The function $V^2 = \frac{1}{n}\sum_{i=1}^n (X_i - \bar{X})^2$ is a consistent, but not unbiased, estimator for the parameter σ^2, since this function tends to underestimating the variance. The (empirical) variance $S^2 = \frac{1}{n-1}\sum_{i=1}^n (X_i - \bar{X})^2$, however, is a consistent and unbiased estimator for the parameter σ^2. (Due to the square root, however, $\sqrt{S^2}$ is *not* an unbiased estimator for the standard deviation.)

Given:　A family of polynomial distributions, that is,

$$f_{X_1,\ldots,X_k}(x_1,\ldots,x_k;\theta_1,\ldots,\theta_k,n) = \frac{n!}{\prod_{i=1}^k x_i!}\prod_{i=1}^k \theta_i^{x_i},$$

where θ_i is the probability that values a_i occurs, and the random variable X_i describes how often the value a_i occurs in the sample.

Desired: Estimators for the unknown parameters $\theta_1,\ldots,\theta_k$.

The relative frequencies $R_i = \frac{X_i}{n}$ of the feature values in the sample are consistent, unbiased, most efficient, and sufficient estimators for the unknown parameters θ_i, $i = 1, \ldots, k$. This is the reason why relative frequencies are used in basically all cases to estimate the probabilities of nominal values.

A.4.2.3 Maximum Likelihood Estimation

Up to now we have simply stated estimation functions for parameters. This is possible because for many standard problems, consistent, unbiased, and efficient estimators are known, so that they can be looked up in standard textbooks. Nevertheless, we consider briefly how one can find estimation functions in principle.

Besides the method of moments, which we omit here, maximum likelihood estimation, as it was developed by R.A. Fisher,[23] is one of the most popular methods for finding estimation functions. The underlying principle is very simple: choose the value of the parameter to estimate (or the set of values of the parameters to estimate if there are several) that renders the given random sample most likely. This is achieved as follows: if the parameter(s) of the true underlying distribution were known, we could easily compute the probability of a random experiment generating the observed random sample. However, this probability can also be written with unknown parameters (though not necessarily be numerically computed). The result is a function that describes the likelihood of a random sample given the unknown parameters. This function is called a **likelihood function**. By taking partial derivatives of this function w.r.t. the parameters to estimate and setting them equal to zero (since the derivative must vanish at a maximum), estimation functions are derived.

A.4.2.4 Maximum Likelihood Estimation Example

Given: A family of normal distributions $N_X(x; \mu, \sigma^2)$, that is,

$$N_X\left(x; \mu, \sigma^2\right) = \frac{1}{\sqrt{2\pi\sigma^2}} \exp\left(-\frac{(x - \mu)^2}{2\sigma^2}\right).$$

Desired: Estimators for the unknown parameters μ and σ^2.

The likelihood function of a simple random sample $x = (x_1, \ldots, x_n)$, which is the realization of a vector $X = (X_1, \ldots, X_n)$ of normally distributed random variables with parameters μ and σ^2, is

$$L\left(x_1, \ldots, x_n; \mu, \sigma^2\right) = \prod_{i=1}^{n} \frac{1}{\sqrt{2\pi\sigma^2}} \exp\left(-\frac{(x_i - \mu)^2}{2\sigma^2}\right).$$

---

[23]However, R.A. Fisher did not invent this method as is often believed. Earlier on C.F. Gauß and D. Bernoulli already made use of it, but Fisher was the first to study it systematically and to establish it in statistics [10].

It describes the probability of the sample (the data set) depending on the parameters μ and σ^2. By exploiting the known rules for computing with exponential functions, this expression can be transformed into

$$L(x_1, \ldots, x_n; \mu, \sigma^2) = \frac{1}{(\sqrt{2\pi\sigma^2})^n} \exp\left(-\frac{1}{2\sigma^2} \sum_{i=1}^{n} (x_i - \mu)^2\right).$$

In order to determine the maximum of this function w.r.t. parameters μ and σ^2, it is convenient to take the natural logarithm (in order to eliminate the exponential function). Since the logarithm is a monotone function, this does not change the location of the maximum. We obtain the **log-likelihood function**

$$\ln L(x_1, \ldots, x_n; \mu, \sigma^2) = -n \ln\left(\sqrt{2\pi\sigma^2}\right) - \frac{1}{2\sigma^2} \sum_{i=1}^{n} (x_i - \mu)^2.$$

In order to find the maximum, we set the partial derivatives w.r.t. μ and σ^2 equal to 0. The partial derivative w.r.t. μ is

$$\frac{\partial}{\partial \mu} \ln L(x_1, \ldots, x_n; \mu, \sigma^2) = \frac{1}{\sigma^2} \sum_{i=1}^{n} (x_i - \mu) \overset{!}{=} 0,$$

from which

$$\sum_{i=1}^{n} (x_i - \mu) = \left(\sum_{i=1}^{n} x_i\right) - n\mu \overset{!}{=} 0,$$

and thus

$$\hat{\mu} = \frac{1}{n} \sum_{i=1}^{n} x_i$$

follows as an estimate for the parameter μ. The partial derivative of the log-likelihood function w.r.t. σ^2 yields

$$\frac{\partial}{\partial \sigma^2} \ln L(x_1, \ldots, x_n; \mu, \sigma^2) = -\frac{n}{2\sigma^2} + \frac{1}{2\sigma^4} \sum_{i=1}^{n} (x_i - \mu)^2 \overset{!}{=} 0.$$

By inserting the estimated value $\hat{\mu}$ for the parameter μ, we obtain the estimator

$$\hat{\sigma}^2 = \frac{1}{n} \sum_{i=1}^{n} (x_i - \hat{\mu})^2 = \frac{1}{n} \sum_{i=1}^{n} x_i^2 - \frac{1}{n^2} \left(\sum_{i=1}^{n} x_i\right)^2$$

for the parameter σ^2. Note that the result is not unbiased. (Recall that, as we mentioned above, the empirical variance with a factor of $\frac{1}{n}$ instead of $\frac{1}{n-1}$ is not unbiased.) This shows that there is no estimator for the variance of a normal distribution that has all desirable properties. Among those that are unbiased, the data is not maximally likely, and the one that makes the data maximally likely is not unbiased.

A.4.2.5 Maximum A Posteriori Estimation

An alternative for maximum likelihood estimation is maximum a posteriori estimation, which rest on Bayes' rule. This estimation method assumes a prior distribution
on the domain of the parameter(s) and computes, with the help of this prior distribution and the given data, a posterior distribution via Bayes' rule. The parameter value
with the greatest posterior probability (density) is chosen as the estimated value.
That is, one chooses the value for θ that maximizes

$$f(\theta \mid D) = \frac{f(D \mid \theta) f(\theta)}{f(D)} = \frac{f(D \mid \theta) f(\theta)}{\int_{-\infty}^{\infty} f(D \mid \theta) f(\theta) d\theta},$$

where D are the given data, and $f(\theta)$ is the assumed prior distribution. Compare
this to maximum likelihood estimation, which chooses the value of the parameter θ
that maximizes $f(D \mid \theta)$, that is, the likelihood of the data.

A.4.2.6 Maximum A Posteriori Estimation Example

In order to illustrate why it can be useful to assume a prior distribution on the possible values of the parameter θ, we consider three situations:

- A drunkard claims to be able to predict the side onto which a tossed coin will
  land (head or tails). On ten trials he always states the correct side beforehand.
- A tea lover claims that she is able to taste whether the tea or the milk was poured
  into the cup first. On ten trials she always identifies the correct order.
- An expert of classical music claims to be able to recognize from a single sheet
  of music whether the composer was Mozart or somebody else. On ten trials he is
  indeed correct every time.

Let θ be the (unknown) parameter that states the probability that a correct prediction
is made. The data is formally identical in all three cases: 10 correct, 0 wrong predictions. Nevertheless we are reluctant to treat these three cases equally, as maximum
likelihood estimation does. We hardly believe that the drunkard can actually predict
the side a tossed coin will land on but assume that he was simply "lucky." The tea
lover we also view sceptically, even though our skepticism is less pronounced as in
the case of the drunkard. Maybe there are certain chemical processes that depend on
the order in which tea and milk are poured into the cup and which change the taste
slightly and thus are noticeable to a passionate tea drinker. We just see this possibility as unlikely. On the other hand, we are easily willing to believe the music expert.
Clearly, there are differences in the style of different composers that may allow a
knowledgeable music expert to see even from a single sheet of music whether it was
composed by Mozart or not.

The three attitudes with which we see the three situations can be expressed by
prior distribution on the domain of the parameter θ. In the case of the drunkard we
ascribe a nonvanishing probability density only to the value 0.5.[24] In the case of the

---

[24]Formally: Dirac pulse at $\theta = 0.5$.

tea lover we may choose a prior distribution, which ascribes values close to 0.5 a high probability density, which quickly declines towards 1. In the case of the music expert, however, we ascribe a significant probability densities also to values closer to 1. In effect, this means that in the case of the drunkard we always estimate θ as 0.5, regardless of the data. In the case of the tea lover only fairly clear evidence in favor of her claim will make us accept higher values for θ. In the case of the music expert, however, few positive examples suffice to obtain a fairly high value for θ.

Obviously, the prior distribution contains background knowledge about the data-generating process and expresses which parameter values we expect and how easily we are willing to accept them. However, how to choose the prior distribution is a tricky and critical problem, since it has to be chosen subjectively. Depending on their experience, different people will choose different distributions.

A.4.2.7 Interval Estimation

A parameter value that is estimated with a point estimator from a data set usually deviates from the true value of the parameter. Therefore it is useful if one can make statements about these unknown deviations and their expected magnitude. The most straightforward approach is certainly to provide a point-estimated value t and the standard deviation $D(T)$ of the estimator, that is,

$$t \pm D(T) = t \pm \sqrt{D^2(T)}.$$

However, a better possibility consists in determining intervals—so-called confidence intervals—that contain the true value of the parameter with high probability.

The boundaries of these confidence intervals are computed by certain rules from the sample values. Hence they are also statistics, and thus, like point estimators, (realizations of) random variables. Therefore they can be treated analogously. Formally, they are defined as follows:

Let $X = (X_1, \ldots, X_n)$ be a simple random vector the random variables of which have the distribution function $F_{X_i}(x_i; \theta)$ with (unknown) parameter θ. Furthermore, let $A = g_A(X_1, \ldots, X_n)$ and $B = g_B(X_1, \ldots, X_n)$ be two estimators defined on X such that

$$P(A < \theta < B) = 1 - \alpha, \qquad P(\theta \leq A) = \frac{\alpha}{2}, \qquad P(\theta \geq B) = \frac{\alpha}{2}.$$

Then the random interval $[A, B]$ (or a realization $[a, b]$ of this random interval) is called a $(1 - \alpha) \cdot 100\%$ **confidence interval** for the (unknown) parameter θ. The value $1 - \alpha$ is called **confidence level**.

Note the term "confidence" refers to the *method* and *not* to the *result* of the procedure (that is, to a realization of the random interval). *Before* data has been collected, a $(1 - \alpha) \cdot 100\%$ confidence interval contains the true parameter value with probability $1 - \alpha$. However, *after* the data has been collected and the interval boundaries have been computed, the interval boundaries are not random variables anymore. Therefore the interval either contains the true value of the parameter θ

or it does not (probability 1 or 0—even though it is not known which of the two possibilities is obtained).

The above definition of a confidence interval is not specific enough to derive a computation procedure from it. Indeed, the estimators A and B are not uniquely determined: the sets of realizations of the random vectors $X_1, \ldots, X_n$ for which $A \geq \theta$ and $B \leq \theta$ hold merely have to be disjoint and must possess the probability $\frac{\alpha}{2}$. In order to derive a procedure to obtain the boundaries A and B of a confidence interval, the estimators are restricted as follows: they are not defined as general functions of the random vector but rather as functions of a chosen point estimators T for the parameter θ. That is,

$$A = h_A(T) \quad \text{and} \quad B = h_B(T).$$

In this way confidence intervals can be determined generally, namely by replacing an investigation of $A < \theta < B$ with the corresponding event w.r.t. the estimator T, that is, $A^* < T < B^*$. Of course, this is only possible if we can derive the functions $h_A(T)$ and $h_B(T)$ from the inverse functions $A^* = h_A^{-1}(\theta)$ and $B^* = h_B^{-1}(\theta)$ that we have to consider w.r.t. T.

$$
\begin{aligned}
\text{Idea:} \quad & P(A^* < T < B^*) = 1 - \alpha \\
\Rightarrow \quad & P(h_A^{-1}(\theta) < T < h_B^{-1}(\theta)) = 1 - \alpha \\
\Rightarrow \quad & P(h_A(T) < \theta < h_B(T)) = 1 - \alpha \\
\Rightarrow \quad & P(A < \theta < B) = 1 - \alpha.
\end{aligned}
$$

Unfortunately, this is not always possible (in a sufficiently simple way).

A.4.2.8 Interval Estimation Examples

Given: A family of uniform distributions on the interval $[0, \theta]$, that is,

$$
f_X(x; \theta) = \begin{cases} \frac{1}{\theta} & \text{if } 0 \leq x \leq \theta, \\ 0 & \text{otherwise.} \end{cases}
$$

Desired: A confidence interval for the unknown parameter θ.

A confidence interval can be computed in this case by starting from the unbiased point estimator $U = \frac{n+1}{n} \max\{X_1, \ldots, X_n\}$:

$$P(U \leq B^*) = \int_0^{B^*} f_U(u; \theta)\, du = \frac{\alpha}{2} \quad \text{and}$$

$$P(U \geq A^*) = \int_{A^*}^{\frac{n+1}{n}\theta} f_U(u; \theta)\, du = \frac{\alpha}{2}.$$

As we know from the section on point estimation,

$$f_U(u; \theta) = \frac{n^{n+1}}{(n+1)^n} \frac{u^{n-1}}{\theta^n}.$$

Thus we obtain

$$B^* = \sqrt[n]{\frac{\alpha}{2}\frac{n+1}{n}}\,\theta \quad \text{and} \quad A^* = \sqrt[n]{1 - \frac{\alpha}{2}\frac{n+1}{n}}\,\theta,$$

that is,

$$P\left(\sqrt[n]{\frac{\alpha}{2}\frac{n+1}{n}}\,\theta < U < \sqrt[n]{1 - \frac{\alpha}{2}\frac{n+1}{n}}\,\theta\right) = 1 - \alpha,$$

from which we can derive easily

$$P\left(\frac{U}{\sqrt[n]{1 - \frac{\alpha}{2}\frac{n+1}{n}}} < \theta < \frac{U}{\sqrt[n]{\frac{\alpha}{2}\frac{n+1}{n}}}\right) = 1 - \alpha.$$

This expression allows us to read the values A and B directly.

Given: A family of binomial distributions, that is,

$$b_X(x; \theta, n) = \binom{n}{x}\theta^x(1-\theta)^{n-x}.$$

Desired: A confidence interval for the unknown parameter θ.

For an exact computation of a confidence interval for the parameter θ, we start in analogy to the above example with

$$P(X \geq A^*) = \sum_{i=A^*}^{n}\binom{n}{i}\theta^i(1-\theta)^{n-i} \leq \frac{\alpha}{2},$$

$$P(X \leq B^*) = \sum_{i=0}^{B^*}\binom{n}{i}\theta^i(1-\theta)^{n-i} \leq \frac{\alpha}{2}.$$

However, these expressions are often difficult to evaluate. Hence one often chooses an alternative approach, namely the approximate computation of a confidence interval based on the central limit theorem (see Theorem A.19 on page 348). This theorem allows us to approximate the binomial distribution by a standard normal distribution:

$$b_X(x; \theta, n) \approx N\left(\frac{x - n\theta}{\sqrt{n\theta(1-\theta)}}; 0, 1\right) = \frac{1}{\sqrt{2\pi n\theta(1-\theta)}}\exp\left(-\frac{(x-n\theta)^2}{2n\theta(1-\theta)}\right).$$

This approximation is fairly good already for $n\theta(1-\theta) > 9$.

Similar to the approach above, we now start from

$$P\left(\frac{X - n\theta}{\sqrt{n\theta(1-\theta)}} \leq B^*\right) = \frac{\alpha}{2} \quad \text{and}$$

$$P\left(\frac{X - n\theta}{\sqrt{n\theta(1-\theta)}} \geq A^*\right) = \frac{\alpha}{2}.$$

Note that this expression does not contain the estimator (here $T = \frac{X}{n}$) for the unknown parameter θ itself but a function of this estimator (because of the approximation used).

Due to the symmetry of the normal distribution, the computations become fairly simple. For example, due to this symmetry, we know that $B^* = -A^*$. Hence we can write

$$P\left(-A^* < \frac{X - n\theta}{\sqrt{n\theta(1 - \theta)}} < A^*\right)$$

$$= \int_{-A^*}^{A^*} \frac{1}{\sqrt{2\pi n\theta(1 - \theta)}} \exp\left(-\frac{(x - n\theta)^2}{2n\theta(1 - \theta)}\right) dx$$

$$= \Phi(A^*) - \Phi(-A^*) = 2\Phi(A^*) - 1 = 1 - \alpha,$$

where Φ is the distribution function of the standard normal distribution. This function cannot be computed analytically but is available in tabulated form, so that one can easily find the value x that corresponds to a given value $\Phi(x)$. Thus we only have to derive an expression $P(A < \theta < B)$ from the above expression. This is done as follows:

$$-A^* < \frac{X - n\theta}{\sqrt{n\theta(1 - \theta)}} < A^*$$

$$\Rightarrow \quad |X - n\theta| < A^*\sqrt{n\theta(1 - \theta)}$$

$$\Rightarrow \quad (X - n\theta)^2 < \left(A^*\right)^2 n\theta(1 - \theta)$$

$$\Rightarrow \quad \theta^2\left(n\left(A^*\right)^2 + n^2\right) - \theta\left(2nX + \left(A^*\right)^2 n\right) + X^2 < 0.$$

From the resulting quadratic equation we easily obtain the values of A and B as

$$A/B = \frac{1}{n + (A^*)^2}\left(X + \frac{(A^*)^2}{2} \mp A^*\sqrt{\frac{X(n - X)}{n} + \frac{(A^*)^2}{4}}\right),$$

where $\Phi(A^*) = 1 - \frac{\alpha}{2}$.

A.4.3 Hypothesis Testing

A hypothesis test is a statistical procedure where a decision is made between two contrary hypotheses about the data generating process. The hypotheses may refer to the value of a parameter *(parameter test)*, to a distribution assumption *(goodness-of-fit test)*, or to the dependence or independence of two quantities *(dependence test)*. One of the two hypotheses is preferred, that is, in case of doubt the decision is made in its favor. The preferred hypothesis is called the **null hypothesis** H_0, and the other is called the **alternative hypothesis** H_a. Only if sufficiently strong evidence is available against the null hypothesis, then the alternative hypothesis is accepted (and thus the null hypothesis is rejected). One also says that the null hypothesis receives the benefit of the doubt.[25]

---

[25] Alternatively, one may say that a court trial is held against the null hypothesis, where the data (sample) act as evidence. In case of doubt the defendant is acquitted (the null hypothesis is accepted). Only if the evidence is sufficiently incriminating, the defendant is convicted (the null hypothesis is rejected).

The test decision is made on the basis of a **test statistic**, that is, a function of the sample values of the given data set. The null hypothesis is rejected if the value of the test statistic lies in the so-called **critical region** C. The development of a statistical test consists in choosing, for a given distribution assumption and a parameter, an appropriate test statistic and then to determine, for a user-specified *significance level* (see the next section), the corresponding critical region C (see the following sections).

A.4.3.1 Error Types and Significance Level

Since the data on which the test decision rests is the outcome of a random process, we cannot be sure that the decision made with a hypothesis test is correct. We may decide wrongly and may do so in two different ways:

- **error of the first kind**:
  The null hypothesis H_0 is rejected, even though it is correct.
- **error of the second kind**:
  The null hypothesis H_0 is accepted, even though it is wrong.

Errors of the first kind are seen as more severe, because the null hypothesis receives the benefit of the doubt and thus is not rejected as easily as the alternative hypothesis. If the null hypothesis is rejected nevertheless, despite being correct, we commit a serious error. Therefore it is tried to limit the probability of an error of the first kind to a certain maximal value. This maximal value α is called the **significance level** of the hypothesis test. It has to be chosen by a user. Typical values of the significance level are 10%, 5%, or 1%.

A.4.3.2 Parameter Test

In a **parameter test** the contrary hypotheses make statements about the values of one or more parameters. For example, the null hypothesis may be that the true value of a parameter θ is at least (or at most) θ_0:

$$H_0: \quad \theta \geq \theta_0, \qquad H_a: \quad \theta < \theta_0.$$

In such a case the test is called **one-sided**. On the other hand, in a **two-sided** test the null hypothesis consists of a statement that the true value of a parameter lies in a certain interval or equals a specific value. Other forms of parameter tests compare the parameters of the distributions that underlie two different samples. Here we only consider a one-sided test as an example.

For a one-sided test, like the one described above, one usually chooses a point estimator T for the parameter θ as a test statistic. In such a case we will reject the null hypothesis H_0 only if the value of the point estimator T has a value c, which does not exceed the **critical value**. Therefore the critical region is $C = (-\infty, c]$. Hence it is clear that the value c must lie to the left of θ_0, because we will not be able to reasonably reject H_0 if even the value of the point estimators T exceeds θ_0. However, even a value that is only slightly smaller than θ_0 will not be sufficient to make the probability of an error of the first kind (the null hypothesis H_0 is rejected even though it is correct) sufficiently small. Therefore c must lie at some distance to the left of θ_0. Formally, the critical value c is determined as follows: We consider

$$\beta(\theta) = P_\theta(H_0 \text{ is rejected}) = P_\theta(T \in C),$$

which can be simplified to $\beta(\theta) = P(T \leq c)$ for a one-sided test. The quantity $\beta(\theta)$ is also called the **power** of the test. It describes the probability of a rejection of H_0 dependently on the value of the parameter θ. For all values θ that satisfy the null hypothesis, the value of $\beta(\theta)$ must be less than the significance level α. The reason is that it the null hypothesis is true, we want to reject it at most with probability α in order to commit an error of the first kind at most with this probability. Therefore we must have

$$\max_{\theta:\theta \text{ satisfies } H_0} \beta(\theta) \leq \alpha.$$

For the test we consider here, it is easy to see that the power $\beta(\theta)$ of the test reaches its maximum for $\theta = \theta_0$: the larger the true value of θ, the less likely it is that the test statistic (the point estimator T) yields a value of at most c. Hence we must choose the smallest value θ that satisfies the null hypothesis $H_0 : \theta \geq \theta_0$. The expression reduces to

$$\beta(\theta_0) \leq \alpha.$$

At this point all that is left to do to complete the test is to determine $\beta(\theta_0)$ from the distribution assumption and the point estimator T.

A.4.3.3 Parameter Test Example

As an example, for a parameter test, we consider a one-sided test of the expected value μ of a normal distribution $N(\mu, \sigma^2)$ with known variance σ^2 [1]. That is, we consider the hypotheses

$$H_0: \quad \mu \geq \mu_0, \qquad H_a: \quad \mu < \mu_0.$$

As a test statistic, we use the standard point estimator for the mean (expected value) of a normal distribution, namely

$$\bar{X} = \frac{1}{n} \sum_{i=1}^{n} X_i,$$

that is, the arithmetic mean of the sample values. (n is the sample size.) As one can easily check, this estimator has the probability density

$$f_{\bar{X}}(x) = N\left(x; \mu, \frac{\sigma^2}{n}\right).$$

Therefore it is

$$\alpha = \beta(\mu_0) = P_{\mu_0}(\bar{X} \le c) = P\left(\frac{\bar{X} - \mu_0}{\sigma/\sqrt{n}} \le \frac{c - \mu_0}{\sigma/\sqrt{n}}\right) = P\left(Z \le \frac{c - \mu_0}{\sigma/\sqrt{n}}\right)$$

with standard normally distributed random variable Z. (The third step in the above transformation served the purpose to obtain a statement about such a random variable.) Thus we have

$$\alpha = \Phi\left(\frac{c - \mu_0}{\sigma/\sqrt{n}}\right),$$

where Φ is the distribution function of the standard normal distribution, which can be found in a tabulated form in many textbooks. From such a table we obtain the value z_α for which $\Phi(z_\alpha) = \alpha$. Then the critical value is

$$c = \mu_0 + z_\alpha \frac{\sigma}{\sqrt{n}}.$$

Note that due to the small value of α, the value of z_α is negative, and therefore c, as already made plausible above, is smaller than μ_0.

In order to give a numeric example, we choose [1] $\mu_0 = 130$ and $\alpha = 0.05$. In addition, let $\sigma = 5.4$, $n = 125$, and $\bar{x} = 128$. From a table of the standard normal distribution we obtain $z_{0.05} \approx -1.645$ and arrive at

$$c_{0.05} \approx 130 - 1.645 \frac{5.4}{\sqrt{25}} \approx 128.22.$$

Since $\bar{x} = 128 < 128.22 = c$, the null hypothesis H_0 is rejected. If we had chosen $\alpha = 0.01$ instead, we would have obtained (with $z_{0.01} \approx -2.326$)

$$c_{0.01} \approx 130 - 2.326 \frac{5.4}{\sqrt{25}} \approx 127.49,$$

and thus H_0 would not have been rejected.

As an alternative, the significance level can be left unspecified. Instead, one provides the value α from which upward the null hypothesis H_0 is rejected. This value α is also called p-**value**. For the above example, it has the value

$$p = \Phi\left(\frac{128 - 130}{5.4/\sqrt{25}}\right) \approx 0.032.$$

That is, the null hypothesis H_0 is rejected for a significance level above 0.032 but accepted for a significance level less than 0.032. Note, however, that one must **not** choose the significance level **after** computing the p-value as this would undermine the validity of the test. The p-value is only a convenience in order to accommodate the different attitudes of users, some of which are more cautious and thus choose lower significance levels α, while other are more daring and thus choose higher significance levels. From the p-value all users can see whether they would reject or accept the null hypothesis and thus need not follow the choice of the writer.

A.4.3.4 Goodness-of-Fit Test

With a goodness of fit test, it is checked whether two distributions, two empirical distributions, or one empirical and one theoretical coincide. Often a goodness-of-fit test is used to check a distribution assumption, as it is needed for parameter estimation. As an example, we consider the χ^2 goodness-of-fit test for a polynomial distribution: let a one-dimensional data set of size n be given for k attribute values $a_1, \ldots, a_k$. In addition, let p_i^*, $1 \leq i \leq k$, be an assumption about the probabilities with which the attribute values a_i occur. We want to check whether the hypothesis fits the data set, that is, whether the actual probabilities p_i coincide with the hypothetical p_i^*, $1 \leq i \leq k$, or not. Thus we contrast the hypotheses

$$H_0 : \forall i, 1 \leq i \leq k : p_i = p_i^* \quad \text{and} \quad H_a : \exists i, 1 \leq i \leq k : p_i \neq p_i^*.$$

An appropriate test statistic can be derived from the following theorem about polynomially distributed random variables, which describe the frequency of the occurrence of the different values a_i in a sample.

Theorem A.20 *Let $(X_1, \ldots, X_k)$ be a k-dimensional polynomially distributed random variable with parameters $p_1, \ldots, p_k$ and n. Then the random variable*

$$Y = \sum_{i=1}^{k} \frac{(X_i - np_i)^2}{np_i}$$

is approximately χ^2-distributed with $k - 1$ degrees of freedom. (In order for this approximation to be sufficiently good, it should be $\forall i, 1 \leq i \leq k : np_i \geq 5$. This can always be achieved by combining attribute values and/or random variables.)

In the expression for calculating the random variable Y, the values of the random variables X_i are compared to their expected values np_i, the deviations are squared (among other reasons, so that positive and negative deviations do not cancel), and summed weighted, with a deviation being weighted the lower, the smaller the expected value is. Since Y is χ^2 distributed, large values are unlikely.

The degrees of freedom result from the number of free parameters of the distribution. The number n is not a free parameter, since it is fixed by the size of the sample. From the k parameters $p_1, \ldots, p_k$ only $k - 1$ can be chosen freely, since it must be $\sum_{i=1}^{k} p_i = 1$. Hence only $k - 1$ of the $k + 1$ parameters of the polynomial distribution remain that determine the degrees of freedom.

By replacing the actual probabilities p_i by the hypothetical p_i^* and replacing the random variables X_i by their realizations (absolute frequency of the occurrence of a_i in the sample), we obtain a test statistic for the goodness-of-fit test, namely

$$y = \sum_{i=1}^{k} \frac{(x_i - np_i^*)^2}{np_i^*}.$$

If the null hypothesis H_0 is correct, that is, if all hypothetical probabilities coincide with the actual ones, it is very unlikely that y takes a large value, since y

is a realization of the random variable Y, which is χ^2 distributed. Therefore the null hypothesis H_0 is rejected if the value of y exceeds a certain critical value c, which depends of the significance level. Hence the critical region is $C = [c, \infty)$. The critical value c is determined from the χ^2 distribution with $k - 1$ degrees of freedom, namely as the value for which $P(Y > c) = \alpha$ (or equivalently, for which $P(Y \leq c) = F_Y(c) = 1 - \alpha$), where F_Y is the distribution function of the χ^2-distribution with $k - 1$ degrees of freedom.

Note that in practice the number k of attribute values may have to be reduced by combining attribute values, in order to ensure that $\forall i, 1 \leq i \leq k : np_i \geq 5$ holds. Otherwise the approximation by the χ^2 distribution is not sufficiently good.

For continuous distributions, the domain of values can be divided into nonoverlapping regions that approximate the hypothetical distribution by a step function (constant function value in each interval). Then the goodness-of-fit test for a polynomial distribution is applied, with each interval yielding an outcome of the experiment. In this case one may either work with a completely determined hypothetical distribution (all parameters being given, test whether a specific distribution fits the data) or one may estimate all or some of the parameters from the data (test whether a distribution of a given type fits the data). In the latter case the degrees of freedom of the χ^2 distribution has to be reduced by 1 for each parameter that is estimated from the data. This is actually plausible, because with each parameter that is estimated from the data, the power of the test should go down. However, this is exactly the effect of reducing the degrees of freedom.

As an alternative of applying the χ^2 goodness-of-fit test to an interval partition, the **Kolmogorov–Smirnov test** may be used for continuous distributions, which does without an interval partition, but directly compares the empirical and hypothetical distribution functions.

A.4.3.5 Goodness-of-Fit Test Example

A die is suspected to be unfair, that is, that when tossed, the die shows the different numbers of pips with different probabilities. In order to test this hypothesis, the die is tossed 30 times and it is counted how frequently the different numbers turn up:

$$x_1 = 2, \ x_2 = 4, \ x_3 = 3, \ x_4 = 5, \ x_5 = 3, \ x_6 = 13.$$

That is, one pip turned up twice, two pips four times, etc. Now we contrast the hypotheses

$$H_0 : \forall i, 1 \leq i \leq 6 : p_i = \frac{1}{6} \quad \text{and} \quad H_a : \exists i, 1 \leq i \leq 6 : p_i \neq \frac{1}{6}.$$

Since $n = 30$, we have $\forall i : np_i = 30\frac{1}{6} = 5$, and thus the prerequisites of Theorem A.20 are satisfied. Hence the χ^2 distribution with 5 degrees of freedom is a good approximation of the random variable Y. We compute the test statistic

$$y = \sum_{i=1}^{6} \frac{(x_i - 30 \cdot \frac{1}{6})^2}{30 \cdot \frac{1}{6}} = \frac{1}{5} \sum_{i=1}^{6} (x_i - 5)^2 = \frac{67}{5} = 13.4.$$

For a significance level of $\alpha_1 = 0.05$ (5% probability for an error of the first kind), the critical value is $c \approx 11.07$, since a χ^2 distributed random variable Y with five degrees of freedom satisfies

$$P(Y \leq 11.07) = F_Y(11.07) = 0.95 = 1 - \alpha_1,$$

as one may easily obtain from tables of the χ^2 distribution. Since $13.4 > 11.07$, the null hypothesis that the die is fair can be rejected on a significance level of $\alpha_1 = 0.05$. However, it cannot be rejected on a significance level of $\alpha_2 = 0.01$, since

$$P(Y \leq 15.09) = F_Y(15.09) = 0.99 = 1 - \alpha_2$$

and $13.4 < 15.09$. The p-value is

$$p = 1 - F_Y(13.4) \approx 1 - 0.9801 = 0.0199.$$

That is, for a significance level of 0.0199 and above, the null hypothesis H_0 is rejected, while for a significance level below 0.0199, however, it is accepted.

A.4.3.6 (In)Dependence Test

With a dependence test, it is checked whether two quantities are dependent. In principle, any goodness-of-fit test can easily be turned into a dependence test: simply compare the empirical joint distribution of two quantities with a hypothetical independent distribution that has the same marginals. In such a case the marginal distributions are usually estimated from the data.

As an example, we consider the χ^2 **dependence test** for two nominal values, which is derived from the χ^2 goodness-of-fit test. Let X_{ij}, $1 \leq i \leq k_1$, $1 \leq j \leq k_2$, be random variables that describe the absolute frequency of the joint occurrence of the values a_i and b_j of two attributes A and B, respectively. Furthermore, let $X_{i.} = \sum_{j=1}^{k_2} X_{ij}$ and $X_{.j} = \sum_{i=1}^{k_1} X_{ij}$ be the marginal frequencies (absolute frequencies of the attribute values a_i and b_j). Then, as a test statistic, we compute

$$y = \sum_{i=1}^{k_1} \sum_{j=1}^{k_2} \frac{(x_{ij} - \frac{1}{n}x_{i.}x_{.j})^2}{\frac{1}{n}x_{i.}x_{.j}} = \sum_{i=1}^{k_1} \sum_{j=1}^{k_2} n \frac{(p_{ij} - p_{i.}p_{.j})^2}{p_{i.}p_{.j}}$$

from the realizations x_{ij}, $x_{i.}$, and $x_{.j}$ of these random variables, which are counted in a sample of size n, or from the estimated joint probabilities $p_{ij} = \frac{x_{ij}}{n}$ and marginal probabilities $p_{i.} = \frac{x_{i.}}{n}$ and $p_{.j} = \frac{x_{.j}}{n}$. The critical value c is determined with the help of the chosen significance level from a χ^2 distribution with $(k_1 - 1)(k_2 - 1)$ degrees of freedom. The degrees of freedom are justified as follows: for the $k_1 \cdot k_2$ probabilities p_{ij}, $1 \leq i \leq k_1$, $1 \leq j \leq k_2$, and for the occurrence of the different combinations of a_i and b_j, it must be $\sum_{i=1}^{k_1} \sum_{j=1}^{k_2} p_{ij} = 1$. Thus $k_1 \cdot k_2 - 1$ free parameters remain. From the data we estimate the k_1 probabilities $p_{i.}$ and the k_2 probabilities $p_{.j}$. However, they must also satisfy $\sum_{j=1}^{k_2} p_{i.} = 1$ and $\sum_{i=1}^{k_1} p_{.j} = 1$, so that the degrees of freedom are reduced by only $(k_1 - 1) + (k_2 - 1)$. In total, we have $k_1 k_2 - 1 - (k_1 - 1) - (k_2 - 1) = (k_1 - 1)(k_2 - 1)$ degrees of freedom.

Appendix B
The R Project

R is an open-source statistics and data analysis software available under the General Public License (GPL). This means especially that R can be downloaded, used, and distributed freely.

R is based on a very simple command-line language that can be used interactively, but also for writing programs in R. The sections in this book referring to R are in no way intended to give a comprehensive introduction to R and do not claim to be complete in anyway. The main purpose of these sections is to enable the reader to apply methods introduced in the "theoretical chapters" directly to their own data. For most of the methods whose usage is explained in R, one or two commands will be sufficient.

This appendix explains how to get started with R a provides quick overview on the very basics of R. More details can be found at the website for R

http://www.r-project.org

B.1 Installation and Overview

R can be downloaded from the above-mentioned website where versions suitable for standard operating systems are available. All is needed to download the corresponding installation file, unzip and start it and then follow the installation instructions. The simplest way is to stick to the proposed default settings.

Once R is installed, double click the R symbol on the desktop that should have been created during installation.

Figure B.1 shows a screenshot of R after the command

```
> plot(iris)
```

has been entered in the console window. Note that the prompt symbol > does not belong to the command. We will always display the prompt symbol in order to distinguish R commands from outputs generated after a command has been entered. Outputs will be shown without the prompt symbol.

M.R. Berthold et al., *Guide to Intelligent Data Analysis*, 369
Texts in Computer Science 42,
DOI 10.1007/978-1-84882-260-3, © Springer-Verlag London Limited 2010

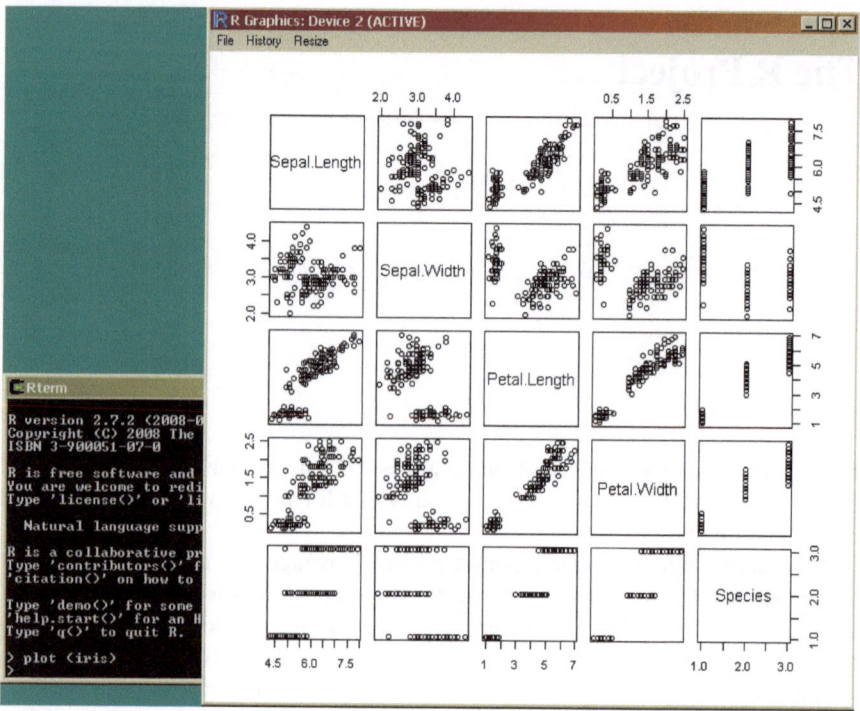

Fig. B.1 A screenshot of R

The graphics displayed on the right in Fig. B.1 is generated by the `plot` command.

B.2 Reading Files and R Objects

R is a type-free language. This means that variables need not to be declared before their use. A variable might be used to store just a single number, but it can also be a complex object with many attributes. In most cases, the objects in this book will contain data sets or analysis results. Assignments in R are denoted by the two symbols <-. Before we can analyze a data set with R, we have to load the data set into R. The easiest way to achieve this uses the function `read.table`:

```
> mydata <- read.table(file.choose(),header=TRUE)
```

This command will open a file chooser window to find and select the file with the data set to be analyzed. After the file has been chosen, it will be stored in the variable `mydata`. The specification `header=TRUE` when calling the function `read.table` tells R that the first line in the file should not be interpreted as data,

but rather as names for the attributes. It is therefore assumed that the structure of the
file looks like

```
x        y      z
1.3      2.8    a
3.4      1.9    b
2.7      4.2    a
...      ...    ...
```

In this case, there are three attributes named x, y, and z given in the first line.
The records then come in the following lines. Because three attribute names have
been given in the first line, each of the following lines must also have three entries
separated by an arbitrary number of blanks. If the values of the attributes are not
separated by blanks but by another symbol, say a comma, then one would have to
write

```
> mydata <- read.table(file.choose(),header=TRUE,sep=",")
```

Now the object named mydata contains the data from the file. Assume that the file
contains only three records and not more as indicated by the dots above.

At least for smaller data sets, one can take a look at the data by simply typing

```
> mydata
    x   y z
1 1.3 2.8 a
2 3.4 1.9 b
3 2.7 4.2 a
```

or print(mydata), which gives the same result.

The summary function gives main properties—in the case of an object representing a data set simple statistics—of an object:

```
> summary(mydata)
       x                y            z
 Min.   :1.300    Min.   :1.900    a:2
 1st Qu.:2.000    1st Qu.:2.350    b:1
 Median :2.700    Median :2.800
 Mean   :2.467    Mean   :2.967
 3rd Qu.:3.050    3rd Qu.:3.500
 Max.   :3.400    Max.   :4.200
```

print and summary can be applied to any R object. If the corresponding R
object is not a data set but a result of some analysis method, then the main properties
of the analysis result will be listed.

One can access a specific attribute of an object by writing the name of the object,
followed by the symbol $ and then by the name of the variable:

```
> mydata$y
[1] 2.8 1.9 4.2
```

In this simple example, we only have three records and therefore only three values for the attribute *y*. If one line is not enough to list all values, R will simply continue in the next line and list the index of the first data entry in each line in square brackets. So if we had 15 records, the result might look like the following one:

```
> mydata$y
 [1] 2.8 1.9 4.2 2.4 3.0 1.7
 [7] 4.1 3.3 2.6 1.8 4.3 3.1
[13] 3.7 2.1 1.8
```

We can also assess the values of an attribute—a column in our data table—by using its index in square brackets:

```
> mydata[2]
    y
1 2.8
2 1.9
3 4.2
```

A specific record, a row in our data table, can be selected in the following way:

```
> mydata[2,]
    x   y z
2 3.4 1.9 b
```

Most of the R code examples given in the "Practical . . ." section at the ends of the corresponding chapters are based on the Iris data set. It is not necessary to first load this data set into R. R provides some simple data sets, and, among them, there is the Iris data set that can be accessed via an object called `iris`. The attribute names are `Sepal.Length`, `Sepal.Width`, `Petal.Length`, `Petal.Width`, and `Species`. So, if we want to know the mean value of the sepal length, we simply need to enter

```
> mean(iris$Sepal.Length)
[1] 5.843333
```

without loading any file in advance.

B.3 R Functions and Commands

Functions in R can have quite a number of parameters, but, when using a function, most of the parameters do not have to be specified, unless one wants to use a value different from the default value of the parameter. Such a parameters can be set inside the call of the function by specifying the parameter name, followed by "=" and the value of the parameter. We have seen examples for such parameters already in the function read.table in the beginning of the previous section. The default value of the parameter header is FALSE, assuming that the file to be read does not contain the names of the attributes. Only if we have a file whose first line defines the names of the attributes, we have to use header=TRUE, as we have done using the function read.table. Another parameter of this function is sep. We do not need to specify the value of this parameter when the attribute values in a row of our file are separated by blanks. If another symbol like comma is used to separate the values, we must assign the corresponding value (symbol) to the parameter sep.

Not all parameters are specified in this form. As in most programming languages, the data set is just handed over to the function as normal argument as in the very first example of this appendix (plot(iris)).

One can browse through the history of commands or functions that have been used in an R session by the key "cursor up" and "cursor down."

B.4 Libraries/Packages

There are various libraries or packages for R for special topics or specialized methods. Some of these libraries come along with R, and other need to be downloaded. Downloading an additional package is very easy. Given that the computer is connected to the Internet, just type

```
> install.packages
```

and wait for the window asking you to choose a mirror site from which you would like to download the package. After you have clicked the mirror site, the packages will be listed in alphabetical order in a new window, and you can choose the package you need by clicking it.

Once a package—for instance, the package cluster—has been downloaded, it can be added to an R session by the following command:

```
> library(cluster)
```

When a package has been downloaded once, it is not necessary to download it again. However, unless the workspace (see next the section) is saved and reloaded, the packages must be added to the R session each time R is restarted.

B.5 R Workspace

The actual R session can be stored, so that it can be reloaded next time, and all the R objects, like the data that had been loaded and the analysis results that have been stored in objects, can be recovered:

```
> save(list = ls(all=TRUE), file="all.Rdata")
```

In order to load a workspace that has been stored before, the command

```
> load("all.Rdata", .GlobalEnv)
```

can be used. Of course, the file does not have to be called all.Rdata.

B.6 Finding Help

If detailed information about an R function or command is needed,

```
> help(...)
```

will provide the description of the R function or command that has been entered in place of the three dots.

If one does not know the command,

```
> help.search("...")
```

will help. It will list all the R functions/commands in which the specified term that should be given in place of the three dots occurs.

Even if you know the correct name of the command or function you are interested in, R will not be able to provide help if the function belongs to a package that has not been included in the corresponding R session. So

```
> help(scatterplot3d)
```

will not give any information on scatterplot3d, unless the package to which the function scatterplot3d belongs has been added to the session. (In this case, the package name is even identical with the function name.)

If you know neither the exact function name nor the corresponding package, you can simply search the R website for the topic or use a search engine and type in R the corresponding term for which you would like to find an R function.

B.7 Further Reading

There are various introductory books on R and numerous books on specialized topics like time series analysis or Bayesian statistics where R code is included. It is impossible to list all of them here. As a starting point for R and how to apply basic statistics with R, we refer to [17]. The book [19] provides details for manipulating data and connecting to databases with R. An introduction to programming with R is given in [16]. Those who are not satisfied with the simple default graphics provided by R might like to take a look at the book [18].

Appendix C
KNIME

KNIME, pronounced *[naim]*, is a modular data exploration platform that enables the user to visually create data flows (often referred to as pipelines), selectively execute some or all analysis steps, and later investigate the results through interactive views on data and models. This appendix will give a short introduction to familiarize the readers of this book with the basic usage of KNIME. Considerably more information regarding the use of KNIME is available online at

http://www.knime.org

C.1 Installation and Overview

In order to install KNIME, download one of the versions suitable for your operating system and unzip it to any directory for which you have write permissions. No other action to install KNIME is required, in particular no setup routine has to be launched. In order to start KNIME for the first time, double click the *knime.exe* file on Windows or on Linux launch *knime*.

KNIME is uninstalled from the system by simply deleting the installation directory. Per default the workspace is also in this directory. If a different location for the workspace was chosen, this directory needs to be deleted manually as well.

When KNIME is started the first time, a welcome screen opens. From here the user can

- **Open KNIME workbench**: opens the KNIME workbench to immediately start exploring KNIME, build own workflows, and explore your data.
- **Get additional nodes**: In addition to the ready-to-use basic KNIME installation, there are additional plug-ins for KNIME, e.g., an R and Weka integration, modules for image and text processing, or the integration of the Chemistry Development Kit with additional nodes for the processing of molecular structures. These features can be downloaded also later from within KNIME itself if you choose to skip this step.

M.R. Berthold et al., *Guide to Intelligent Data Analysis,*
Texts in Computer Science 42,
DOI 10.1007/978-1-84882-260-3, © Springer-Verlag London Limited 2010

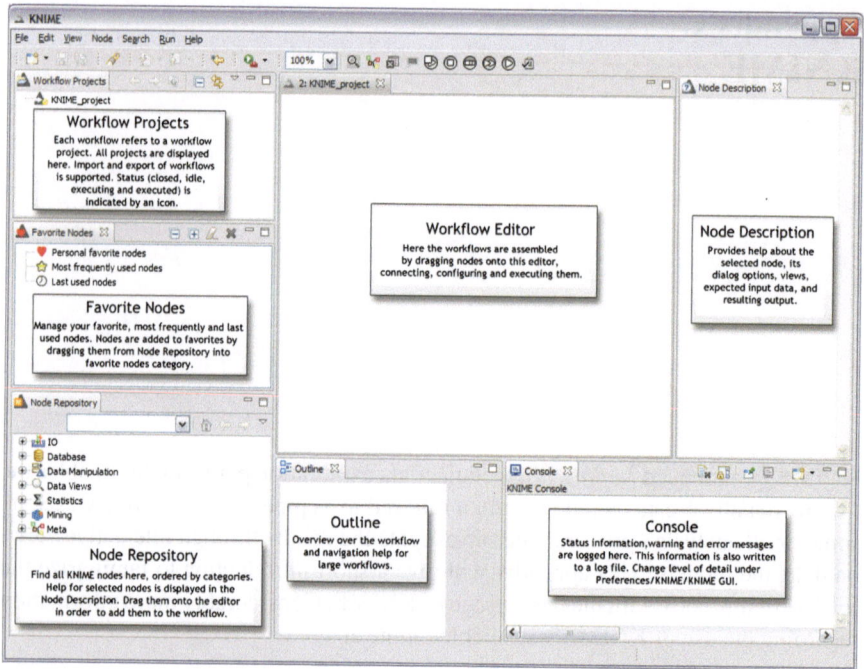

Fig. C.1 The standard outlay of the KNIME workbench

The KNIME Workbench is organized as depicted in Fig. C.1. It consists of different areas:

- Workflow Projects: All KNIME workflows are displayed in the Workflow Projects view. The status of the workflow is indicated by an icon showing whether the workflow is closed, idle, executing, or if execution is complete.
- Favorite Nodes: The Favorite Nodes view displays your favorite, most frequently used, and last used nodes. A node is added to the favorites by dragging it from the node repository into the personal favorite nodes category. Whenever a node is dragged onto the workflow editor, the last used and most frequently used categories are updated.
- Node Repository: The node repository contains all KNIME nodes ordered in categories. A category can contain another category, for example, the Read category is a subcategory of the IO category. Nodes are added from the repository to the workflow editor by dragging them to the workflow editor. Selecting a category displays all contained nodes in the node description view; selecting a node displays the help for this node. If the user knows the name of a node, you can enter parts of the name into the search box of the node repository. As you type, all nodes are filtered immediately to those that contain the entered text in their names.
- Outline: The outline view provides an overview over the whole workflow even if only a small part is visible in the workflow editor (marked in gray in the outline view). The outline view can also be used for navigation: the gray rectangle can be

moved with the mouse, which causes the editor to scroll so that the visible part matches the gray rectangle.

- Console: The console view prints out error and warning messages in order to give you a clue of what is going on under the hood. The same information is written to a log file, which is located in the workspace directory.
- Node Description: The node description displays information about the selected node (or the nodes contained in a selected category). In particular, it explains the dialog options, the available views, the expected input data, and resulting output data.
- Workflow Editor: The workflow editor is used to assemble workflows, configure and execute nodes, inspect the results, and explore your data. This section describes the interactions possible within the editor.

C.2 Building Workflows

A workflow is built by dragging nodes from the *Node Repository* onto the *Workflow Editor* and connecting them there. Nodes are the basic processing units of a workflow. Each node has a number of input and/or output ports. Data (or a model) is transferred over a connection from an out-port to the in-port(s) of other nodes.

When a node is dragged onto the workflow editor, the status light is usually red, which means that the node has to be configured in order to be able to be executed. A node is configured by right clicking it, choosing *Configure*, and adjusting the necessary settings in the node's dialog (see Fig. C.2 on the left).

When the dialog is closed by pressing the *OK* button, the node is configured, and the status light changes to yellow: the node is ready to be executed. A right-click on the node again shows an enabled *Execute* option; pressing it will execute the node,

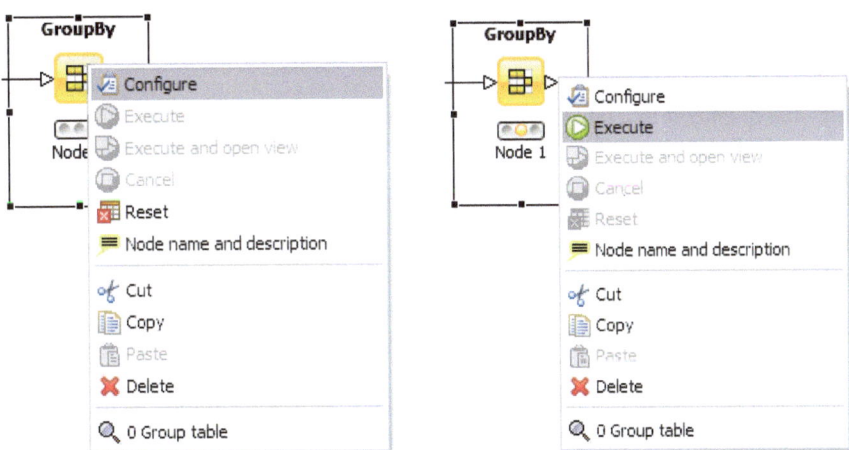

Fig. C.2 The node dialog allows one to configure (*left*) and later execute (*right*) individual nodes

Fig. C.3 Different port
types: data, database, PMML,
and unspecified ports

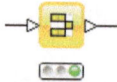

and the result of this node will be available at the out-port (see Fig. C.2 on the right). After a successful execution the status light of the node is green, indicating that the processed data is now available on the outports. The result(s) can be inspected by exploring the out-port view(s): the last entries in the context menu open them.

Ports on the left are input ports, where the data from the outport of the predecessor node is fed into the node. Ports on the right are outgoing ports. The result of the node's operation on the data is provided at the out-port to successor nodes. A tooltip gives information about the output of the node.

Nodes are typed such that only ports of the same type can be connected; Fig. C.3 shows the corresponding symbols for the following, most prominently encountered port types:

- Data Ports: The most common type is the data port (a white triangle) which transfers flat data tables from node to node.
- Database Ports: Nodes executing commands inside a database can be identified by their port color and shape (brown square):
- PMML Ports: Data Mining nodes learn a model which is passed to a model writer or predictor node via a blue squared PMML port:
- Other Ports: Whenever a node provides data which does not fit a flat data table structure, a general purpose port for structured data is used (dark cyan square). Ports that are not data, database, PMML, or ports for structured data are displayed as unknown types (gray square):

C.3 Example Flow

This section demonstrates the basic process of building a small, simple workflow: read data from an ASCII file, assign colors based on certain properties, cluster the data, and display it in a table and a scatter plot.

Start KNIME with an empty workflow and create new empty workflow by right clicking in the *Project Repository* and selecting New Project.

In the Node Repository expand the *IO* category and the *Read* subcategory as shown in Fig. C.4. Drag&drop the *File Reader* node onto the *Workflow Editor* window.

The next node will be a learning node, implementing the well-known k-means clustering algorithm. Expand the *Mining* category followed by the *Clustering* category and then drag the *K-Means* node onto the flow.

We will find the third node by using the convenient search node facility. In the search box above the *Node Repository*, enter "color" and press *Enter*. This limits the nodes shown to the ones with "color" in their name (see Fig. C.5). Pull the *Color Manger* node onto the workflow (this node will be used to define the color in the

Fig. C.4 Locating the file
reader in the node repository

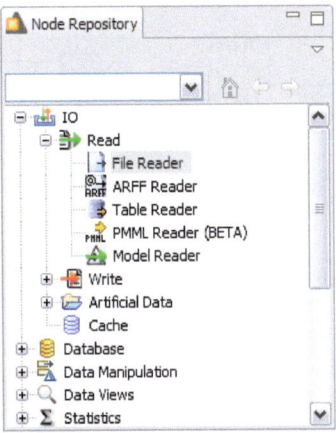

Fig. C.5 Search for nodes in
the node repository

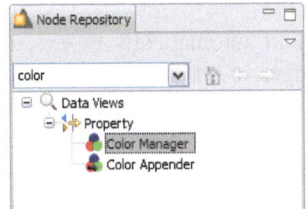

data views later). To again see all nodes in the repository, press ESC or Backspace
in the search field of the *Node Repository*.

Finally, drag the *Interactive Table* and the *Scatter Plot* from the *Data Views* category to the *Workflow Editor* and position them to the right of the *Color Manager*
node.

After placing all nodes, we can now connect them (one can, of course, also later
drag new nodes onto the workbench). Click one output (right) port of the file reader
and drag the connection to the input port of the k-means node. Then continue to
connect the ports as shown in Fig. C.6. (Note that your nodes will not show a green
status, as long as they are not configured and executed.)

Some of the now connected nodes may still show a red status icon, indicating
that it must be configured in order to produce meaningful results. Right click the
File Reader and select *Configure* from the menu. Navigate to the *IrisDataSet* directory located in the KNIME installation directory. Select the *data.all* file from this
location. The *File Reader*'s preview table shows a sample of the data, which should
match the structure of the data file correctly. Click *OK* to confirm this configuration.
Once the node has been configured correctly, it switches to yellow (indicating that
it is ready for execution). After that, the *K-Means* node will also turn yellow, since
its default settings can be applied. To be sure that the default settings fit your needs,
open the dialog and inspect the default settings.

In order to configure the *Color Manager* node, you must first execute the *K-
Means* node by right clicking the node and selecting *Execute*. Note how the File

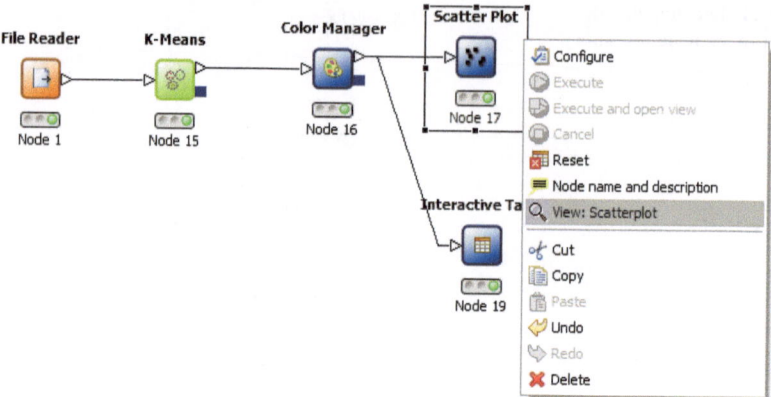

Fig. C.6 The example flow to cluster and visualize the sample data set

Reader node will automatically be executed as well. After execution all nominal values and ranges of all attributes are known at the outport of the executed node: this meta information is propagated to the successor nodes. The *Color Manager* needs this data before it can be configured. Once the *K-Means* node is executed, open the configuration dialog of the *Color Manger* node. The node will suggest to color the rows in our table based on the clustering results. Accept these default settings by clicking *OK*.

Finally, execute the *Scatter Plot*. In order to examine the data and the results, open the nodes' views. In our example, the *K-Means*, the *Interactive Table*, and the *Scatter Plot* have views. Open them from the nodes' context menus.

Select some points in the scatter plot and choose "Hilite Selected" from the "Hilite" menu. The hilited points are marked with an orange border. You will also see the hilited points in the table view. The propagation of the hilite status works for all views in all branches of the flow displaying the same data. Figure C.7 shows an example of the views with a couple of highlighted points.

C.4 R Integration

One of the nice features of KNIME is the modular, open API which allows one to easily integrate other data processing or analysis projects. From the KNIME web-page one can already download a number of such integrations of third party libraries and projects, most notable the statistical data analysis package R, and the machine learning library Weka. In addition, a number of external contributors are providing nodes integrating their own projects into KNIME.

The Weka integration is fairly straightforward to use, one simply drags the node corresponding to the desired learning algorithm onto the workbench, connects it, and opens the configuration dialog which then provides access to all appropriate parameters. If views are available, the KNIME–Weka nodes allow one to open those

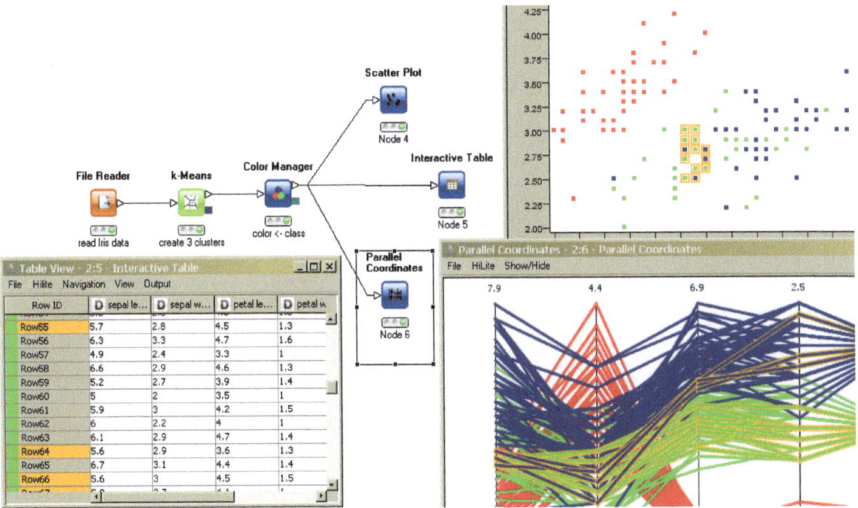

Fig. C.7 A number of open views with highlights patterns

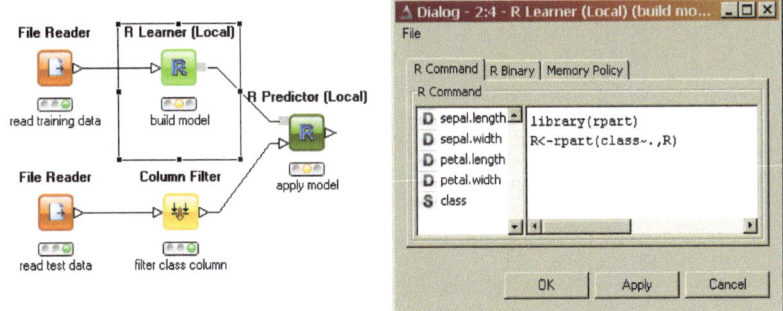

Fig. C.8 A small example workflow using an R code fragment

just like other KNIME views. Weka models can also be fed into a predictor node, similar to KNIME models and hence applied to other data.

The R integration is a bit different, however. Since R really provides more of a statistical programming language, it would require thousand of nodes to cover all the possibilities hidden within the language. KNIME therefore offer nodes which allow one to call small fragments of R code instead—which allows one to use the power of R when needed, e.g., for sophisticated statistical analyses and rely on KNIME's strengths for data loading, integration, and transformation and some of the built-in analysis routines. KNIME offers to point to a local R installation (one can actually download an integrated R installation together with the corresponding KNIME nodes), and it also allows one to access an R installation residing on a server. Dif-

ferent R nodes allow one to execute an R script on incoming data and produce again
a data table, a view, or a model. The latter can then be used in the R predictor node
and applied to other data. Figure C.8 shows a small example flow and the dialog for
an R snippet node.

References

Appendix A

1. Berthold, M., Hand, D.: Intelligent Data Analysis. Springer, Berlin (2009)
2. Buffon, G.-L.L.: Mémoire sur le Jeu Franc-Carreau, France (1733)
3. Everitt, B.S.: The Cambridge Dictionary of Statistics, 3rd edn. Cambridge University Press, Cambridge (2006)
4. Freedman, S., Pisani, R., Purves, R.: Statistics, 4th edn. Norton, London (2007)
5. Friedberg, S.H., Insel, A.J., Spence, L.E.: Linear Algebra, 4th edn. Prentice Hall, Englewood Cliffs (2002)
6. Huff, D.: How to Lie with Statistics. Norton, New York (1954)
7. Kolmogorow, A.N.: Foundations of the Theory of Probability. Chelsea, New York (1956)
8. Krämer, W.: So lügt man mit Statistik, 7 Auflage. Campus-Verlag, Frankfurt (1997)
9. Landau, L.D., Lifshitz, E.M.: Mechanics, 3rd edn. Butterworth-Heinemann, Oxford (1976)
10. Larsen, R.J., Marx, M.L.: An Introduction to Mathematical Statistics and Its Applications, 4th edn. Prentice Hall, Englewood Cliffs (2005)
11. Lay, D.C.: Linear Algebra and Its Applications, 3rd edn. Addison Wesley, Reading (2005)
12. von Mises, R.: Wahrscheinlichkeit, Statistik und Wahrheit. Berlin (1928)
13. Press, W.H., Teukolsky, S.A., Vetterling, W.T., Flannery, B.P.: Numerical Recipes in C—The Art of Scientific Computing, 2nd edn. Cambridge University Press, Cambridge (1992)
14. Sachs, L.: Angewandte Statistik—Anwendung statistischer Methoden, 11 Auflage. Springer, Berlin (2003)
15. Wichura, M.J.: Algorithm AS 241: the percentage points of the normal distribution. Appl. Stat. **37**, 477–484 (1988)

Appendix B

16. Chambers, J.: Software for Data Analysis: Programming with R. Springer, New York (2008)
17. Dalgaard, P.: Introductory Statistics with R, 2nd edn. Springer, New York (2008)
18. Murrell, P.: R Graphics. Chapman & Hall/CRC, Boca Raton (2006)
19. Spector, P.: Data Manipulation with R. Springer, New York (2008)

M.R. Berthold et al., *Guide to Intelligent Data Analysis*,
Texts in Computer Science 42,
DOI 10.1007/978-1-84882-260-3, © Springer-Verlag London Limited 2010

Index

χ^2 dependence test, 367
χ^2 distribution, 349
χ^2 measure, 190
χ^2-goodness-of-fit test, 365
σ-algebra, 326
0–1 loss, 224

A

a priori property, 181
absolute frequency, 305, 324
absolute scale, 37
accuracy, 37, 224
accurate predictor, 285
acyclic directed graph, 227
AdaBoost, 288
additivity, 326
agglomerative hierarchical clustering, 147
AIC, 110
Aikake information criterion, 168
Akaike's information criterion, 110
algorithmic error, 101
alternating optimization, 93
alternative hypothesis, 361
alternatives, 305
antecedent, 180, 245
antimonotone, 181
Apriori, 180, 184
architecture, 82
area chart, 307
area under curve, 98
artificial neural networks, 290
association analysis, 11
association rule, 180
  χ^2 measure, 190
  confidence, 186

  minimum, 186
  information gain, 190
  lift value, 190
  mutual information, 190
  support, 186
    minimum, 186
association rule induction, 186
association rules, 146
assumption, 29
attribute, 34, 304
attribute type, 304
attribute value, 304
attributed graph, 191
AUC, 98
automorphism, 192

B

backward elimination, 120
backward feature elimination, 139
bag-of-words, 134, 158
bagging, 105, 287
bar chart, 40, 307
batch training, 276
Bayes classifier, 208
  full, 219, 222
    classification formula, 222
  idiot's, 219
  mixed, 222
  naive, 219, 221
    classification formula, 221
Bayes error, 94
Bayes network, 226
Bayes' rule, 220, 331
Bayesian information criterion, 110

M.R. Berthold et al., *Guide to Intelligent Data Analysis,*
Texts in Computer Science 42,
DOI 10.1007/978-1-84882-260-3, © Springer-Verlag London Limited 2010